Encyclopaedia of Mathematical Sciences

Volume 14

Editor-in-Chief: R.V. Gamkrelidze

R. V. Gamkrelidze (Ed.)

Analysis II

Convex Analysis and Approximation Theory

With 21 Figures

Springer-Verlag
Berlin Heidelberg New York
London Paris Tokyo
Hong Kong

Title of the Russian edition:
Itogi nauki i tekhniki, Sovremennye problemy matematiki,
Fundamental'nye napravleniya, Vol. 14, Analiz 2
Publisher VINITI, Moscow 1987

Mathematics Subject Classification (1980):
26Axx, 26Bxx, 26Cxx, 30Exx, 41-xx, 42-xx, 52-xx

Library of Congress Cataloging-in-Publication Data
Analiz. English. Analysis: Convex Analysis and Approximation Theory.
(Encyclopaedia of mathematical sciences; v. 13–)
Translation of: Analiz, issued as part of the serial: Itogi nauki i tekhniki. Serii͡a sovremennye problemy matematiki. Fundamental'nye napravlenii͡a.
Vol. 2. Convex Analysis and Approximation Theory.
Includes bibliographies and indexes.
1. Operational calculus. 2. Integral representations. 3. Integral transforms. 4. Asymptotic expansions.
I. Gamkrelidze, R.V. II. Title. III. Series.
IV. Series: Encyclopaedia of mathematical sciences; v. 13, etc.
QA432.A6213 1990 515'.72 89-6163
ISBN-13: 978-3-642-64768-0 e-ISBN-13: 978-3-642-61267-1
DOI: 10.1007/978-3-642-61267-1

Softcover reprint of the hardcover 1st edition 1990
Typesetting: Asco Trade Typesetting Ltd., Hong Kong
2141/3140-543210 – Printed on acid-free paper

List of Editors, Contributors and Translators

Editor-in-chief

R.V. Gamkrelidze, Academy of Sciences of the USSR, Steklov Mathematical Institute, ul. Vavilova 42, 117966 Moscow, Institute for Scientific Information (VINITI), Baltiiskaya ul. 14, 125219 Moscow, USSR

Consulting Editor

R.V. Gamkrelidze, Academy of Sciences of the USSR, Steklov Mathematical Institute, ul. Vavilova 42, 117966 Moscow, Institute for Scientific Information (VINITI), Baltiiskaya ul. 14, 125219 Moscow, USSR

Contributor

V.M. Tikhomirov, Department of Mathematics, Moscow State University, Leninskie Gory, 119899 Moscow, USSR

Translator

D. Newton, University of Sussex, Falmer, Brighton BN1 9QH, Great Britain

Contents

I. Convex Analysis
V.M. Tikhomirov
1

II. Approximation Theory
V.M. Tikhomirov
93

Author Index
245

Subject Index
250

I. Convex Analysis

V.M. Tikhomirov

Translated from the Russian
by D. Newton

Contents

Preface 3
Introduction 4
1. What is Studied in Convex Analysis? 4
2. Duality and the Basic Operators of Convex Analysis 4
3. Finite-Dimensional Convex Geometry 11
4. Convex Calculus 15
5. A Brief Historical Sketch 19
List of Notations 22
Chapter 1. The Basic Ideas of Convex Analysis 24
§1. Convex Sets and Functions 24
1.1. Subspaces, Convex Cones, Affine Manifolds, Convex Sets 24
1.2. Linear, Convex Conical, Affine and Convex Hulls 26
1.3. Convex Functions 27
1.4. Operations Over Convex Objects 29
§2. Duality of Linear Spaces. Dual Operators of Convex Analysis 31
2.1. The Definition of Duality 31
2.2. Elementary Dual Relations 32
2.3. The Basic Operators of Convex Analysis 33
2.4. Examples 34
§3. Topological Properties of Convex Sets and Functions 37
3.1. Topology of Duality 37
3.2. Topological Properties of Convex Sets 38
3.3. Topological Properties of Convex Functions 38
3.4. Topological Properties of Finite-Dimensional Convex Objects 39
§4. Basic Theorems 40
4.1. The Hahn-Banach Theorem and Separation 40
4.2. The Krejn-Mil'man Theorem 42
4.3. The Banach-Alaoglu-Bourbaki Theorem 42
4.4. Appendix 42

Chapter 2. Convex Calculus 42
§ 1. Theorems on Involutiveness 43
1.1. Statement of the Theorems 43
1.2. The Proof of the Fenchel-Moreau Theorem 43
1.3. Some Relations Between the Basic Operators 45
1.4. The Proofs of Theorems 1 and 2 46
§ 2. The Legendre-Young-Fenchel Transform 46
2.1. Statements of the Theorems 46
2.2. Proofs 47
2.3. Some Properties of the Legendre-Young-Fenchel Transform 48
2.4. The Conjugate Function of a Convolution Integral 48
§ 3. The Calculus of Convex Sets and Sublinear Functions 49
3.1. Tables of Basic Formulae 49
3.2. Proofs 50
§ 4. Subdifferential Calculus 51
4.1. The Simplest Properties of Subdifferentials 51
4.2. The Fundamental Formulae of Subdifferential Calculus 52
4.3. The Refinement Theorem 52
§ 5. Convex Analysis in Spaces of Measurable Vector-Functions 53
5.1. Introduction 53
5.2. Subdifferentials of Convex Integral Functionals and Their Applications 54
5.3. The Calculation of the Adjoint Functional I_f^* 54
Chapter 3. Some Applications of Convex Analysis 55
§ 1. Linear and Convex Programming 56
1.1. Statements of the Problems 56
1.2. The Method of Duality in Convex Programming 57
1.3. The Fundamental Theorems of Linear Programming 58
1.4. Convex Programming 58
§ 2. Convexity in Geometry 59
2.1. Extremal Geometric Inequalities 59
2.2. Convex Geometry and the Approximation Theory 61
§ 3. Convex Analysis and Variational Problems 64
3.1. Lyapunov's Theorem and its Generalisations 64
3.2. The Existence of Solutions and Extensions 65
3.3. Duality of Variational Problems 67
3.4. Duality in Many-dimensional Problems 69
3.5. Appendices 72
Chapter 4. Extensions of the Sphere of Convex Analysis and Generalizations of Convexity 74
§ 1. Non-smooth Analysis 74
1.1. Introduction 74
1.2. Convexity and Differentiability 75

1.3. The Origin of Non-smooth Analysis. Clarke's Approach 76
1.4. Other Approaches 78
§2. Convexity and Order 79
2.1. Basic Definitions 79
2.2. Theorems on Extensions and Their Corollaries 80
2.3. The Subdifferential Calculus of Convex Operators 81
2.4. The Method of General Position 81
2.5. The Legendre-Young-Fenchel Transform 82
§3. Choquet Theory and Generalisations of Convexity 82
3.1. Introduction 82
3.2. Existence and Uniqueness Theorems 83
3.3. Other Generalizations of Convexity 84
3.4. Supremal Generators 84
Summary of the Literature 85
References 86

Preface

Convex analysis is that special branch of mathematics which directly borders onto classical (smooth) analysis on the one side and geometry on the other. Almost all mathematicians (and very many practitioners) must have the skills to work with convex sets and functions, and extremal problems, since convexity continually crops up in the investigation of very diverse problems in mathematics and the natural sciences. It seems that some of the elements of convex analysis must occupy a place in mathematical education at any level. Bearing all of this in mind, this article was written to take account of a very broad circle of readers. In particular, this is true of the introduction, where an attempt has been made to explain the fundamental concepts, ideas, statements of problems and the essence of the matter to any interested person.

Chapters 1–3 are addressed to those who wish to become acquainted with the foundations of convex analysis and those areas of mathematics where convex analysis is applied. Chapter 4 is devoted to a brief survey of the areas bordering convex analysis which have been intensively developed in recent times.

The author has not set himself the aim of encompassing the whole of convex analysis; his endeavour has been to discuss the sources, the fundamental concepts and results, to relate the most important applications of convexity and to mention some of the most significant and beautiful theorems.

In conclusion the author expresses his thanks to V.I. Blagodatskikh , A.D. Ioffe, S.V. Konyagin, A.G. Kusraev, V.L. Levin, G.G. Magaril-Il'yaev, B.Sh. Mordukhovich and A.F. Filippov who participated in discussions about this article.

Introduction

1. What is Studied in Convex Analysis? We have become familiar with the concept of convexity in geometry lessons at school. The notion itself arose in ancient times. This is what Archimedes said (Werke, 1963): "I give the name convex on one side to those surfaces for which the line joining any two of its points will... lie on one side of the surface." This definition has remained virtually unchanged up to the present day. We recall that a figure (that is, a closed bounded set) is called *convex* if together with any two of its points it contains the whole segment joining them. The boundary of a planar convex figure is called a *convex curve*; the boundary of a convex figure of higher dimension is called a *convex surface*. A function is called *convex* if its graph lies no higher than the chord joining any pair of points on the graph. Figures 1 and 2 depict convex and non-convex figures and functions. The intersection of all the convex sets containing a given set A is called the *convex hull* of A and denoted co A.

Amongst all figures the convex ones form a fairly narrow special class. The same can be said of the convex functions. However, convexity plays a very important role in mathematics and its applications. The interest in the ideas associated with convexity, and the wealth of applications, led to the creation of a separate direction in mathematics, which has been given the name "Convex Analysis". In convex analysis we study the properties of convex sets and functions, and extremal problems.

2. Duality and the Basic Operators of Convex Analysis. The most important phenomenon associated with convex analysis is *duality*. The fundamental thesis of convex analysis can be formulated as follows: convex objects have a dual description.[1] Let us clarify this with some examples.

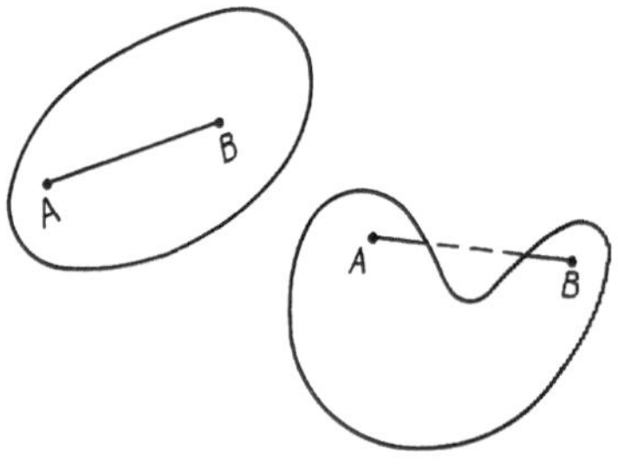

Fig. 1

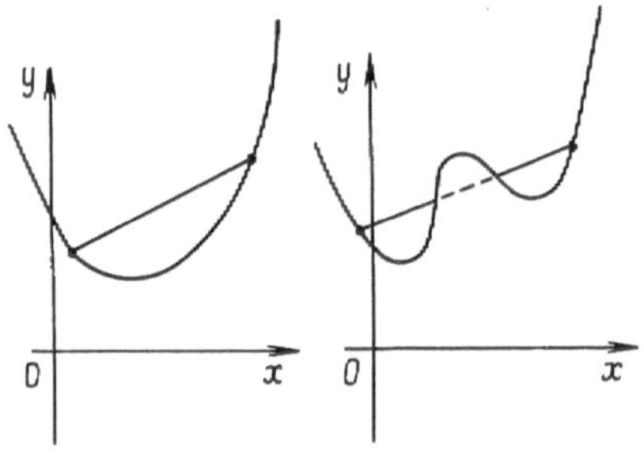

Fig. 2

[1] Corresponding to each convex object A there is a dual object B. If B is given then A can be uniquely recovered.

A segment $\Delta x = [0, x]$ in the plane (or in a higher-dimensional space) joining the origin to a point x admits a dual description: it uniquely determines a half-plane (half-space) Π, containing the origin and bounded by the line (plane) $\partial\Pi$, any vector y of which has scalar product with x equal to 1. In other words

$$\Pi = \Pi x = \{y: \langle y, x\rangle \leqslant 1\}, \qquad \partial\Pi = \{y: \langle y, x\rangle = 1\}$$

where the scalar product $\langle y, x\rangle$ in the plane is $y_1x_1 + y_2x_2$, and in a higher-dimensional space is $\sum_{i=1}^{n} y_ix_i$. Thus there is an association segments $\leftrightarrow$ half-planes: $\Delta x \leftrightarrow \Pi x$.

Any closed convex set A in the plane (or in a higher-dimensional space) containing the origin is a union of the segments Δx which lead from the origin to its points. The intersection of the corresponding family of half-planes (half-spaces) Πx forms a convex set, which we denote by A° and call the *polar* of A. Following from what has been said we have the definition.

$$A^\circ = \{y: \langle y, x\rangle \leqslant 1 \text{ for any } x \text{ in } A\}.$$

In this way we obtain one of the most important dualities, a set and its polar: $A \to A^\circ$.

A ray λ from the origin uniquely determines a half-plane (half-space) $\Pi_0\lambda = \{y: \langle y, x\rangle \geqslant 0\}$ (for any x in λ). The boundary $\partial(\Pi_0\lambda)$ of this set is the line (plane) perpendicular to λ and passing through the origin. Thus $\lambda \leftrightarrow \Pi_0\lambda$ (see Fig. 4a).

Any *convex cone* K (that is, any convex set which is homothety invariant; in the plane, any convex angle) is a union of the rays λ which it contains. The intersection of the corresponding family of half-planes (half-spaces) $\Pi_0\lambda$ is itself a convex cone. It is denoted K^* and called the *conjugate cone*. Here is its analytic definition:

$$K^* = \{y: \langle y, x\rangle \geqslant 0 \text{ for any } x \text{ in } K\}.$$

A correspondence $K \leftrightarrow K^*$ has been obtained (see Fig. 4b).

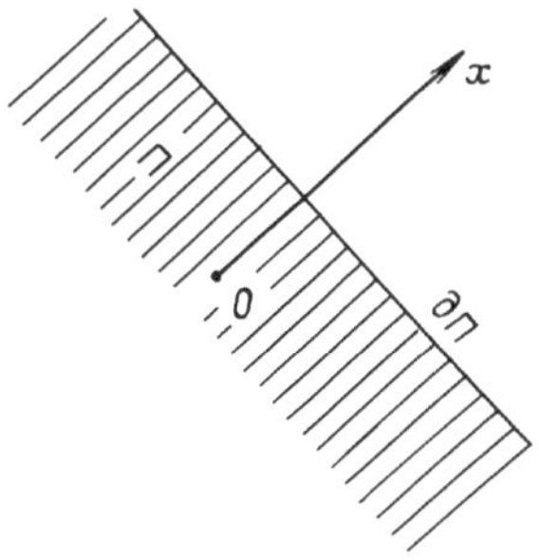

Fig. 3

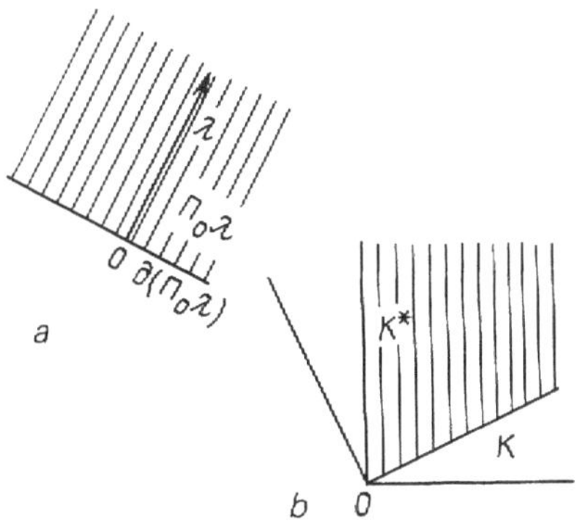

Fig. 4

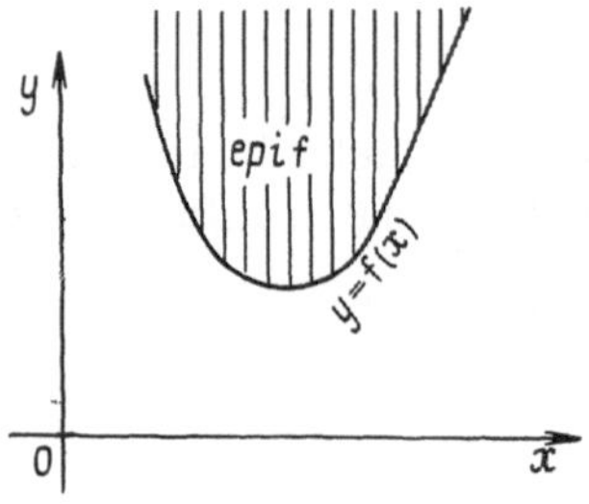

Fig. 5

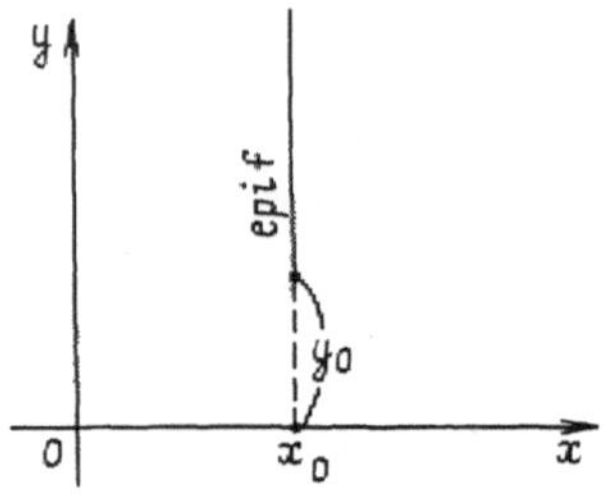

Fig. 6

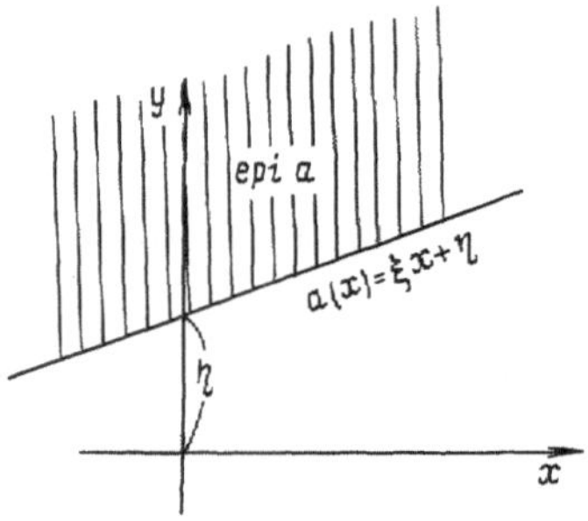

Fig. 7

The next example is concerned with functions. Let $y = f(x)$ be a function of one or several variables. From each point $(x, f(x))$ of the graph of this function draw a ray upwards (in the direction of the Oy axis). The union of these rays forms a set called the *epigraph of the function* and denoted epi f (Fig. 5). Any set which contains the whole of a ray from some point, parallel to the Oy axis in the direction of increasing y, or the whole vertical, is the epigraph of some function (provided we agree to allow the function to take the values $+\infty$ and $-\infty$). In particular, a ray by itself issuing from a point (x_0, y_0) is the epigraph of an "elementary" function which takes the value $+\infty$ everywhere except at $x = x_0$, where $f(x_0) = y_0$ (Fig. 6).

Let us now consider a function $a(x) = \xi x + \eta$ (for the many-variable case $a(x) = \langle \xi, x \rangle + \eta$). Such a function is called *affine* (Fig. 7). The function a admits a dual description: it uniquely determines an "elementary" function $f_a(x) = -\eta$ for $x = \xi$ and $+\infty$ for $x \neq \xi$. Thus a $\leftrightarrow f_a$.

Consider the function which is the least upper bound of the affine functions associated with the family of elementary functions which make up the epigraph of a convex function f. It is denoted f^* and called the *conjugate function*, or *Legendre-Young-Fenchel transform*, of f. Here is its analytic description:

$$f^*(y) = \sup_x (\langle x, y\rangle - f(x)).$$

We have obtained a correspondence $f \to f^*$.

And finally, one last correspondence. Here we associate a function with a set. A vector y generates a linear function $l_y(x) = \langle x, y\rangle$ and, conversely, a linear function $l(x)$ determines a vector y for which $l(x) = \langle x, y\rangle$. Thus $y \leftrightarrow l_y$.

It can be shown that any convex closed function p which is *homogeneous of first degree* (that is, $p(x_1 + x_2) \leqslant p(x_1) + p(x_2)$, $p(\alpha x) = \alpha p(x)$ if $\alpha \geqslant 0$) is the least upper bound of a family of linear functions (they have been called *sublinear*). The collection of vectors corresponding to this family of linear functions forms a convex set. It is called a *subdifferential* of p and denoted ∂p. The analytic definition is:

$$\partial p = \{y: \langle y, x\rangle \leqslant p(x) \text{ for any } x\}.$$

A correspondence $p \leftrightarrow \partial p$ is obtained.

Relative to a convex homogeneous function we have learned how to construct a set—its subdifferential. One of the possibilities for an inverse association is the following. Let A be a set. Put

$$sA(y) = \sup\{\langle x, y\rangle: \text{for all } x \text{ in } A\}.$$

It is easy to see that $sA(y)$ is a sublinear function. It is called a *support function*. As a result we obtain a correspondence $A \to sA$. We note that if $O \in A$, then $A^\circ = \{y: sA(y) \leqslant 1\}$.

We have introduced the basic operators of convex analysis: $A \to A^\circ$, $K \to K^*$, $f \to f^*$, $p \to \partial p$, $A \to sA$. We remark that all of these operators have been constructed by a single recipe from the "elementary" duality relations $\Delta x \leftrightarrow \Pi x$, $\lambda \leftrightarrow \Pi_0 \lambda$, $\operatorname{epi} f_a \leftrightarrow a$, $y \leftrightarrow l_y$.

In what follows, for convenience, alongside the standard notations we will introduce others, namely we will denote A° by πA, K^* by σK, f^* by $\mathit{l}\!f$.

What is duality in convex analysis? It is this; that it may actually be possible to find "inverse" operators to the operators of polarity, conjugation for functions, subdifferentials and supports which rigorously allow us to discuss a single object from two points of view.

Namely, there are the following theorems.

Theorem 1 (on the bipolar). *In order that $\pi^2 A = A$ ($\Leftrightarrow A^{\circ\circ} = A$) it is necessary and sufficient that A be convex, closed and contain the origin.*

Thus the polar operator is involutive. Here the bipolar $A^{\circ\circ} = \{x: \langle x, y\rangle \leqslant 1$ for all y in $A^\circ\}$, is the "polar of the polar". The remaining theorems have a similar meaning.

Theorem 2 (on the conjugate cone). *In order that $\sigma^2 K = K$ ($\Leftrightarrow K^{**} = K$) it is necessary and sufficient that K be a closed convex cone.*

This follows from Theorem 1.

Theorem 3 (Fenchel-Moreau). *Suppose there is an affine function a such that* $a \leqslant f$. *In order that* $l^2 f = f$ ($\Leftrightarrow f^{**} = f$) *it is necessary and sufficient that* f *be closed and convex.*

Thus the conjugation operators for cones and functions are involutive on the sets of convex and closed cones and functions.

Theorem 4 (on the inversion of the operator ∂). *In order that* $s\partial p = p$ *it is necessary and sufficient that p be sublinear and closed.*

Theorem 5 (on the inversion of the operator s). *In order that* $\partial s A = A$ *it is necessary and sufficient that A be a closed convex set.*[2]

Thus the operators ∂ and s are invertible on the classes of objects described above. Here we approach the interpretation of the thesis via the twofold description of convex objects. We say that a closed convex set containing zero is described as the union of the segments going from the origin to its boundary and in the dual form as the intersection of a family of half-spaces. The same can be said of the other objects.

We have quoted five duality formulae above. They all result from one cause—*separation.*

Duality in convex analysis is a consequence of the possibility of separating disjoint convex sets by the level lines (surfaces) of linear functionals, and this is the second most important thesis of convex analysis.

Let $l_y(x) = \langle x, y \rangle$ be a linear functional, $y \neq 0$. Its level surfaces (in the plane, its level lines) are the sets $\{x: \langle x, y \rangle = c\} =: \Gamma$. These are also called *hyperplanes* (in the plane, lines). Let A_1 and A_2 be sets. We say that a hyperplane Γ *separates* A_1 from A_2, if A_1 lies in one of the half-spaces bounded by Γ, and A_2 lies in the other. Or, analytically, if $\inf\{\langle x, y \rangle: x \in A_1\} \geqslant \sup\{\langle x, y \rangle: x \in A_2\}$. We will say that Γ *properly separates* A_1 and A_2 if one of the sets does not wholly lie inside Γ, and *strictly separates* them if for some $\varepsilon > 0$ $\inf\{\langle x, y \rangle: x \in A_1\} \geqslant \sup\{\langle x, y \rangle: x \in A_2\} + \varepsilon$.

Theorem 6 (First separation theorem). *Two disjoint convex sets can be separated by a hyperplane.*

Theorem 7 (Second separation theorem). *A non-empty closed convex set can be strictly separated from a point which does not belong to it.*

Here it is appropriate to say that in this introduction we have considered only *finite-dimensional* convex analysis. But in fact the majority of the theorems in

[2] In the statements of Theorems 1–5 we had to repeat the same words: convex set, closed set, convex function, etc. For brevity we will sometimes use the following notations:
Co(X)—the set of all convex subsets of X;
Cl(X)—the set of all closed subsets of X;
Comp(X)—the set of all compact subsets of X;
Co(X, $\mathbf{R}$), (Co(X, $\bar{\mathbf{R}}$))—the set of all convex finite
(convex with values in $\bar{\mathbf{R}}$) functions on X, etc.

question have infinite-dimensional generalizations. In the finite-dimensional case it is possible to regard direct and dual objects as belonging to the same space. In the infinite-dimensional case this, in general, is impossible. There we must consider *spaces in duality*, that is, different spaces X and Y connected by a bilinear form $\langle x, y\rangle$ which satisfies certain non-degeneracy requirements (see Chap. 1, § 1). A bilinear form $\langle x, y\rangle$ allows us to introduce *topologies* in X and Y, that is, to define what are to be the open and closed sets in X and Y. Here the polar operator associates a set in X to a set in Y, etc. Despite all of this the statements of Theorems 1–5 do not change (see Chap. 2). This is also the case for Theorem 7 (see Chap. 1). But Theorem 6 is no longer valid in the infinite-dimensional case. There we must impose additional conditions on one of the convex sets in order that the conclusion of the Theorem should be true. It suffices, for example, to assume that the interior of one of them is nonempty.

We return to the finite-dimensinal case. The second separation theorem is very simple to prove. We must find a point y of A which is closest to x and through some point of the segment $[x, y]$, not x or y, draw a hyperplane perpendicular to this segment (Fig. 8). For an elementary proof of Theorem 6 see V.M. Alekseev, V.M. Tikhomirov, S.V. Fomin [1987].

Following directly from the second separation theorem is the result.

Theorem 8 (Minkowski, on the intersection of half-spaces). *Any non-empty closed convex set is the intersection of the half-spaces (half-planes in the case of the plane) which contain it.*

In fact, if we assume that the set A', which is the intersection of the half-spaces containing A, contains a point $x \in A$ (clearly $A \subset A'$), then we can strictly separate x from A and immediately obtain a contradiction (Fig. 9).

In fact, we have also proved *Hörmander's theorem on sublinear functions.*

A closed convex sublinear function is the least upper bound of the linear functions which do not exceed it.

A closed convex (proper) function is the least upper bound of the family of affine functions which do not exceed it (Minkowski).

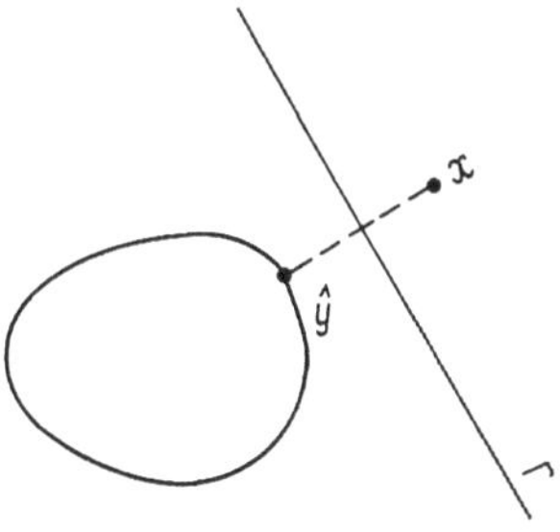

Fig. 8

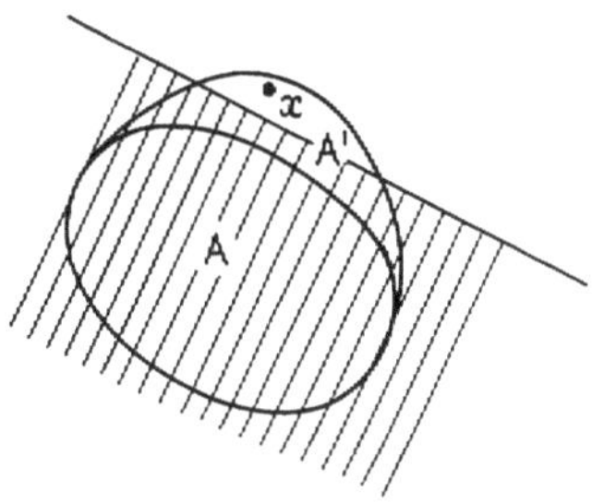

Fig. 9

All the duality theorems (1–5) are proved in a similar way. We will prove, as an example, the theorem on the bipolar. It is clear that $A^{\circ\circ}$ is a closed convex set containing the origin (as the intersection of the family of closed half-spaces $\langle x, y\rangle \leqslant 1$, $y \in A^\circ$). In addition, we obtain directly from the definition that $A \subset A^{\circ\circ}$. If we suppose that there is an $\hat{x} \in A^{\circ\circ}$ but $\hat{x} \notin A$, then, by the second separation theorem (a corollary of the assumptions of convexity and closedness of A), we can find a $\hat{y}$ such that $\langle x, \hat{y}\rangle \leqslant 1$ for all $x \in A$ and $\langle \hat{x}, \hat{y}\rangle > 1$. It follows from the first inequality that $\hat{y} \in A^\circ$ and from the second that $\hat{x} \notin A^{\circ\circ}$—a contradiction.

Thus duality is a consequence of separation.

The third thesis of convex analysis is that there exist (and sometimes significantly "less massive") subsets of a convex set which carry all the information on the convex set itself.

If we draw a convex polygon on the blackboard then we can erase everything except its vertices; knowing the vertices we can recover the whole polygon (Fig. 10). Thus we arrive at the concept of an *extreme point.* A point of a convex set is called an *extreme* point if it is not the midpoint of a segment whose endpoints belong to the convex set. The set of extreme points of the set is denoted ext A.

Theorem 9 (Minkowski, on extreme points). *A closed bounded convex set (in a finite-dimensional space) is the convex hull of its extreme points* ($\Leftrightarrow A \in \operatorname{Comp}\operatorname{Co}(\mathbf{R}^n) \Rightarrow A = \operatorname{co}\operatorname{ext} A$).

Minkowski's theorem played a prominent role in the development of functional analysis. In Chap. 1 we will discuss the Krejn-Mil'man theorem—the infinite-dimensional generalization of Minkowski's theorem, and in Chap. 4 we will discuss Choquet's theorem, where the Krejn-Mil'man theorem is generalized.

We will sketch the proof of Minkowski's theorem in the case of the plane. First we establish that extreme points exist. For this we take a line, intersecting A, and "move" it parallel to itself up to the "last moment" at which at has no points in common with A (Fig. 11). At this "last moment" A lies on one side of the line and intersects the line in a point or a segment. This point or a boundary point of the segment (in Fig. 11, x_1 or x_2) will, as is easy to see, be extreme.

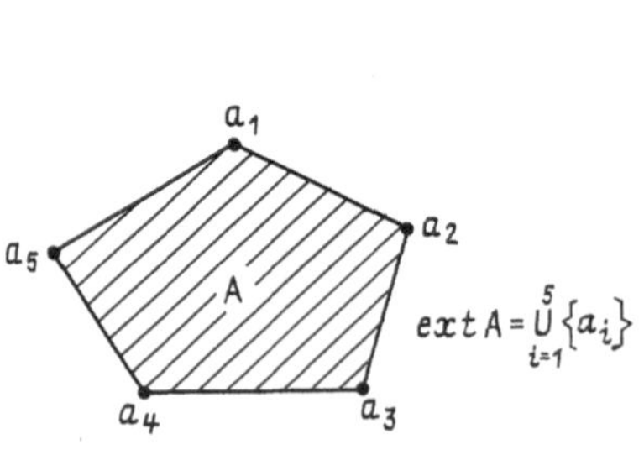

Fig. 10

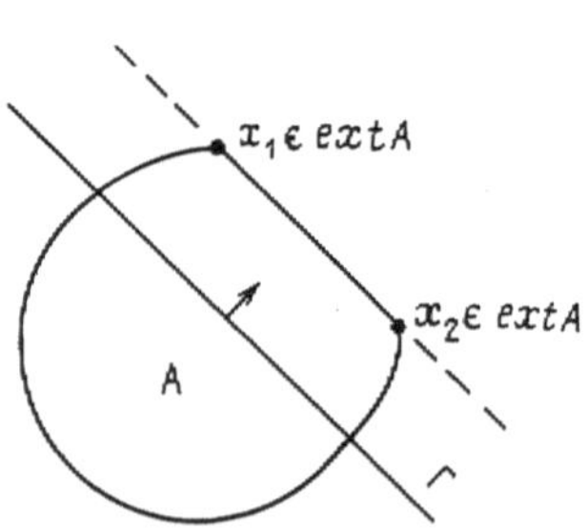

Fig. 11

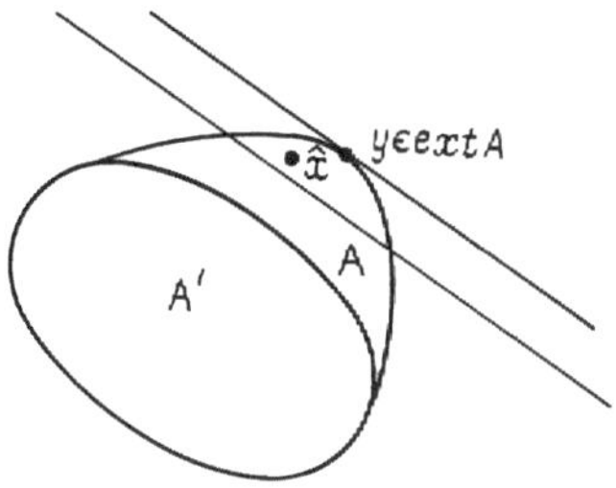

Fig. 12

Now take the set $A' = \operatorname{co}\operatorname{ext} A$. It can be shown to be convex and closed.

If we suppose that there is a point $\hat{x}$ of A, not belonging to A', then by the separation theorem we can separate it from A'. Move this line "on the side of $\hat{x}$" and further ("away from A'") again up to the last moment when it no longer intersects A (Fig. 12). At this last moment on this line, as was said earlier, there will be an extreme point of A. By construction it does not belong to A', and by definition does belong. This contradiction proves the theorem.

The finite-dimensional Minkowski theorem is easily derived from the plane case by induction.

3. Finite-Dimensional Convex Geometry. Initially the properties of finite-dimensional convex sets were studied. We will talk about this briefly in this subsection. Let us agree on some notations. We will denote by $\mathbf{R}^n$ the n-dimensional vector space ($\mathbf{R}$ the number line, $\mathbf{R}^2$ the plane, $\mathbf{R}^n = \{x = (x_1, \ldots, x_n): x_i \in \mathbf{R}\}$). The collection of convex sets in $\mathbf{R}^n$ will be denoted $\operatorname{Co}(\mathbf{R}^n)$, the closed sets by $\operatorname{Cl}(\mathbf{R}^n)$, the closed convex sets by $\operatorname{Cl}\operatorname{Co}(\mathbf{R}^n)$, the convex closed and bounded (compact) sets by $\operatorname{Comp}\operatorname{Co}(\mathbf{R}^n)$.

We will continue from the point at which the last subsection finished. Minkowski's Theorem in $\mathbf{R}^n$ can be strengthened.

Theorem 10 (Carathéodory). *Let $A \in \operatorname{Co}(\mathbf{R}^n)$ and $x \in A$. Then x belongs to the convex hull of not more than $(n + 1)$ points of* $\operatorname{ext} A$.

For the proof of this theorem see, for example, Danzer, Grünbaum & Klee [1963], Leichtweiss [1980].

Let $A \in \operatorname{Cl}\operatorname{Co}(\mathbf{R}^n)$. It may contain lines (that is to say, subspaces or half-spaces). Denote by $L_A = \{x \in \mathbf{R}^n: x + A = A\}$ (Fig. 13).

Theorem 11 (A criterion for $\operatorname{ext} A$ to be non-empty). $\operatorname{ext} A \neq \varnothing \Leftrightarrow L_A = \{0\}$ (*extreme points exist if and only if L_A consists only of zero*).

Let $A \in \operatorname{Co}(\mathbf{R}^n)$ not contain lines. A direction x is called *recessive* if $A + tx \subset A$ for any $t \geqslant 0$. The collection of recessive directions forms a convex cone, called the *recessive cone* and denoted $\operatorname{rec} A$ (Fig. 14). A ray $\lambda = \{x: x = \xi + t\eta, t \geqslant 0\} \subset$

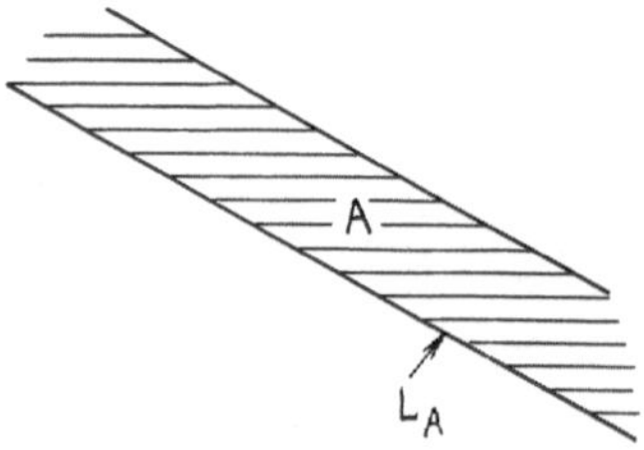

Fig. 13

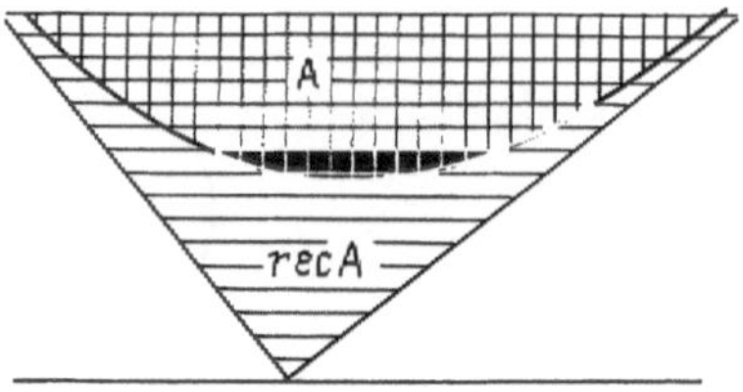

Fig. 14

A is called *extreme*, if $x \in A$, $y \in A$, $\alpha x + (1 - \alpha)y \in \lambda$ for $0 < \alpha < 1$, implies that $x \in \lambda$, $y \in \lambda$. The collection of extreme rays is denoted rext A.

Theorem 12 (Klee, see Leichtweiss [1980]). $A \in \mathrm{Cl\,Co}(\mathbf{R}^n)$ & $L_A = \{0\} \Rightarrow A = \mathrm{co}(\mathrm{ext}\, A \cup \mathrm{rext}\, A)$.

And, finally, we give a general theorem on the representation of a convex closed set. Denote by $L^\perp$ the orthogonal complement of the subspace L, $L^\perp = \{x: \langle x, y\rangle = 0, \forall y \in L\}$.

Theorem 13 (Holmes [1975]). $A \in \mathrm{Cl\,Co}(\mathbf{R}^n) \Rightarrow A = L_A + \mathrm{rec}(A \cap L_A^\perp) + \mathrm{co\,ext}(A \cap L_A^\perp)$

And now one of the most beautiful (and with significant applications) theorems of finite-dimensional analysis.

Theorem 14 (Helly). *Let* $\mathscr{A} = \{A_\alpha\}_{\alpha \in \mathfrak{A}}$, $A_\alpha \subset \mathbf{R}^n$, *be a family of convex closed sets in* $\mathbf{R}^n$, *of which at least one is bounded. Suppose that each* $(n + 1)$ *sets from this family have a common point. Then there is a common point of all the sets.*

We will derive this theorem from the following result.

Theorem 15 (Radon). *Every finite set in* $\mathbf{R}^n$ *containing* $\geqslant (n + 2)$ *points can be divided into two non-empty disjoint subsets whose convex hulls have at least one point in common.*

In fact, let the points be $\{x_1, \ldots, x_m\}$, $m \geqslant n + 2$. Then the vectors $\{x_1 - x_m, \ldots, x_{m-1} - x_m\}$ are linearly dependent

$$\Rightarrow \exists \lambda_1, \ldots, \lambda_{m-1}: \sum_{i=1}^{m-1} |\lambda_i| \neq 0 \quad \text{and} \quad \sum_{i=1}^{m-1} \lambda_i(x_i - x_m) = 0$$

$$\Rightarrow \sum_{i=1}^{m} \alpha_i x_i = 0,$$

where $\alpha_i = \lambda_i$, $i = 1, \ldots, m - 1$; $\alpha_m = -(\lambda_1 + \cdots + \lambda_{m-1})$. Enumerate x_i so that $\alpha_1 \geqslant 0, \ldots, \alpha_k \geqslant 0$, $\alpha_{k+1} < 0, \ldots, \alpha_m < 0$ ($\sum \alpha_i = 0$), and put $A_1 = \{x_1, \ldots, x_k\}$, $A_2 = \{x_{k+1}, \ldots, x_m\}$. It is clear that $A_1 \cap A_2 = 0$. On the other hand

$$y = \left(\left(\sum_{i=1}^{k} \alpha_i x_i\right) \Big/ \sum_{i=1}^{k} \alpha_i\right) = \left(\left(\sum_{i=k+1}^{m} (-\alpha_i) x_i\right) \Big/ \sum_{i=k+1}^{m} (-\alpha_i)\right), \quad \text{i.e.}$$

$$y \in \operatorname{co} A_1 \cap \operatorname{co} A_2.$$

We will now prove Helly's theorem for a finite system using induction over the number of sets. The result is obvious if the number of sets is equal to $(n + 1)$ (by assumption). Suppose the theorem has been proved for subsets consisting of $m - 1$ sets, where $m \geqslant n + 2$: and let $A_{\alpha_1}, \ldots, A_{\alpha_m}$ be sets such that each $m - 1$ of them have a common point. Let $x_i \in \bigcap_{j=1, j\neq i}^{m} A_{\alpha_j}$. By Radon's lemma, $\{x_1, \ldots, x_m\}$ can be divided into two sets $C_1 = \{x_i : i \in I_1\}$ and $C_2 = \{x_i : i \in I_2\}$, $I_1 \cap I_2 = \emptyset$, $I_1 \cup I_2 = \{1, \ldots, m\}$ so that $\operatorname{co} C_1 \cap \operatorname{co} C_2 \neq \emptyset$ and $\exists y \in \operatorname{co} C_1 \cap \operatorname{co} C_2$. In view of the convexity of A_{α_i}, and the fact that $x_i \in A_{\alpha_j}$, (if $j \in I_1, i \in I_2$) it follows that $y \in \bigcap_{i \in I_1} A_{\alpha_i}$; similarly $y \in \bigcap_{i \in I_2} A_{\alpha_i} \Rightarrow y \in \bigcap_{i=1}^{m} A_i$. The result of the theorem now follows from a lemma on centred systems, Kolmogorov & Fomin [1981], according to which, if a system of compact subsets has the property that any finite number of them has a non-empty intersection, then the whole family has a non-empty intersection.

Let us say a few words about the theory of convex figures and polyhedra, one of the most ancient and most beautiful branches of geometry. It takes its history from the theory of *platonic solids* and *Cauchy's theorems on polyhedra.* The initial objects of the theory are compact convex sets, that is, $\operatorname{Comp}\operatorname{Co}(\mathbf{R}^n)$. Otherwise called *convex figures.*

We introduce a distance in $\operatorname{Comp}\operatorname{Co}(\mathbf{R}^n)$ by putting $h(A_1, A_2) = \inf\{\varepsilon \geqslant 0: A_1 + \varepsilon B \supset A_2, A_2 + \varepsilon B \supset A_1\}$, where B is the unit sphere in $\mathbf{R}^n$, $B := \{x \in \mathbf{R}^n: \langle x, x\rangle \leqslant 1\}$. The quantity $h(A_1, A_2)$ is called the *Hausdorff distance* between A_1 and A_2. We note further that the set $\operatorname{Comp}\operatorname{Co}(\mathbf{R}^n)$ itself is a "convex cone", since convex sets can be added and multiplied by non-negative numbers:[3]

$$A_1 + A_2 = \{x: x = x_1 + x_2, x_1 \in A_1, x_2 \in A_2\}, \quad \alpha A = \{x: x = \alpha y, y \in A\}.$$

There is the important

Theorem 16 (Blaschke, on compactness). *The set* $\operatorname{Comp}\operatorname{Co}(\mathbf{R}^n)$ *is a locally compact cone (relative to the topology induced by the Hausdorff metric).*

In many cases this theorem allows us to prove existence theorems (see, for example, Chap. 3, § 2.2).

By Minkowski's theorem $A \in \operatorname{Comp}\operatorname{Co}(\mathbf{R}^n) \Rightarrow A = \operatorname{co}\operatorname{ext} A$. If $\operatorname{ext} A$ is a finite set then A is called a *convex polyhedron.* We denote the collection of convex polyhedra by $\mathfrak{M}(\mathbf{R}^n)$. We have

---

[3] This cone is naturally embedded in the linear "space of convex sets" in the same way as a semigroup is embedded in a group, etc.

Theorem 17 (on approximation). *Let $A \in \text{Comp Co}(\mathbf{R}^n)$ and $\varepsilon > 0$. Then there are $A', A'' \in \mathfrak{M}(\mathbf{R}^n)$ such that $A' \subset A \subset A''$ and $h(A', A'') < \varepsilon$, that is, a compact convex set can be arbitrarily closely approximated by polyhedra.*

This theorem allows many facts of the geometry of polyhedra to be translated into convex surfaces.

The study of convex polyhedra began in ancient times. Book XIII of Euclid's "Elements" is devoted to the five regular convex polyhedra (the platonic solids): the regular tetrahedron, cube, octahedron, dodecahedron and icosahedron. In $\mathbf{R}^4$ there are 6 regular convex polyhedra, and in $\mathbf{R}^n$, for $n \geqslant 5$ there are only three: the regular simplex, cube and octahedron.

An important role in various questions in the natural sciences is played by the regular convex polyhedra which regular decompose space (see Fedorov, 1949, where there is an elucidation of the history and the following results).

A figure, formed by the intersection of a finite number of half-spaces, is called *polyhedral*. We denote the collection of such figures by $\mathscr{P}(\mathbf{R}^n)$.

Theorem 18 (polyhedral criterion). *$P \in \mathscr{P}(\mathbf{R}^n)$ if and only if P is finitely generate, that is, there are $\xi_i \in \mathbf{R}^n$, $i = 1, \ldots, m$, such that for any $x \in P$ there are numbers $\lambda_i \geqslant 0$, $i = 1, \ldots, m$, such that $x = \sum_{i=1}^{k} \lambda_i \xi_i + \sum_{i=k+1}^{m} \lambda_i \xi_i$, $\sum_{i=1}^{k} \lambda_i = 1$.*

Hence, in particular, it follows that a bounded polyhedral set is a convex polyhedron, and a finitely generated cone $\{x \colon x = \sum_{i=1}^{k} \lambda_i \xi_i, \lambda_i \geqslant 0\}$ is closed, since it is polyhedral.

For further details see Rockafellar [1970].

The first results, relating to the theory of convex surfaces were obtained in the nineteenth century.

Theorem 19 (Cauchy, on polyhedra). *Two convex polyhedra, formed in the same way from correspondingly congruent faces, are congruent.*

In other words, convex polyhedra are rigid. For non-convex polyhedra this is not so: Fig. 15 depicts an example of two different polyhedra formed in the same way from congruent faces.

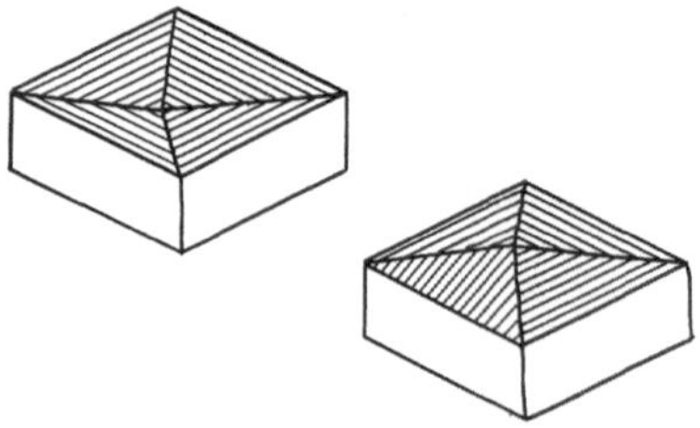

Fig. 15

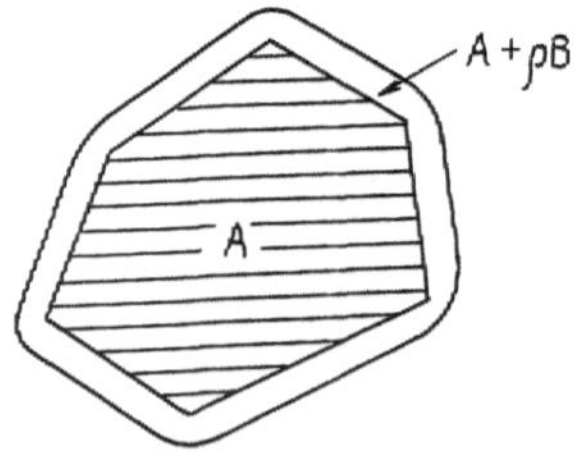

Fig. 16

At the beginning of this century Minkowski proved another theorem on convex polyhedra.

Theorem 20 (Minkowski, on polyhedra). *Let m unit vectors* $\omega_1, \ldots, \omega_m$ *and m positive numbers* S_i *be connected by the relation* $\sum_{i=1}^m S_i \omega_i = 0$. *Then there is a unique convex polyhedron for which the* ω_i *are the normals and the* S_i *the areas of the corresponding faces.*

The theorems of Cauchy and Minkowski have reached a definitive conclusion in the works of the Soviet mathematicians A.D. Alexandrov and A.V. Pogorelov (see Aleksandrov [1950]; Pogorelov [1952, 1969].

After Cauchy the research of Brunn and Steiner played an outstanding role in the geometry of convex sets. Let $V(A)$ denote the volume of a bounded convex set $A \subset \mathbf{R}^n$.

Theorem 21 (Brunn-Minkowski, Hadwiger [1957], Leichtweiss [1980]). *If* A_0 *and* A_1 *are convex bodies in* $\mathbf{R}^n$, $A_\alpha = \alpha A_0 + (1-\alpha)A_1$, $0 \leqslant \alpha \leqslant 1$, $v(\alpha) = (V(A_\alpha))^{1/n}$, *then*

$$v(\alpha) \geqslant \alpha v(0) + (1-\alpha)v(1),$$

and equality holds if and only if A_0 *and* A_1 *are homothetic.*

Let $A_i \in \operatorname{Comp}\operatorname{Co}(\mathbf{R}^n)$, $i = 1, \ldots, m$. It is easy to show that the volume $V(\sum_{i=1}^m \alpha_i A_i)$ can be written in the form

$$\sum_{i_1=1}^{m} \cdots \sum_{i_n=1}^{m} V(A_{i_1}, \ldots, A_{i_n}) \alpha_{i_1} \ldots \alpha_{i_n}.$$

The coefficients $V(A_{i_1}, \ldots, A_{i_n})$ are called the *mixed volumes*.

Let A be a convex body, $A_\rho = A + \rho B$, $\rho \geqslant 0$.

Theorem 22 (Steiner). *The function* $\rho \to V(A_\rho)$ *is a polynomial*:

$$V(A_\rho) = \sum_{k=0}^{n} W_k(A) \rho^k.$$

$W_k(A) = V(\underbrace{A, \ldots, A}_{n-k}, B, \ldots, B)$, where $B = \{x \in \mathbf{R}^n : |x| \leqslant 1\}$. In the plane, when $n = 2$, $W_0(A)$ is the area of A, $W_1(A)$ is its length and $W_2(A) = \pi$ is the area of the unit circle (Fig. 16).

On the geometry of convex figures and surfaces see the monographs Aleksandrov [1950]; Berger M. [1977]; Bonnesen & Fenchel [1934]; Busemann [1956] and others.

4. Convex Calculus. In the introduction we have already used the terminology "convex calculus". It is not usual, but appears to us to be apt. Firstly, the combination of words reminds us of the differential calculus, and in fact there are fruitful analogies in the treatment of differentiable and convex functions.

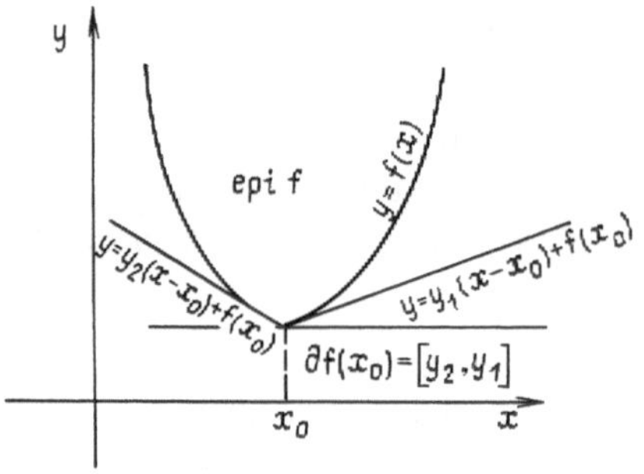

Fig. 17

Secondly, the very word "calculus", in its original meaning, fits very well into what we have to say, since the question will be as to how one calculates with the operators of convex analysis (polar, adjoint, etc.) in combination with various operations over convex sets (say, how "to calculate" the polar of the sum of two sets).

First let us give the definition of a recent important idea of convex analysis—the concept of the subdifferential of a function at a point. The *subdifferential of a function f at a point x_0* is the set

$$\partial f(x_0) = \{y: f(x) - f(x_0) \geqslant \langle y, x - x_0 \rangle\}.$$

For a function of one variable this is the collection of angular coefficients y for which the lines $f(x_0) + y(x - x_0)$ lie under the graph of f (Fig. 17).

The subdifferential is a concept which generalizes that of a derivative for convex functions: if f is differentiable at x_0, then $\partial f(x_0) = f'(x_0)$. There is one further important resemblance. The notion of differential for a smooth function is related to the approximation of $f(x) - f(x_0)$ in a neighbourhood of x_0 by a linear function. For a convex function the function itself is approximable. This takes the form:

$$f'(x_0, x) = \lim_{\alpha \downarrow 0} (f(x_0 + \alpha x) - f(x_0)) \cdot \alpha^{-1}.$$

The function $f'(x_0, x)$ is convex and homogeneous in x. Its subdifferential, in the sense of subsection 2.2, coincides with $\partial f(x_0)$.

An essential part of differential calculus is the set of rules of the type "the differential of a sum", "the differential of a product", etc.

The same is true of convex analysis. The same circle of ideas is studied—how are the various operators of the type l, π, σ, ∂ and s and the operations which take convex objects to convex objects related to each other. We quote two formulae of convex calculus:

$$\partial(f_1 + f_2) = \partial f_1 + \partial f_2, \qquad l(f_1 + f_2) = lf_1 \oplus lf_2.$$

These formulae are valid under certain hypotheses known as "*general position*" (regarding this see Chap. 4). The first of these (called the *Moreau-Rockafellar formula*) is concerned with "*subdifferential calculus*". It is analogous to the formula of differential calculus on the derivative of a sum and requires no explanation: the subdifferential of the sum of two functions at a point x is the sum of the two convex sets $\partial f_1(x)$ and $\partial f_2(x)$. In the second formula there is a new operation:

$$(\varphi_1 \oplus \varphi_2)(x) = \inf\{\varphi_1(x) + \varphi_2(x): x_1 + x_2 = x\},$$

called *infimal convolution*. Combining the various operations and operators, we obtain an extensive collection of formulae which are given in a table in Chap. 2.

From Hörmander's theorem, a convex function with a closed epigraph is nothing but the least upper bound of a family of affine functions. Thus often it is sufficient to be able to understand a single function—the maximum of several or the least upper bound of a whole family of convex functions. More will be said about this in Chap. 4, but here we will give an important formula of subdifferential calculus regarding the subdifferential of a maximum. Denote by $(f_1 \vee f_2)(x)$ the function $\max(f_1(x), f_2(x))$—the maximum of the two functions f_1 and f_2 (there is no such operation in smooth analysis). The following *formula* due to *Dubovitskij-Milyutin* holds (again under the hypothesis of general position): for $f_1(x) = f_2(x)$

$$\partial(f_1 \vee f_2)(x) = \operatorname{co}(\partial f_1(x) \cup \partial f_2(x)).$$

This is true, in particular, if we assume that both functions are continuous and convex at x.

The Dubovitskij-Milyutin formula has far reaching generalizations, which form the basis for the whole of the calculus of convex functions. One such is, for example, the "*refinement theorem*" (see Chap. 2). This is the name given to the formula in subdifferential calculus for the subdifferential of the function $f(x) = \max_{t \in T} f(t, x)$, where $f(t, \cdot)$ is a family of convex functions depending on a parameter t.

In the formulae used in the calculus of convex sets a fundamental rôle is played by Minkowski's theorem, proved above, on the description of a convex set as an intersection of half-spaces. §§ 2–5 of Chap. 2 are devoted to convex calculus.

Recently an attempt has been made to unite differential and convex calculus and to lay the foundation for non-smooth analysis. We will speak on this in Chap. 4.

We propose the question: where does one meet convexity? Where is convex calculus applied?

Let us have no illusions: convex analysis occupies, and will occupy, a rather modest place in analysis. The object of this study is a comparatively narrow class of functions, sets and extremal problems. However, convexity does arise naturally in very different areas of the natural sciences, engineering and economics. The attractive peculiarity of the theory of convexity for mathematicians is the aesthetic element in this theory of much beauty.

Let us try here to outline the circle of applications of convex analysis.

Convexity in mathematics. It was mentioned above that geometry is the cradle of convex analysis: in the first half of the nineteenth century the foundations of the theory of convex surfaces and convex polyhedra were laid. Many of the extremal problems of geometry were convex (the most ancient being the classical isoperimetric problem). One of the most important examples of an extremal geometric problem is the problem of the distance of a point from a subspace. But this is related to the theory of approximations. This means that the methods of convex analysis find very broad application in this science. We note that the first works of Minkowski on convex geometry arose from the requirements of number theory. It must also be said that convex analysis is one of the fundamental scientific directions which led to the creation of functional analysis.

Later we will attempt to show that convexity is present in all the problems of classical variational calculus and optimal control. Thus, thanks to this genetic connection of these disciplines with convexity, many conditions in variational calculus and optimal control take the form of maximum principles (the Weierstrass necessary condition, the Pontryagin maximum principle, etc.). One could even advance the thesis that: the theory of variational problems and optimal control is the union of smooth and convex analysis.

To a large extent the role of convexity in the theory of differential equations follows from here. We can also mention duality in the theory of characteristics. For example, the theory of elliptic problems is very closely bound up with the ideas of convexity Courant & Hilbert [1962]. Many applications of convex analysis can be found in the theory of functions and complex analysis. In particular, the theory of embeddings of function spaces is interlaced with the ideas of convexity. This is because to a large extent it is based upon interpolation theorems, which are sometimes called convexity theorems. Recently convex geometry, and particularly the geometry of convex polyhedra, has found numerous applications in algebra, singularity theory and differential equations theory (Newton polyhedra). Convexity plays a very large rôle in functional analysis.

Convexity in the natural sciences. The Legendre transform finds application in very diverse areas of the natural sciences. In classical mechanics the Legendre transform is connected with the Hamiltonian and the Lagrangian, which, ultimately, connect Lagrangian and Hamiltonian mechanics.

In thermodynamics the Legendre transform provides a passage from some functions of states to others, for example, from specific volume and entropy to temperature and pressure.

Every time that the Legendre transform is applied, it is possible to say with some confidence that the corresponding problem contains, perhaps hidden, convexity. Therefore we can see the versatility of the applications of the conception of convexity in classical and statistical mechanics.

For the same reasons we can see the connections with convex analysis, which arise in the theory of plasticity and other areas of the natural sciences.

Convexity in economics. The role of convexity in modern mathematical economics is huge. This area has for a long time been one of the basic providers of new problems and ideas in convex analysis. We will speak later on the rôle of convexity in economic problems, in illustrating the problems of convex programming.

Convexity in engineering arises fundamentally through extremal problems.

Here we have sketched some of the outline of the theme, which illuminates this work. Now it seems appropriate to speak on the basic historical landmarks, connected with the creation and development of convex analysis.

5. A Brief Historical Sketch. The two principal branches of convex analysis—geometric and analytic—arose at different times.

We have already mentioned above the origin of the idea of convexity and of the concept of convex geometry.

The Legendre transform in embryo was already available to Leibniz. In approximately 1673 he discovered the so called "theorem of transmutation", in the proof of which he made the transition from a function f to the function $x \to xf'(x) - f(x) = f^*(f'(x))$. In fact this transform was found by Euler in 1776, although it was explicitly defined by Legendre in 1789. In the nineteenth century this idea was very widely used in classical mechanics and variational calculus, although the theory of convex functions was not yet properly created. Certain very preliminary concepts and facts, concerning convex functions, were introduced and studied at the end of the nineteenth and the beginning of the twentieth centuries (Stolz [1893], Jensen [1906] and others). The foundation of the theory of convex functions in its modern form was laid only comparatively recently—on the boundary of the forties and fifties in this century in the work of Fenchel [1949, 1951].

After these papers the definitions of the basic concepts of the theory were formulated—the Legendre-Young-Fenchel transform and subdifferentials. A subdifferential, as was said before, has a clear connection with the derivative. After this had been introduced, a "subdifferential calculus" was created, similar to the differential calculus but adapted for convex functions. The theory of convex functions then took its place as a specific division of mathematical analysis.

The theory of convex sets arose in some sense both earlier and later than the theory of convex functions. Its foundation was constructed at the end of the nineteenth century by Minkowski. Minkowski immediately gave the theory a fairly definitive form and took it approximately to the level, to which the theory of convex functions was lifted after the work of Fenchel.

Minkowski's contribution to the theory of convexity was exceptional. He introduced the most important ideas—the polar, the support function, an extreme point—proved a number of fundamental results, concerning the description of convex sets (in fact, he discovered the phenomenon of duality for convex objects), proved a deep and fundamental theorem on polyhedral, see Leichtweiss [1980], and generalized the Brunn inequality.

The beginning of this century was celebrated by the stormy development of the geometry of convex sets. Of the results obtained then, we recall Helly's theorem, Helly [1923], Radon [1921], playing a very major role in what followed, and the theorems of Carathéodory [1907], Radon [1921] and others.

At the same time there arose a new direction in convex geometry, where extremal problems in the collection of convex sets were solved. We mention the problems, solved by Young [1901], Blaschke [1914] and Bieberbach [1915]. It was then that the discussion of the properties of a "cone of convex sets" was initiated and, in particular, its local compactness was proved (Blaschke, see Leichtweiss [1908]). The results of the early period of the theory of convex sets were given in the well-known monograph of Bonnesen and Fenchel [1934].

The beginning of this century saw the birth of functional analysis, in which we can find the completion of many of the concepts of finite-dimensional convex geometry. Minkowski's theorem on separation was a predecessor of theorems of Hahn-Banach type and general results on separation in the theory of linear topological spaces, his theorem on the representation of the points of a polyhedron via the extreme points was extended to the infinite-dimensional case by M.G. Krejn and D.P. Mil'man [1940]. These and many other facts of convex geometry, in the infinite-dimensional version, became in powerful tools in all of modern analysis. Duality theory for linear spaces (Chapter I) was constructed in the forties.

In the thirties and forties there were several important events concerning convexity. F. Riesz [1928], and also L.V. Kantorovich [1935] and their followers had created the theory of ordered spaces. Such a space, above all, was adapted for various generalizations of the concept of convexity. In fact the primitive concepts—of convex figure and convex function—were most closely related to the fact that the number line has an "order", that is, it is possible to say which number is smaller and which greater than the other, or whether a given number lies between two others. To be exact, it is the order on the number line which allows the definition of the concept of segment, chord and, by the same token, of convex set and convex function. The ordered spaces are precisely those linear spaces which are like the number line in relation to order.

Subsequently the concepts of convex analysis found completion in the theory of ordered spaces (see Chapter 4).

Here we must name M.G. Krejn and his followers, who developed the theory of convex cones and also the theory of duality (in particular, in relation to the problem of moments).

In addition, at the end of the thirties, the attention of mathematicians turned to certain specific problems, which arose in the formalization of a number of concrete problems of economics. It became clear that many such problems reduced to the finding of extrema of linear functions under linear constraints.

These problems were first investigated by L.V. Kantorovich [1939]. Then, in the forties, a new direction in the theory of extremal problems was formalized, which became known as linear programming.

Yet another event of this same period. A.A. Lyapunov [1940] proved that the set of values of a continuous vector measure was convex and closed. This work then remained practically unnoticed. Its role in convex analysis and the theory of extremal problems came to be understood later. It must be remarked that even in the nineteenth century, in the works of Weierstrass on the formulation of necessary conditions, there appeared relations very reminiscent of the inequality which characterises convex functions (the Weierstrass necessary condition). At that time the relation between these conditions and convexity was not totally clear. It became clear when N.N. Bogolyubov [1969] proved that from the theoretical point of view one need only consider those variational problems, for which the Weierstrass necessary condition was satisfied everywhere, that is, those, for which the integrand was convex with respect to derivatives. It turned out, that convexity was intimately connected with the semi-continuity from below of the functionals of classical variational calculus.

The creation of the theory of optimal control influenced the formulation of convex analysis in a very essential way. The maximum principle of L.S. Pontryagin [1976]—the most important tool of optimal control—was discovered at the crossroads of classical and convex analysis (see Chap. 3).

This all served as a precursor to the fact that in the fifties and sixties convex analysis acquired a completely distinct outline.

In the new discipline all the aspects listed above were united: the geometric, the analytic and the extremal.

The works of John [1948], Kuhn-Tucker [1951], and Slater laid the foundation of nonlinear and convex programming. In the above mentioned works of Fenchel and his followers and then in Hörmander [1955]; Moreau [1962]; A. Ya. Dubovitskij and A.A. Milyutin [1965], Rockafellar, Brønstead [1964]; B.N. Pshenichnyj [1980, 1982]; E.G. Gol'shtein [1971]; A.D. Ioffe and V.M. Tikhomirov [1968a, 1968b, 1969, 1974], V.L. Levin [1969, 1972, 1975, 1985] and others, the calculus of convex functions and sets in finite- and infinite-dimensional spaces was created. The results of this period have been given in the monograph of Rockafellar [1970]. The name for the new division (more precisely for Rockafellar's book) was given by Tucker. In the forties and fifties geometry was also intensively developed. In particular we must mention the classical results of A.D. Aleksandrov [1958] and A.V. Pogorelov [1952, 1969]. In these same years we also see the development of Choquet theory, originating from Minkowski's theorems on extreme points, Choquet [1962], Phelps [1978].

Let us touch briefly on later events. In the sixties the close connection between convexity and variational calculus and optimal control was revealed and the general theory of integral functionals in finite-dimensional spaces was created. In this connection the relationship between the Weierstrass necessary condition, the theorems of N.N. Bogolyubov and the theorems of A.A. Lyapunov (see Chap.

3) was clarified. In our day convex analysis in spaces of measurable vector-valued functions has been created (see § 5, Chapter 2).

In the seventies numerous attempts have been made to create a single smooth-convex calculus, that is, ultimately a calculus of non-smooth functions. The pioneering work of Clarke [1975] found many followers and served as a starting point for the creation of a new direction—non-smooth analysis. At the same time there has been intensive research into the relations between convexity and smoothness, the starting point for which was the work of Asplund [1968].

In the seventies and eighties convex calculus was extended to ordered linear spaces (V.L. Levin, Valadier, Castaing, A.G. Kusraev, S.S. Kutateladze). Noticeable progress has been made in Choquet theory and the axiomatic theory of convexity. We will deal with these problems in later chapters.

List of Notations

$\forall$	universal quantifier "for all"
$\exists$	existential quantifier "there exists"
$\Rightarrow$	implication sign "form... follows"
$\Leftrightarrow$	equivalence sign
$=:$ or $\stackrel{\text{def}}{=}$	equal by definition
$\stackrel{(i)}{=}$ ($\stackrel{(i)}{\geqslant}$ etc.)	equal (greater or equal etc.) by (i)
$\{x: P(x)\}$	set of elements having the property P
$\varnothing$	empty set
$x \in A$ ($x \notin A$)	x belongs (does not belong) to the set A
$A \cup B$	union of the sets A and B
$A \cap B$	intersection of the sets A and B
$\{x_1, \ldots, x_n, \ldots\}$	set comprising the elements $x_1, \ldots, x_n, \ldots,$
$\{A_\alpha\}_{\alpha \in \mathscr{A}}$	a family of sets parametrised by a set $\mathscr{A}$
$\Lambda^{-1}y$	inverse image of y under the mapping Λ
ΛA	image of the set A under the mapping Λ
$\mathbf{R}$	set of real numbers, the number line
$\bar{\mathbf{R}} = \mathbf{R} \cup \{\pm\infty\}$	extended number line
$\mathbf{R}^n$	arithmetic n-dimensional space (usually provided with Euclidean structure), $\mathbf{R}^n_+ = \{x = (x_1, \ldots, x_n) \in R^n: x_i \geqslant 0\}$
$\lvert x \rvert$	Euclidean norm in $\mathbf{R}^n$, $S^{n-1} = \{x \in \mathbf{R}^n: \lvert x \rvert = 1\}$
$h(A_1, A_2)$	Hausdorff distance from A_1 to A_2
$\mathrm{Aff}(X)$	collection of all affine manifolds in X
$\mathrm{Cl}(X)$	collection of all closed sets in X
$\mathrm{Co}(X, \mathbf{R})$ ($\mathrm{Co}(X, \bar{\mathbf{R}})$)	collection of all convex finite (with values in $\bar{\mathbf{R}}$) functions on X
$\mathrm{Co}(X)$	collection of all convex sets in X
$\mathrm{Co}(X)_0$	collection of all convex sets in X containing zero
$\mathrm{Comp}(X)$	collection of all compact sets in X
$\mathrm{Comp\,Co}(X)$	collection of all convex compact sets in X
$\mathrm{Cone\,Co}(X)$	collection of all convex cones in X

$\operatorname{Lin}(X)$	collection of all subspaces on X
$\mathscr{L}(X, Y)$	set of all linear continuous mappings of X to Y
$S\mathscr{L}(X)$	set of all sublinear functions on X
$\operatorname{Cl}\operatorname{Co}(X)$ ($\operatorname{Cl} S\mathscr{L}(X)$ etc.)	set of closed convex sets (closed sublinear functions etc.) in X
$\operatorname{aff} A$	affine hull of the set A
$\operatorname{cl} A$	closure of the set A
$\operatorname{co} A$	convex hull of the set A
$\operatorname{cone} A$	convex conical hull of the set A
$\operatorname{lin} A$	linear hull of the set A
$\operatorname{cl}\operatorname{co} A$	convex closure of the set A
$\operatorname{ext} A$	set of extreme points of the set A
$\operatorname{int} A$	interior of the set A
$\operatorname{ri} A$	relative interior of the set A
$\operatorname{cl} f$	closure of the function f
$\operatorname{dom} f$	effective set of the function f
$\operatorname{epi} f$	epigraph of the function f
$+$	operation of sum for sets and functions
$\oplus$	infimal convolution
$\vee$	operation of maximum for functions
$\operatorname{co} \wedge$	convex hull minimum
fA	inverse image of the function f under a linear mapping A
Af	image of the function f under a linear mapping A
$\boxplus$	convolution (Helly sum) of sets
$\operatorname{co} \cup$	convex hull union
$\oint$	convolution integral
δA	indicator function of a set A
μA	Minkowski function of the A
sA	support function of the set A
ρA	function of distance to the set A
$X^{\#}(X')$	algebraic (topological) adjoint space of X
X^*	adjoint of a normed space
$x^{\#}(x', x^*)$	elements of $X^{\#}(X', X^*)$
$X \leftarrow d \rightarrow Y$	spaces X and Y which are in duality
$\langle \cdot, \cdot \rangle$	canonical bilinear form
$\sigma(X, Y), \sigma(Y, X)$	weak topologies in spaces in duality
lf or f^*	Young or Young-Fenchel or Legendre-Young-Fenchel transform of f
σK or K^*	conjugate cone of K
P	subdifferential of the sublinear function p
$f(x)$	subdifferential of the function f at the point x
πA or A^0	polar of the set A
$L^{\perp}$	annihilator of the subspace L

$\pi^2 A(l^2 f, \sigma^2 K, L^{\perp\perp}$ etc.$) = \pi(\pi A)\ (l(lf), \sigma(\sigma K), (L^{\perp})^{\perp}$ etc.$)$

$\mu A(\delta A$ etc.$) = (\mu A)\ ((\delta A)$ etc.$)$

$f(x) \rightarrow \inf(\sup), x \in C,$	problem of finding the greatest lower (least upper) bound of the functional f under the constraint C
$f'(x_0, x)$	directional derivative
$\partial_C f(x)$	Clarke subdifferential
$\partial_A f(x)$	approximative subdifferential

Abbreviations:

l.s.	linear space
n.s.	normed space
l.c.s.	locally convex space

Chapter 1
The Basic Ideas of Convex Analysis

In this chapter we concentrate the information needed in the construction of convex analysis.

In the first section we introduce the most important concepts; in the second, spaces in duality and duality operators. In the third section we discuss the topological properties of convex objects; in the fourth, the fundamental theorems of functional analysis, needed in the later chapter.

§ 1. Convex Sets and Functions

In what follows we will only consider linear spaces over the field of real numbers and this situation will not be expressly specified.

In this section we give the definitions of the fundamental objects of convex analysis.

1.1. Subspaces, Convex Cones, Affine Manifolds, Convex Sets. Let X be a linear ($\Leftrightarrow$ vector) space, $A \subset X$ a set in X. We denote by l the line in the plane $\{(\alpha_1, \alpha_2) \in \mathbf{R}^2 : \alpha_1 + \alpha_2 = 1\}$, and by Δ the segment $\{(\alpha_1, \alpha_2) \in \mathbf{R}^2 : \alpha_1 \geqslant 0, \alpha_2 \geqslant 0, \alpha_1 + \alpha_2 = 1\}$.

A set A is called:

—*a subspace*, if together with any two of its points x_1, x_2 it contains all their *linear combinations*, that is, all expressions of the form $\alpha_1 x_1 + \alpha_2 x_2$, $\alpha_1, \alpha_1 \in \mathbf{R}$

$$\Leftrightarrow \{A : \alpha_1 A + \alpha_2 A \subset A, \forall (\alpha_1, \alpha_2) \in \mathbf{R}^2\};$$

—*a convex cone*, if with any two of its points x_1, x_2 it contains all their *conical combinations*, that is, expressions of the form $\alpha_1 x_1 + \alpha_2 x_2$, $\alpha_1, \alpha_2 \in \mathbf{R}_+$

$$\Leftrightarrow \{A : \alpha_1 A + \alpha_2 A \subset A \;\; \forall (\alpha_1, \alpha_2) \in \mathbf{R}^2_+\};$$

—*an affine manifold*, if together with any two of its points x_1, x_2 it contains all *affine combinations*, that is, expressions of the form $\alpha_1 x_1 + \alpha_2 x_2$, $\alpha_1, \alpha_2 \in \mathbf{R}$, $\alpha_1 + \alpha_2 = 1$,

$$\Leftrightarrow \{A : \alpha_1 A + \alpha_2 A \subset A \;\; \forall (\alpha_1, \alpha_2) \in l\};$$

—*a convex set*, if together with any two of its points x_1, x_2 it contains all *convex combinations*, that is, expressions of the form $\alpha_1 x_1 + \alpha_2 x_2$, $\alpha_1, \alpha_2 \in \mathbf{R}_+$, $\alpha_1 + \alpha_2 = 1$

$$\Leftrightarrow \{A : \alpha_1 A + \alpha_2 A \subset A \;\; \forall (\alpha_1, \alpha_2) \in \Delta\};$$

We have said already that a *cone* is a set which is invariant under homotheties, that is, A is a cone if $x \in A \Rightarrow \alpha x \in A\ \forall \alpha \geqslant 0$. Therefore, convex cones are those cones which are convex sets.

To put it another way, subspaces (of dimension $\geqslant 2$) are subsets which together with any two of their points contain the two-dimensional plane generated by them, cones contain the planar angle on whose sides the given points lie, affine manifolds contain the line passing through the points, and convex sets contain the segment joining them. Clearly all these sets are convex.

These definitions can be amalgamated and this gives us the possibility of using the same scheme to describe other objects that we shall need. Let $\mathscr{T} \subset \mathbf{R}^2$ be some subset of pairs (α_1, α_2), $\alpha_1, \alpha_2 \in \mathbf{R}$. We call $A \subset X$ a *$\mathscr{T}$-set*, and write $A \in \mathscr{T}(X)$, if $\alpha_1 A + \alpha_2 A \subset A\ \forall (\alpha_1, \alpha_2) \in \mathscr{T}$.

Thus, subspaces are $\mathbf{R}^2(X)$-sets, convex cones are $\mathbf{R}^2_+(X)$-sets, affine manifolds and convex sets are respectively in $l(X)$ and $\Delta(X)$ and, for example, *absolutely convex* sets are in $\mathscr{T}(X)$, where $\mathscr{T} = \{(\alpha_1, \alpha_2) \in \mathbf{R}^2 \colon |\alpha_1| + |\alpha_2| \leqslant 1\}$, etc.

The collections introduced above, in particular the collections of convex sets and convex cones, are the most important geometric objects of convex analysis. We will denote the collection of convex sets in X by $\mathrm{Co}(X)$, the convex cones by $\mathrm{Cone\,Co}(X)$, the convex sets containing zero by $\mathrm{Co}(X)_0$, the affine manifolds by $\mathrm{Aff}(X)$ and the subspaces by $\mathrm{Lin}(X)$.

We give some examples of convex sets and convex cones in $\mathbf{R}^n$ to which we will repeatedly return.

Included in $\mathrm{Co}(\mathbf{R}^n)$ are:
the *p-sphere*

$$B_p^n(a) = \left\{x \in \mathbf{R}^n \colon \left(\sum_{i=1}^n |x_i/a_i|^p\right)^{1/p} \leqslant 1\right\}, \qquad 1 \leqslant p < \infty,$$

$$a = (a_1, \dots, a_n), \qquad a_i \geqslant 0,$$

if $a = (1, \dots, 1)$ then this sphere is denoted simply by B_p^n, and if $p = 2$ then by B^n or B;
the *parallelopiped*:

$$B_\infty^n(a) = \{x \in \mathbf{R}^n \colon |x_i| \leqslant a_i,\ 1 \leqslant i \leqslant n\};$$

the *octahedron*:

$$B_1^n(a) = \left\{x \in \mathbf{R}^n \colon \sum_{i=1}^n |x_i/a_i| \leqslant 1\right\} \qquad \text{(Fig. 18);}$$

Amongst the convex cones in $\mathbf{R}^n$ we mention the *right circular cones*:

$$K_a(\alpha) = \{x \in \mathbf{R}^n \colon \langle x, a\rangle \geqslant |x| \cos\alpha\}, \qquad 0 < \alpha < \pi/2,$$

where a is a vector in $\mathbf{R}^n$ of unit length. Let us give several more examples of the description of finite-dimensional convex sets and convex cones.

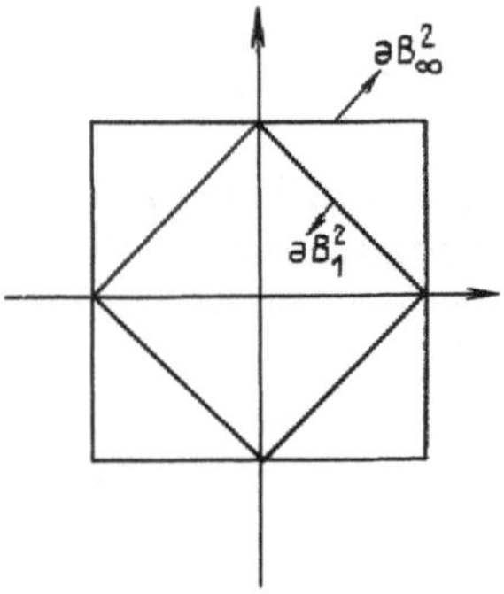

Fig. 18

Subspaces in $\mathbf{R}^n$ are described by a homogeneous system of equations:

$$L = \left\{x \in \mathbf{R}^n \colon \sum_{j=1}^{n} a_{ij}x_j = 0, i = 1, \dots, s\right\} = \operatorname{Ker} \Lambda,$$

$\Lambda \in \mathscr{L}(\mathbf{R}^n, \mathbf{R}^s)$. If $\operatorname{rank}(a_{ij}) = k$, then the dimension of L is equal to $n - k$ (that is, $\dim L = n - \operatorname{rank} \Lambda$).

Affine manifolds are described by a system of inhomogeneous equations:

$$M = \left\{x \in \mathbf{R}^n \colon \sum_{j=1}^{n} a_{ij}x_j = y_i, i = 1, \dots, s\right\} = \Lambda^{-1}(y),$$

where $\Lambda \in \mathscr{L}(\mathbf{R}^n, \mathbf{R}^s)$, $y \in \mathbf{R}^s$, $\Lambda^{-1}y$ is the inverse image of y. An affine manifold is a shift of a subspace. The affine manifolds of the form $\Gamma(a, \beta) = \{x \in \mathbf{R}^n \colon \sum_{i=1}^{n} a_i x_i = \beta\}$, $a \neq 0$, are called *hyperplanes*. They have dimension $n - 1$.

Sets of the form

$$\Pi_+(a, \beta) = \left\{x \in \mathbf{R}^n \colon \sum_{i=1}^{n} a_i x_i \geqslant \beta\right\},$$

$$\Pi_-(a, \beta) = \left\{x \in \mathbf{R}^n \colon \sum_{i=1}^{n} a_i x_i \leqslant \beta\right\},$$

are called *half-spaces*. Their boundary is the hyperplane $\Gamma(a, \beta)$.

We remark that the set $\operatorname{Co}(X)$ is itself a convex cone relative to the operations of sum and multiplication by scalars introduced above.

1.2. Linear, Convex Conical, Affine and Convex Hulls. Again let A be a subset of X, $\mathscr{T} \subset \mathbf{R}^2$. The intersection of all the $\mathscr{T}$-sets containing A is called the $\mathscr{T}$*-hull* of A and denoted $H_{\mathscr{T}}(A)$. For the four cases introduced above the $\mathscr{T}$-hulls have special names and notations:

$H_{\mathbf{R}_2}(A)$ is called the *linear hull* and denoted $\operatorname{lin} A$;
$H_{\mathbf{R}^2_+}(A)$ is called the *convex conical hull* and denoted $\operatorname{cone} A$;
$H_l(A)$ is called the *affine hull* and denoted $\operatorname{aff} A$;
$H_\Delta(A)$ is called the *convex hull* and denoted $\operatorname{co} A$.

Points $\{x_1, \dots, x_k\}$ are called *linearly (affinely) independent* if no one of them is a linear (affine) combination of the remainder.

Here:

$$\operatorname{lin} A = \bigcap\{L\colon L \in \operatorname{Lin}(X), A \subset L\},$$

$$\operatorname{aff} A = \bigcap\{M\colon M \in \operatorname{Aff}(X), A \subset M\},$$

$$\operatorname{cone} A = \bigcap\{K\colon K \in \operatorname{Cone}(X), A \subset K\},$$

$$\operatorname{co} A = \bigcap\{B\colon B \in \operatorname{Co}(X), A \subset B\}.$$

The last relation means that the convex hull of A is the intersection of all convex sets containing A; the rest are interpreted in a similar way.

It is easy to demonstrate, that a) $H_{\mathscr{T}}(A)$ is the smallest $\mathscr{T}$-set containing A, b) if $A_1 \subset A_2$, then $H_{\mathscr{T}}(A_1) \subset H_{\mathscr{T}}(A_2)$, c) $A \in \mathscr{T}(X) \Rightarrow H_{\mathscr{T}}(A) = A$.

Let us write down the $\mathscr{T}$-hulls of finite sets (for the $\mathscr{T}$'s considered above). Let $A_0 = \{x_1, \dots, x_m\}$. Then

$$H_{\mathscr{T}}(A_0) = \left\{x\colon x = \sum_{i=1}^{m} \alpha_i x_i\right\},$$

where a) $\alpha_i \in R$ in the case $\mathscr{T} = \mathbf{R}^2$, b) $\alpha_i \geqslant 0$ in the case $\mathscr{T} = \mathbf{R}^2_+$ c) $\sum_{i=1}^n \alpha_i = 1$, if $\mathscr{T} = l$ and d) $\alpha_i \geqslant 0$, $\sum_{i=1}^n \alpha_i = 1$ in the case $\mathscr{T} = \Delta$ Here $H_{R^2}(A_0)$ is called the *subspace*, $H_{R^2_+}(A_0)$ the *convex polyhedral cone*, $H_l(A_0)$ the *affine manifold* generated by A_0, and $H_\Delta(A_0)$ is called an *$(m - 1)$-dimensional simplex* if the x_i are affinely independent, and a *convex polyhedron* in the general case.

The associated "$\mathscr{T}$-combinations" $\sum_{i=1}^m \alpha_i x_i$ are called *linear*, *conical*, *affine* and *convex* combinations. In subsection 1.1 these objects were introduced only for $m = 2$. It is easy to prove

Lemma 1. *The $\mathscr{T}$-hull of a set A is made up of the $\mathscr{T}$-hulls of the finite subsets of A.*

In other words:

$$H_{\mathscr{T}}(A) = \bigcup\{H_{\mathscr{T}}(A_0)\colon A_0 = \{x_1, \dots, x_m\}, x_i \in A\}, \quad m = 1, 2, \dots$$

1.3. Convex Functions. Again let X be a linear space. We will now consider a function $f\colon X \to \bar{\mathbf{R}}$, where $\bar{\mathbf{R}} = R \cup \{-\infty\} \cup \{+\infty\}$ is the *extended real line*. Associated with each function are two sets:

$$\operatorname{dom} f = \{x \in X\colon f(x) < +\infty\},$$

$$\operatorname{epi} f = \{(x, \alpha)\colon \alpha \geqslant f(x), x \in \operatorname{dom} f\}$$

called respectively the *effective set* and the *epigraph* of f.

A function f for which $\operatorname{dom} f \neq \emptyset$ and where $f(x) > -\infty$, is called *proper*, the remainder are called *improper*. If f is proper, then $\operatorname{epi} f \neq \emptyset$. A function f is called *convex* if $\operatorname{epi} f$ is a convex subset of $X \times \mathbf{R}$, and *convex homogeneous* if f is convex and positively homogeneous of degree one: $f(\alpha x) = \alpha f(x)\ \forall x \in X$, $\alpha > 0$. Convex homogeneous functions will, as a rule, be denoted by the letter p. They have been called *sublinear functions* (or functionals). The reason for this name was seen above: sublinear functions are the suprema of linear functions.

Convex functions and convex homogeneous functions form the most important function classes of convex analysis: we will them denote $\mathrm{Co}(X, \mathbf{R})$ and $\mathrm{S}\mathscr{L}(X, \mathbf{R})$ respectively.

Included in $\mathrm{Co}(\mathbf{R}, \mathbf{R})$ are the powers $|x|^\alpha$, $\alpha \geqslant 1$, the exponentials a^x, $a > 0$; the functions $f(x) = \sum_{i=1}^n a_i |x_i|^{p_i}/p_i$, $a_i \geqslant 0$, belong to $\mathrm{Co}(\mathbf{R}^n, \mathbf{R})$ for $p_i \geqslant 1$. Convex homogeneous functions on the line have a very simple structure (depending on two parameters a and b):

$$f(x; a, b) = \begin{cases} bx, & x \geqslant 0 \\ ax, & x \leqslant 0,\ a < b. \end{cases}$$

Amongst the functions from $\mathrm{S}\mathscr{L}(\mathbf{R}^n, \mathbf{R})$ we mention

$$N_p(x) = \left(\sum_{k=1}^n |x_k|^p\right)^{1/p}.$$

We give some other important examples.

A *linear function on* $\mathbf{R}^n$: $f(x) = \langle a, x\rangle =: \sum_{k=1}^n a_k x_k$ is convex homogeneous.

An *affine function* on R^n: $f(x) = \langle a, x\rangle + a_0 =: \sum_{k=1}^n a_k x_k + a_0$ is a convex function.

The *indicator function of a set* $A \subset X$ is defined by the relation:

$$\delta A(x) = \begin{cases} 0, & x \in A, \\ +\infty, & x \notin A. \end{cases}$$

It belongs to $\mathrm{Co}(X, \overline{\mathbf{R}})$ if and only if $A \in \mathrm{Co}(X)$, and belongs to $\mathrm{S}\mathscr{L}(X, \overline{\mathbf{R}})$ if and only if $A \in \mathrm{Cone}(X)$.

The *Minkowski function of a set* $A \subset X$ is defined as follows:

$$\mu A(x) = \begin{cases} 0, & \text{if } x = 0, \\ \inf\{t > 0: x/t \in A\}, & \\ \infty, & \text{if } x/t \notin A\ \forall t > 0,\ x \neq 0. \end{cases}$$

The Minkowski function has been called a *gauge* and denoted $\gamma(x|A)$ (Rockafellar, 1970). It is easy to see that if $A \in \mathrm{Co}(X)$, then $\mu A \in \mathrm{S}\mathscr{L}(X, \overline{\mathbf{R}})$.

Let $f \in \mathrm{Co}(X, R)$ and $f(x) > -\infty\ \forall z \in X$. It is not difficult to show that f is a convex function if and only if it satisfies *Jensen's inequality*, that is,

$$f(\alpha_1 x_1 + \alpha_2 x_2) \leqslant \alpha_1 f(x_1) + \alpha_2 f(x_2)$$

for any $x_1, x_2 \in X$, $\alpha_1 > 0$, $\alpha_2 > 0$, $\alpha_1 + \alpha_2 = 1$.

Moreover, the convex homogeneous functions can be described as the functions for which epi p is a convex cone.

Finally we establish an important connection between convex and convex homogeneous functions. Let f be a convex proper function, $x \in \operatorname{dom} f$ and for any $h \in X$ let there be an $\varepsilon(h) > 0$ for which $x + dh \in \operatorname{dom} f$, $|d| < \varepsilon(h)$, then we will write $x \in \operatorname{core}\operatorname{dom} f$.

Lemma 2. *Let $f \in \operatorname{Co}(X, R)$ and $x \in \operatorname{core}\operatorname{dom} f$. Then for every $h \in X$ the directional derivative*

$$f'(x, h) = \lim_{\alpha \downarrow 0} \alpha^{-1}(f(x + \alpha h) - f(x)),$$

exists and the mapping $h \to f'(x, h)$ is a convex homogeneous function.

The largest convex function not exceeding f is called its *convex hull* and denoted $\operatorname{co} f$.

1.4. Operations Over Convex Objects. Here we we list some of the basic operations over convex objects. We begin with convex functions. On $\operatorname{Co}(X, \bar{\mathbf{R}})$ we can introduce the following operations:

1) *the sum*:

$$(f_1 + f_2)(x) = f_1(x) + f_2(x);$$

2) *the convolution*:

$$(f_1 \oplus f_2)(x) = \inf_{x = x_1 + x_2} (f_1(x_1) + f_2(x_2));$$

3) *the maximum*:

$$(f_1 \vee f_2)(x) = \max(f_1(x), f_2(x));$$

4) *the convex hull minimum*:

$$(f_1 \operatorname{co}\wedge f_2)(x) = \operatorname{co}(\min\{f_1(\cdot), f_2(\cdot)\})(x);$$

5) *the inverse image of a function under a linear mapping*:

$$(fA)(x) = f(Ax);$$

6) *the image of a function under a linear mapping*:

$$(Af)(y) = \inf\{f(x) \colon x \in A^{-1}y\},$$

where $A^{-1}y$ is the inverse image of y.

In 5) and 6) A is a linear mapping of X to a linear space Y, $f \in \operatorname{Co}(Y, \bar{\mathbf{R}})$ in 5) and $f \in \operatorname{Co}(X, \bar{\mathbf{R}})$ in 6).

All of these operations take convex functions to convex functions. The odd-numbered operations are quite standard and are "local", that is, for the calculation of $f_1 + f_2$, $f_1 \vee f_2$, and fA at a point x it is only required to know f_1, f_2 and f at x. The even-numbered operations are quite unusual and are "non-local".

They require a "global" knowledge of the functions f_1 and f_2, in 2) and 4), and f in 6). The role and significance of these operations will become clear when we write down the formulae of convex calculus. We note that 1) and 2) have natural extensions to any finite, and operations 3) and 4) to any, number of functions:

$$\left(\bigvee_{\alpha\in\mathscr{A}} f_\alpha\right)(x) = \sup_{\alpha\in\mathscr{A}} f_\alpha(x), \qquad \operatorname{co}\bigwedge_{\alpha\in\mathscr{A}} (f_\alpha)(x) = \operatorname{co}\left(\inf_{\alpha\in\mathscr{A}} f_\alpha(x)\right).$$

Here, if $f_\alpha \in \operatorname{Co}(X, \bar{\mathbf{R}})$, $\alpha \in \mathscr{A}$, then the functions $\bigvee_\alpha f_\alpha$ and $\operatorname{co}\bigwedge f_\alpha$ are convex.

On $\operatorname{Co}(X)$ we introduce the following operations:

1′) *the sum*:

$$A_1 + A_2 = \{x: x = x_1 + x_2, x_1 \in A_1, x_2 \in A_2\};$$

2′) *the convolution or Kelley sum*:

$$A_1 \boxplus A_2 = \bigcup\{\alpha_1 A_1 \cap \alpha_2 A_2: (\alpha_1, \alpha_2) \in \Delta\};$$

3′) *the convex hull union*:

$$A_1 \operatorname{co}\bigcup A_2 =: \operatorname{co}(A_1 \cup A_2) = \{x: x = \alpha_1 x_1 + \alpha_2 x_2, x_i \in A_i, i = 1, 2, (\alpha_1, \alpha_2) \in \Delta\};$$

4′) *the intersection*:

$$A_1 \cap A_2 = \{x: x \in A_1, x \in A_2\};$$

5′) *the inverse image under a linear mapping*: $A\Lambda =: \Lambda^{-1}A$

6′) *the image under a linear mapping*: ΛA.

In 5′) and 6′) Λ is a linear mapping from X to a linear space Y and $A \in \operatorname{Co}(X)$ in 6′) and $\operatorname{Co}(Y)$ in 5′).

All these operations take convex sets to convex sets. Here, for convex sets, only the Kelley sum is somewhat unusual. Its role becomes clear when we introduce the formulae of convex calculus. The even numbered operations here are local.

Operations 1′) and 2′) admit natural extensions to any finite number, and operations 3′) and 4′) to any number, of elements:

$$\operatorname{co}\bigcup_{\alpha\in\mathscr{A}} A_\alpha =: \operatorname{co}\left(\bigcup_{\alpha\in\mathscr{A}} A_\alpha\right); \qquad \bigcap_{\alpha\in\mathscr{A}} A_\alpha =: \bigcap\{A_\alpha: \alpha \in \mathscr{A}\}.$$

Here, if the A_α are convex, then $\operatorname{co}\bigcup_\alpha A_\alpha$ and $\bigcap_\alpha A_\alpha$ will be convex.

On $\mathrm{S}\mathscr{L}(X, \bar{\mathbf{R}})$ there are the same operations 1) $(+)$, 2) $(\bigvee)$, 3) $(\vee)$, 4) $(\operatorname{co}\bigwedge)$, 5) $p \to p\Lambda$, as on $\operatorname{Co}(X, \bar{\mathbf{R}})$. In addition to these we introduce one more (whose role will be revealed later):

$$p_1 \nabla p_2 = \bigvee\{\alpha_1 p_1 + \alpha_2 p_2: (\alpha_1, \alpha_2) \in \Delta\}.$$

Here $p_1, p_2 \in \mathrm{S}\mathscr{L}(X, \bar{\mathbf{R}}) \Rightarrow p_1 \nabla p_2 \in \mathrm{S}\mathscr{L}(X, \bar{\mathbf{R}})$.

On $\operatorname{Cone}(X)$, $\operatorname{Aff}(X)$, $\operatorname{Lin}(X)$ there are the same operations as on $\operatorname{Co}(X)$, but in this connection it is necessary to keep in view that $\boxplus$ here is $\cap$, and $\operatorname{co}\bigcup$ is $+$.

§2. Duality of Linear Spaces. Dual Operators of Convex Analysis

Duality in convex analysis is based on the duality of linear spaces. Here we will define duality of spaces first, and then define dual operators. This material was presented in a descriptive manner in section 2 of the introduction.

2.1. The Definition of Duality. The basic phenomenon connected with convexity, *duality*, assumes, first of all, the presence of a dual structure for the linear spaces themselves. This subsection is devoted to the description of that structure.

Let X and Y be linear spaces and let $B: X \times Y \to R$ be a bilinear form on $X \times Y$, that is, a function $(x, y) \to B(x, y)$ linear in each argument separately.

We say that the *bilinear form B places X* and *Y* in *duality*, or that X and *Y are in duality relative to B*, and we write $X \leftarrow d \rightarrow Y$, if

$$B(x, y) = 0 \ \forall x \in X \Rightarrow y = 0 \quad \text{and} \quad B(x, y) = 0 \ \forall y \in Y \Rightarrow x = 0.$$

This can be put another way. Let $X^{\#}$ and $Y^{\#}$ denote the algebraic dual spaces of X and Y respectively, that is, the spaces of all linear functionals on X and Y. We denote by $\langle x^{\#}, x\rangle$ $(\langle y^{\#}, y\rangle)$ the action of the element $x^{\#} \in X^{\#}$ $(y^{\#} \in Y^{\#})$ on x (y). The forms $\langle x^{\#}, x\rangle$ and $\langle y^{\#}, y\rangle$ are called the *canonical forms* on $X^{\#} \times X$ and $Y^{\#} \times Y$. From the definition of a bilinear form, for any $y \in Y$ the functional $l_y(x) = B(x, y)$ belongs to $X^{\#}$, and for any $x \in X$ the functional $l_x(y) = B(x, y)$ belongs to $Y^{\#}$. Thus there are mappings $yB: y \to l_y(\cdot)$ and $Bx: x \to l_x(\cdot)$. The requirements of the definition of duality reduce to these mappings being monomorphisms, that is, that they have null kernel.

Therefore the mapping which associates to each $y \in Y$ the linear functional $x^{\#}(y) \in X^{\#}$, by the formula $\langle x^{\#}(y), x\rangle = B(x, y)$, maps Y isomorphically (in the sense of isomorphism of linear spaces) to some linear subspace $Y_1 \subset X^{\#}$. The form B then becomes the canonical bilinear form and the spaces X and Y_1 are dual relative to the restriction of the canonical bilinear form to $X \times Y_1$. Thus later on we will always consider Y as a linear subspace in $X^{\#}$ are take duality relative to the canonical form $\langle \cdot, \cdot \rangle$.

The relation of duality becomes symmetric if in the presence of a duality $X \leftarrow d \rightarrow Y$ relative to a bilinear form $(x, y) \to \langle x, y\rangle$ we define a duality $Y \leftarrow d \rightarrow X$ by the form $(y, x) \to \langle x, y\rangle$. Therefore when speaking of a duality $X \leftarrow d \rightarrow Y$ relative to $\langle \cdot, \cdot \rangle$, we use various expressions of the type: X and Y are in duality, X is dual with Y, Y is dual with X. etc.

In this way each of the spaces X and Y appears in a dual role both as itself and as the space of linear functionals over the other space. Duality is illustrated most simply with a Hilbert space, which can be identified with its dual. Therefore an object and its dual may be "represented" in one and the same space. This situation was utilised in the introduction where a pair of dual objects were simultaneously represented in the plane.

We will mention some important examples of spaces occurring in duality.

Finite-dimensional spaces. It is not difficult to see that if $\mathbf{R}^n \leftarrow d \rightarrow \mathbf{R}^m$, then $n = m$. If $n = m$ then the spaces $X = \mathbf{R}^n$ and $Y = \mathbf{R}^m$ are in duality with respect to any bilinear form

$$B(x, y) = \sum_{i=1}^{n} \sum_{j=1}^{n} b_{ij} x_i y_j,$$

with a nondegenerate matrix (b_{ij}). Later, for $X = Y = \mathbf{R}^n$, $\langle \cdot, \cdot \rangle$ will denote the standard bilinear form $\langle x, y \rangle = \sum_{i=1}^{n} x_i y_i$.

Any linear space (l.s.) X is in duality with its algebraic dual $X \leftarrow d \rightarrow X^{\#}$. Denote by $\langle x^{\#}, x \rangle$ the bilinear form which associates the pair $(x^{\#}, x)$, $x^{\#} \in X^{\#}$, $x \in X$, to the number $\langle x^{\#}, x \rangle$ which is the value of $x^{\#}$ at x. This form is also called canonical. The canonical form places X and X^* in duality.

Any locally convex linear topological space (l.c.s) (in particular a normed space) is in duality with its (topological) dual $X'(X^*)$ relative to the canonical bilinear form $(x', x) \rightarrow \langle x', x \rangle$ $((x^*, x) \rightarrow \langle x^*, x \rangle)$.[1]

The proof of this fact (which follows at once from the Hahn-Banach theorem) will be given later.

2.2. Elementary Dual Relations. In this subsection we return once again to the theme touched on in the introduction.

Let X and Y be two spaces in duality.

Associated with each $y \in Y$ there is a linear function $\langle \cdot, y \rangle$. So there is a dual relation

$$y \Leftrightarrow \langle \cdot, y \rangle, \tag{1}$$

(and similarly, $x \Leftrightarrow \langle x, \cdot \rangle$).

Let $y \neq 0$. Then we can associate with this element the segment $\Delta y = [0, y]$ in Y and a halfspace containing zero in X: $\Pi y = \{x \in X : \langle x, y \rangle \leqslant 1\}$. So there is a dual relation

$$\Delta y \Leftrightarrow \Pi y, \tag{2}$$

(and similarly, $\Delta x \Leftrightarrow \Pi x$).

Associated with a ray $\lambda y = \{\alpha y : \alpha \geqslant 0\}$, $y \neq 0$, there is a half-space, bounded by a hyperplane passing through zero: $\Pi_0 y = \{x \in X : \langle x, y \rangle \geqslant 0\}$. So there is a dual relation

$$\lambda y \Leftrightarrow \Pi_0 y \tag{3}$$

(and similarly $\lambda x \Leftrightarrow \Pi_0 x$).

For each $\eta \in Y$ and $\alpha \in \mathbf{R}$ there is an affine function $a(\eta, \alpha) = \langle x, \eta \rangle + \alpha$. So there is a dual relation

$$a(\eta, \beta) \Leftrightarrow \delta\eta - \beta \in \mathrm{Co}(Y, \bar{\mathbf{R}}), \tag{4}$$

(and similarly $a(\xi, \alpha) \Leftrightarrow \delta\xi - \alpha \in \mathrm{Co}(X, \bar{\mathbf{R}})$).

---

[1] X' denotes the dual of a linear topological space, whereas for the dual of a normed space we preserve the usual notation X^*; all l.c.s are assumed to be separable.

Here $(\delta\eta - \beta)(y)$ is equal to $-\beta$ if $y = \eta$, and $+\infty$ if $y \neq \eta$. We call functions of the type $\delta\eta - \beta$ *elementary*.

The duality (2) is at the foundation of the correspondence set $\Leftrightarrow$ polar for convex sets, (3) is at the foundation of the correspondence cone $\Leftrightarrow$ conjugate cone, (4) is at the foundation of the correspondence function $\Leftrightarrow$ Legendre-Young-Fenchel transform, and, in the final analysis, they are all based on (1). All of this was briefly mentioned in subsection 0.2.

Two more definitions. We denote by $\Pi_{\pm}(y, \beta)$ the *half-spaces*

$$\Pi_{\pm}(y, \beta) = \{x: \langle x, y\rangle \geqslant \beta \text{ for } +, \leqslant \beta \text{ for } -\},$$

by $\Gamma(y, \beta)$ the *hyperplane*

$$\Gamma(y, \beta) = \{x: \langle x, y\rangle = \beta\}.$$

In this connection we will say that $\Gamma(y, \beta)$ bounds the half-spaces $\Pi_{+}(y, \beta)$ and $\Pi_{-}(y, \beta)$. Thus $\Pi y = \Pi_{-}(y, 1)$, $\Pi_0 y = \Pi_{-}(y, 0)$.

The sets $\{x: \langle x, y\rangle > \beta\} =: \mathring{\Pi}_{+}(y, \beta)$ or $\{x: \langle x, y\rangle < \beta\} =: \mathring{\Pi}_{+}(y, \beta)$ are called *open half-spaces*. Sets $\Pi_{\pm}(x, \alpha)$, $\Gamma(x, \alpha)$ in Y are defined in a similar way.

2.3. The Basic Operators of Convex Analysis. Let X and Y be spaces in duality, f a function on X, A a subset of X, K a cone in X, p a homogeneous (of first degree) function on X. Here we introduce the most important operators of convex analysis (they have already been described in subsection 0.2).

The function $lf(y) = \sup_x(\langle x, y\rangle - f(x))$ is called the *Legendre transform* (or *Young*, *Young-Fenchel* or *Legendre-Young-Fenchel transform*) of f.

The set

$$\pi A = \{y \in Y: \langle x, y\rangle \leqslant 1 \ \forall x \in A\}$$

is called the *polar* of A. The polar of a subspace L is called the *annihilator* of L and denoted $L^{\perp}$. If $X = Y$ is finite-dimensional Euclidean, or Hilbert, then $L^{\perp}$ is the orthogonal complement of L.

The *cone*

$$\sigma K = \{y \in Y: \langle x, y\rangle \geqslant 0 \ \forall x \in K\}$$

is called the *conjugate* to K.

The set

$$\partial p = \{y \in Y: \langle x, y\rangle \leqslant p(x) \ \forall x \in X\}$$

is called the *subdifferential of the homogeneous function p*.

The function

$$sA(y) = \sup_{x \in A} \langle x, y\rangle$$

is called the *support function of A*.

The set

$$\partial f(\hat{x}) = \{y \in Y: f(x) - f(\hat{x}) \geqslant \langle x - \hat{x}, y\rangle \ \forall x \in X\}$$

is called the *subdifferential of f at $\hat{x}$*. Although the subdifferential of a sublinear function and the subdifferential of a convex function at a point have similar notation this will not lead to any confusion.

It is clear that if f is homogeneous then $\partial f = \partial f(0)$.

All of these operators are defined by similar formulae for a function g, a subset B, a cone L and a convex homogeneous function q on Y:

$$lg(x) = \sup_y (\langle x, y\rangle - g(y)), \qquad \pi B = \{x: \langle x, y\rangle \leqslant 1\ \forall y \in B\},$$

$$\sigma L = \{x: \langle x, y\rangle \geqslant 0\ \forall y \in L\}, \qquad \partial q = \{x: \langle x, y\rangle \leqslant q(y)\ \forall y \in Y\},$$

$$sB(x) = \sup_{y \in B} \langle x, y\rangle, \qquad \partial g(y) = \{x: g(y) - g(\hat{y}) \geqslant \langle x, y - \hat{y}\rangle\ \forall y \in Y\}.$$

Here we will write $l^2 f$, $\pi^2 f$, $\sigma^2 K$, $\partial s A$, $s\partial p$, etc., meaning $l(lf)$, $\pi(\pi A)$, $\sigma(\sigma K)$, $\partial(sA)$, $s\partial p)$.

Our notation simplifies many of the formulae of convex analysis. But it is must be said that there are traditional notations for all the operators introduced above. These were given in the introduction. We compare once again the traditional and the new notations:

$$lf(y) \text{ denotes } f^*(y) (lf \Leftrightarrow f^*), \qquad \pi A \Leftrightarrow A^\circ, \qquad \sigma K \Leftrightarrow K^*,$$

$\partial p \Leftrightarrow \partial p(0)$, $\mu A(\cdot) \Leftrightarrow \mu(\cdot|A)$, $sA(y)$ is denoted variously by: $s(y|A)$, $s_A(y)$, $c(A, y)$. The notation $\partial f(x_0)$ is standard.

We resolved to change the traditional notations, because we wished to make use later on of the operator properties of these mappings. For example, we will write $A_1 \boxplus A_2 = \pi\partial(\mu A_1 + \mu A_2)$. whereas in the usual notations we would have had to write

$$A_1 \boxplus A_2 = (\partial(\mu(\cdot|A_1) + \mu(\cdot|A_2))(0))^\circ.$$

The operators that we have introduced here will play a basic role in what follows. If you are not familiar with them it is reasonable to spend some time calculating several examples. Say, those which we give next. All the formulae, proved below, are easily deduced from the definitions by simple calculations.

2.4. Examples. 1) *Legendre-Young-Fenchel transforms.* In a)–c) $X = Y = \mathbf{R}$.

a) $f(x) = |x|^p/p, \quad p > 1 \Rightarrow lf(y) = |y|^{p'}/p^1,$

$$(p')^{-1} + p^{-1} = 1;$$

b) $f(x) = |x| \Rightarrow lf(y) = \delta[-1, 1](y)$;

c) $$f(x) = \exp(x) \Rightarrow lf(y) = \begin{cases} y(\log y - 1), & y > 0 \\ 0, & y = 0 \\ +\infty, & y < 0. \end{cases}$$

In d), e) $X = Y = \mathbf{R}^n$, $x = (x_1, \ldots, x_n)$, $X \to d \leftarrow Y$, relative to the standard form.

d) $f(x) = \sum_{i=1}^{n} |x|^{p_i}/p_i, \quad p_i > 1 \Rightarrow f(y) = \sum_{i=1}^{n} |y_i|^{p_i'}/p_i',$

$$(p_i')^{-1} + p_i^{-1} = 1;$$

e) $f(x) = \max\{x_1, \dots, x_n\} \Rightarrow f(y) = \delta\Sigma^{n-1}(y),$

$$\Sigma^{n-1} = \left\{y \in \mathbf{R}_+^n : \sum_{i=1}^{n} y_i = 1\right\}.$$

In f) $X = Y$ is a Hilbert space and $\langle x, y\rangle$ is the scalar product of x and y.

f) $lf(x) = f(x) \Leftrightarrow f(x) = \langle x, x\rangle/2.$

In g) X is a normed space in duality with its dual X^*.

g) $f(x) = N(x) = \|x\| \Rightarrow (lN)(y^*) = \delta B^*(y^*),$

$$B^* = \{x^* \in X^* : \|x^*\| \leqslant 1\}.$$

2) *Polars*. Let $f: X \to \mathbf{R}^n$ be a convex, smooth, positive function and $A = \{x: f(x) \leqslant 1\}$. To find the polar we proceed as follows. Let $x_0 \in \partial A$ ($\Leftrightarrow f(x_0) = 1$). The equation of the tangent hyperplane to ∂A at x_0 is:

$$\langle f'(x_0), x - x_0\rangle = 0 \Leftrightarrow \langle f'(x_0), x\rangle = \langle f'(x_0), x_0\rangle.$$

Dual to this tangent plane (by the general rule that says the plane $\langle a, x\rangle = 1$ is dual to the point $y = a$) is the point $y = f'(x_0)/\langle f'(x_0), x_0\rangle$. The equation of the polar is obtained by eliminating x_0 from the system of equations $f(x_0) = 1$, $y = f'(x_0)/\langle f'(x_0, x_0\rangle$.

In particular, if $f(\cdot)$ is homogeneous of degree m, then, by Euler's theorem, $\langle f'(x_0), x_0\rangle = mf(x_0) = m$ and the polar is obtained by eliminating x_0 from the system of equations $f(x_0) = 1$, $my = f'(x_0)$. As an example, if $f(x) = \sum_{i-1}^{n} b_{ij}x_ix_j = \langle x, Bx\rangle$, $B > 0$, then $my = f'(x_0) \Leftrightarrow y = Bx_0 \Rightarrow x_0 = B^{-1}y$ and the equation of the polar is obtained in the form $\langle B^{-1}y, y\rangle = 1$. In a)–c) $X = Y = \mathbf{R}^n$, $X \leftarrow d \rightarrow Y$ relative to the standard form

a) $A = B_p^n =: \left\{x: \sum_{i=1}^{n} |x_i|^p \leqslant 1\right\}, \quad p > 1 \Rightarrow \pi A = B_{p'}^n,$

$$(p')^{-1} + p^{-1} = 1;$$

b) $A = B_\infty^n(a) =: \{x: |x_i| \leqslant a_i\} \Rightarrow \pi A = B_1^n(a^{-1}) =: \left\{y: \sum_{i=1}^{n} |a_i y_i| \leqslant 1\right\};$

c) $A = B_1^n(a) =: \left\{x: \sum_{i=1}^{n} |x_i/a_i| \leqslant 1\right\} \Rightarrow \pi A = B_\infty^n(a^{-1}) =: \{y: |y_i| \leqslant a_i^{-1}\}.$

In d) $X = Y$ is a Hilbert space and $\langle x, y\rangle$ is the scalar product of x and y.

d) $\pi A = A \Leftrightarrow A = \{x: \langle x, x\rangle = 1\}$ – the unit sphere.

In e) X is a normed space in duality with its dual X^*, B is the unit sphere in X and B^* is the unit sphere in X^*.

e) $\pi B = B^*$.

3) *Conjugate cones.* In a) $X = Y = \mathbf{R}^n$, $X \leftarrow d \rightarrow Y$ relative to the standard form, in b) $X = Y$ is a Hilbert space $X \leftarrow d \rightarrow Y$ relative to the scalar product.

a) $K = \mathbf{R}^n_+ \Rightarrow \sigma K = \mathbf{R}^n_+$;

b) $K = K_a(\alpha) = \{x: \langle x, a\rangle \geqslant \|x\| \cos\alpha, \|a\| = 1, 0 < \alpha < \pi/2\}$

$\Rightarrow \sigma K = K_a(\pi/2 - \alpha)$.

In c) X is a normed space, X^* its dual.

c) $K = \Pi_0\xi^* =: \{x: \langle \xi^*, x\rangle \leqslant 0\} \Rightarrow \sigma K = \lambda\xi^*$

$= \{x^*: x^* = \alpha\xi^*, \alpha \geqslant 0\}$.

4) *Subdifferentials.* In a) $X = Y = \mathbf{R}$, in b) $X = Y = \mathbf{R}^n$, in c) $X \leftarrow d \rightarrow X^*$, where X is a normed space, X^* its dual.

a) $p(x) = p(x; a, b) = \{bx, x \geqslant 0, ax, x \leqslant 0, a < b\} \Leftrightarrow \partial p = [a, b]$;

b) $p(x) = \max\{x_1, \ldots, x_n\} \Rightarrow \partial p = \Sigma^{n-1} = \left\{y \in R^n: \sum_{i=1}^{n} y_i = 1\right\}$;

c) $p(x) = N(x) = \|x\| \Rightarrow \partial N = B^*, B^* = \{x^*: \|x^*\| \leqslant 1\}$.

5) *Support functions.* In a)–c) $X = Y = R^n$, in d) $X \leftarrow d \rightarrow X^*$, where X is a normed linear space X^* its dual.

a) $A = B^n_p \Rightarrow sA(y) = \|y\|_{p'} =: \left(\sum_{i=1}^{n} |y_i|^{p'}\right)^{1/p'}$,

$1/p' + 1/p = 1 (p > 1)$;

b) $A = B^n_\infty(a) \Rightarrow sA(y) = \sum_{i=1}^{n} |a_i y_i|$;

c) $A = B^n_1(a) \Rightarrow sA(y) = \max_{1 \leqslant i \leqslant n} |a_i y_i|$;

6) *Subdifferentials of functions at a point.* In a)–b) $X = Y = \mathbf{R}$, in c) $X = Y$ is a Hilbert space.

a) $f(x) = |x| \Rightarrow \partial f(x) = \begin{cases} -1, & x < 0, \\ [-1, 1], & x = 0, \\ 1, & x > 0; \end{cases}$

b) $f(x) = \max\{e^x, 1 - x\} \Rightarrow \partial f(x) = \begin{cases} e^x, & x > 0, \\ [-1, 1], & x = 0, \\ -1, & x < 0; \end{cases}$

c) $f(x) = |x| \Rightarrow \partial f(x) = \begin{cases} x(\|x\|)^{-1}, & x \neq 0, \\ \{y: \|y\| \leqslant 1\}, & x = 0; \end{cases}$

d) The function $f(x) = \begin{cases} -\sqrt{(1 - x^2)}, & |x| \leqslant 1 \\ +\infty, & |x| > 1 \end{cases}$

has the property that $f(\pm 1) = 0$, $\partial f(\pm 1) = \varnothing$.

For other examples of subdifferentials at a point see Chap. 2, § 4.

§ 3. Topological Properties of Convex Sets and Functions

It can already be seen in the introduction that the conjunction of the concepts of convexity and topology turns out to be beneficial in convex analysis. That is the topic of this section.

3.1. Topology of Duality. Let X and Y be two spaces in duality. The duality relation in a natural way provides topologies in each of the spaces. These topologies (usually called *weak*) are defined as the weakest of the topologies in which all the linear forms $x \to \langle x, y\rangle$ on X and $y \to \langle x, y\rangle$ on Y are continuous. They are denoted by $\sigma(X, Y)$ and $\sigma(Y, X)$ respectively. Fundamental systems of neighbourhoods of zero in these topologies are formed by the finite intersections of the sets

$$U_y(\varepsilon) = \{x \in X: \langle x, y\rangle < \varepsilon\} \text{ in } X \quad \text{and} \quad V_x(\varepsilon) = \{y \in Y: \langle x, y\rangle < \varepsilon\}$$

in Y. The spaces $(X, \sigma(X, Y))$ and $(Y, \sigma(Y, X))$ are locally convex.

Examples. 1. It is easy to see that $\sigma(\mathbf{R}^n, \mathbf{R}^n)$ is the standard topology on $\mathbf{R}^n$, given by the Euclidean norm $|x|$.

2. Let X be a normed space, X^* its dual. The topology $\sigma(X, X^*)$ is called the *weak topology* on X, the topology $\sigma(X^*, X)$ the *weak* topology* on X^*.

The mappings $\beta(x)$: $y \to \langle x, y\rangle$ and $\kappa(y)$: $x \to \langle x, y\rangle$ map X and Y, respectively, onto subspaces $\beta(X) \subset Y^\#$ and $\kappa(Y) \subset X^\#$. It is not difficult to see that $\beta(X)$ is isomorphic to X, $\kappa(Y)$ is isomorphic to Y and $\beta(X)$ and $\kappa(Y)$ are weakly conjugate to each other.

Now suppose there is given a topology τ on X which turns X into a locally convex space. τ is said to be *compatible with duality*, if $(X, \tau)'$ is isomorphic to $\kappa(Y)$. Compatibility with duality for a topology on Y is defined similarly. It is obvious that $\sigma(X, Y)$ and $\sigma(Y, X)$ are compatible with duality.

For a pair $X \leftarrow d \rightarrow Y$ it is possible to define the strongest locally convex topology, compatible with duality. We denote it by $m(X, Y)$. $m(X, Y)$ is called the *Mackey topology*.

Lemma 3.1 (Bourbaki [1953], Robertson & Robertson [1964]). *A locally convex topology τ is compatible with the duality $X \leftarrow d \rightarrow Y$ if and only if*

$$(X, \tau(X, Y))^* = Y.$$

2. If X is a normed space, then the Mackey topology coincides with the original topology generated by the norm on X.

3. All topologies which are compatible with duality have the same supply of:

a) bounded sets (*Mackey's theorem*);

b) closed convex sets (*Mazur's theorem*).

3.2. Topological Properties of Convex Sets. Let X be a locally convex linear topological space and $A \subset X$. As usual $\operatorname{int} A$ denotes the *interior* of A, and $\operatorname{cl} A$ denotes its *closure*.

Lemma 4. *Let A be a convex set. Then:*

a) $x_1 \in \operatorname{int} A, x_2 \in A \Rightarrow [x_1, x_2) \subset \operatorname{int} A$;

b) $\operatorname{int} A \in \operatorname{Co}(X)$, $\operatorname{cl} A \in \operatorname{Co}(X)$;

c) $\operatorname{int} A \neq \varnothing \Rightarrow \operatorname{cl} A = \operatorname{cl} \operatorname{int} A$.

Proof. a) Let $V \in \mathcal{O}(x_1, X) \cap \operatorname{Co}(X)$, $V \subset A$, and $x \in [x_1, x_2) \Rightarrow x = \alpha x_1 + (1-\alpha)x_2$, $0 < \alpha \leqslant 1 \Rightarrow W = \alpha V + (1-\alpha)x_2 \in \mathcal{O}(x, X)$, $W \subset A$ (because of the convexity of A), that is $x \in \operatorname{int} A$.

b) If $x_i \in \operatorname{int} A$, $i = 1, 2$, then a) $\Rightarrow [x_1, x_2) \subset \operatorname{int} A \Rightarrow \operatorname{int} A \in \operatorname{Co}(X)$. Now let $x_i \in \operatorname{cl} A$, $i = 1, 2$, and $V \in \mathcal{O}(0, X)$, $V \in \operatorname{Co}(X)$. By the definition of closure $\exists x' \in (x_i + V) \cap A$, $i = 1, 2$. Let $x_\alpha = \alpha x_1 + (1-\alpha)x_2$, $0 \leqslant \alpha \leqslant 1$. Put $x'_\alpha = \alpha x'_1 + (1-\alpha)x'_2 \Rightarrow x'_\alpha \in A \cap (\alpha(x_1 + V) + (1-\alpha)(x_2 + V) = x_\alpha + V$, that is, $x \in \operatorname{cl} A \Rightarrow \operatorname{cl} A \in \operatorname{Co}(X)$. Further, from a) $\Rightarrow A \subset \operatorname{cl} \operatorname{int} A$ (if $\operatorname{int} A \neq \varnothing$), that is, $\operatorname{cl} A \subset \operatorname{cl} \operatorname{int} A$. The converse inclusion is obvious.

The intersection of all the convex closed sets containing A is called the *convex closure* of A and denoted $\operatorname{cl} \operatorname{co} A$ or $\overline{\operatorname{co}}\, A$. We denote the collection of all closed subsets of X by $\operatorname{Cl}(X)$.

Lemma 5. a) $\operatorname{cl} \operatorname{co} A = A \Leftrightarrow A \in \operatorname{Cl} \operatorname{Co}(X)$;
b) $\operatorname{cl} \operatorname{co} A = \operatorname{cl}(\operatorname{co} A)$.

3.3. Topological Properties of Convex Functions. In the following theorem X is a linear topological (not necessarily locally convex) space.

Theorem 1. *Let f be a proper convex function on X. Then the following are equivalent:*

a) *f is bounded above in the neighbourhood of some point x;*

b) *f is continuous at some point x;*

c) $\operatorname{int} \operatorname{epi} f \neq \varnothing$;

d) $\operatorname{int}(\operatorname{dom} f) \neq \varnothing$ *and f is continuous on* $\operatorname{int}(\operatorname{dom} f)$. *Here*

$$\operatorname{int}(\operatorname{epi} f) = \{(\alpha, x) \colon x \in \operatorname{int}(\operatorname{dom} f), \alpha > f(x)\}.$$

Let us prove this important theorem.

A) f is continuous at $x \Rightarrow f$ is bounded in some neighbourhood of x. This means a) $\Leftarrow$ b). Now let $\exists U \in \mathcal{O}(x_0, X)$ and $f(x) \leqslant c < +\infty$, $c > 0$, $\forall x \in U$. Without loss of generality, we can assume that $x_0 = 0$ and $f(0) = 0$ (replacing, if necessary, $f(x)$ by $f(x+x_0)-f(x_0)$ and U by $U-x_0$). Taking $0<\varepsilon\leqslant c$, put $V_\varepsilon=(\varepsilon/c)U\cap(-\varepsilon/c)U$. It is clear that $V_\varepsilon\in\mathcal{O}(0,X)$. Let $x\in V_\varepsilon\Rightarrow x\in(\varepsilon/c)U\Rightarrow(c/\varepsilon)x\in U\Rightarrow$ (Jensen's inequality for the convex function f): $f(x)=f\left(\frac{\varepsilon}{c},\frac{c}{\varepsilon},x+\left(1-\frac{\varepsilon}{c}\right)0\right)\leqslant\frac{\varepsilon}{c}f\left(\frac{c}{\varepsilon}x\right)+\left(1-\frac{\varepsilon}{c}\right)f(0)\leqslant\varepsilon$. On the other hand, $x\in(-\varepsilon/c)U\Rightarrow(-c/\varepsilon)x\in U\Rightarrow$ (since $0=(x/(1+(\varepsilon/c)))+(\varepsilon/c)(-c/\varepsilon x)/((1+(\varepsilon/c)))\Rightarrow 0=f(0)\leqslant\frac{1}{1+(\varepsilon/c)}f(x)+\frac{\varepsilon/c}{1+(\varepsilon/c)}f((-c/\varepsilon)x)\Rightarrow$ $f(x)\geqslant-\varepsilon$. The b) $\Leftarrow$ a), that is a) $\Leftrightarrow$ b).

B) a) $\Rightarrow$ c) by definition. Let us prove c) $\Rightarrow$ d). Let $(\hat{\alpha},\hat{x})\in\operatorname{int}(\operatorname{epi}f)\Rightarrow\exists V\in\mathcal{O}(\hat{x},X)$ and a $\delta>0$: $\{(\alpha,x):|\alpha-\hat{\alpha}|<\delta,x\in V\}\subset\operatorname{int}(\operatorname{epi}f)\Rightarrow f$ is bounded in $V\Rightarrow f$ is continuous at $\hat{x}$, and together with $V\subset\operatorname{dom}f\Rightarrow\operatorname{int}(\operatorname{dom}f)\neq\varnothing$. The implication d) $\Rightarrow$ a) is trivial. We suggest that you think over the final formula regarding $\operatorname{int}(\operatorname{epi}f)$ by yourself.

Corollary 1. *Let X be a locally convex space and A a convex neighbourhood of zero. Then the Minkowski function of A is continuous.*

It follows from the definition that $x\in A\Rightarrow\mu A(x)\leqslant 1$.

A major role in convex analysis is played by the closed functions, that is, functions whose epigraphs are closed in $X\times\mathbf{R}$. The following is true

Lemma 5. *Let X be a topological space. Then the following assertions are equivalent:*

a) f is closed;

b) *f is lower semi-continuous;*

c) *for any $\alpha\in\mathbf{R}$ the set $\{x\in X: f(x)\leqslant\alpha\}$ is closed in X;*

d) *for any $\alpha\in\mathbf{R}$ the set $\{x\in X: f(x)>\alpha\}$ is open in X.*

Let $f\in\operatorname{Co}(X,\bar{\mathbf{R}})$. The closure of the epigraph of f is the epigraph of a convex closed function called the *closure* of f and denoted $\operatorname{cl}f$.

For example, the function $y(x)=x\log x$ for $x>0$, $+\infty$ for $x<0$ and any positive number for $x=0$, is convex but not closed. Its closure takes the value zero at $x=0$.

The collection of all convex closed functions on X is denoted $\operatorname{Cl}\operatorname{Co}(X,\bar{\mathbf{R}})$, the collection of all convex closed sublinear functions is denoted $\operatorname{cl}S\mathcal{L}(X,\bar{\mathbf{R}})$.

We remark that a convex continuous function, defined on an open set in a Banach space, is locally Lipshitz.

3.4. Topological Properties of Finite-Dimensional Convex Objects. In the finite-dimensional case a number of theorems can be made more precise.

Theorem 2. $/A, A_1, A_2\in\operatorname{Co}(\mathbf{R}^n)/\Rightarrow$

α) $\operatorname{cl}A$ *and* $\operatorname{ri}A\in\operatorname{Co}(\mathbf{R}^n)$,

β) $\operatorname{cl}\operatorname{ri}A=\operatorname{cl}A$, $\operatorname{ri}(\operatorname{cl}A)=\operatorname{ri}A$,

γ) $\operatorname{ri}(\lambda A)=\lambda\operatorname{ri}A$, $\operatorname{cl}(A_1+A_2)\supset\operatorname{cl}A_1+\operatorname{cl}A_2$, $\operatorname{ri}(A_1+A_2)=\operatorname{ri}A_1+\operatorname{ri}A_2$,

δ) *if* $\Lambda \in \mathscr{L}(\mathbf{R}^n, \mathbf{R}^m)$, *then* $\mathrm{ri}(\Lambda A) = \Lambda\, \mathrm{ri}\, A$, $\Lambda^{-1}\, \mathrm{ri}\, A \neq \varnothing \Rightarrow \mathrm{ri}(\Lambda^{-1} A) = \Lambda^{-1}\, \mathrm{ri}\, A$, $\mathrm{cl}(\Lambda^{-1} A) = \Lambda^{-1}\, \mathrm{cl}\, A$.

It is clear from the following theorem that an improper function in R^n has a rather special character.

Theorem 3. *If f is an improper convex function, then for any $x \in \mathrm{ri}\,\mathrm{dom}\, f \Rightarrow f(x) = -\infty$. If f is lower semi-continuous, then it cannot have finite values.*

Theorem 4. *Let f be a proper convex function. Then $\mathrm{cl}\, f$ coincides with f everywhere except, possibly, at points of the relative boundary of $\mathrm{dom}\, f$. Here, if $x \in \mathrm{ri}(\mathrm{dom}\, f)$, then $\mathrm{cl}\, f(y) = \lim_{\lambda \uparrow 1} f((1-\lambda)x + \lambda y)$ for any y.*

The topological properties of finite-dimensional convex sets and functions have been expounded in detail in Rockafellar [1970]. The duality of linear spaces has been described in many books, first of all, in the classical memoir Bourbaki [1953], and also in Kutateladze [1983], Robertson & Robertson [1964].

§4. Basic Theorems

In this section we have grouped the basic theorems, necessary for the construction of convex analysis.

4.1. The Hahn-Banach Theorem and Separation

Theorem 5 (Hahn-Banach). *Let X be a linear space, p a convex homogeneous function on X and $l\colon X_0 \to \mathbf{R}$ a linear functional on a subspace X_0 of X such that*

$$\langle l, x\rangle \leqslant p(x), \qquad \forall x \in X_0.$$

Then there exists a linear functional Λ, defined on the whole of X (that is, $\Lambda \in X^{\#}$), which is an extension of l (that is, $\langle l, x\rangle = \langle \Lambda, x\rangle\ \forall x \in X_0$) and satisfies the inequality

$$\langle \Lambda, x\rangle \leqslant p(x) \qquad \forall x \in X.$$

This is a well-known theorem (for the proof see Kolmogorov & Fomin [1981] or any other text on functional analysis).

Using the notation of the previous section we can give this result in another form.

Theorem 6 (The Hahn-Banach in subdifferential form). $/X$ a l.s., $p \in S\mathscr{L}(X, \mathbf{R})$, $X_0 \in \mathrm{Lin}(X)/ \Rightarrow$

$$\partial(p + \delta X_0) = \partial p + \partial \delta X_0.$$

In fact the inclusion of the right hand side in the left hand side follows from the definition. Now let $l \in \partial(p + \delta X_0)$. This means, by definition, that $\langle l, x\rangle \leqslant p(x)\ \forall x \in X_0$. By theorem 5 there is an $x^{\#} \in X^{\#}$, for which $\langle l, x\rangle = \langle x^{\#}, x\rangle$

$x \in X_0$ and $\langle x^\#, x\rangle \leqslant p(x)\ \forall x \in X$. But then $l = x_1^\# + x_2^\#$, $x_i^\# \in X^\#$, $i = 1, 2$, $x_1^\# = x^\#$, $x_2^\# = l - x^\#$. Here $x_1^\# \in \partial p$ (by construction) $x_2^\# \in X_0^\perp \subset \partial\delta X_0$, as required.

Corollary 1. *Let X be a locally convex space and $x \neq 0$. Then there exists $x' \in X'$ such that $\langle x', x\rangle \neq 0$.*

By definition there is a convex neighbourhood of zero V, $x \notin V$. By Corollary 1 of Theorem 1, μV is continuous. Consider the subspace $X_0 = \operatorname{lin}\{x\} =: \{\xi: \xi = \alpha x, \alpha \in \mathbf{R}\}$ and put $\langle l, \xi\rangle = \alpha$, if $\xi = \alpha x$. Then from the definitions it is clear that $\langle l, \xi\rangle \leqslant \mu V(\xi)\ \forall \xi \in X_0$. By the Hahn-Banach theorem there is a $\Lambda \in X^\#$ for which $\langle \Lambda, x\rangle \leqslant \mu V(x)$ and, consequently, by Theorem 1, Λ is a continuous functional, that is, $\Lambda = x' \in X'$ and moreover $\langle x', x\rangle = 1$.

It follows at once from this that: a) the canonical form $(x', x) \to \langle x', x\rangle$ places X in duality with X'; b) that the dual of a locally convex space (normed space) is nontrivial. These facts were mentioned in § 2.

Corollary 2 (First separation theorem). *A convex set with a non-empty interior, lying in a linear topological space, can be separated by a hyperplane from any point which does not belong to it.*

We give an analytic formulation:

$$/X \text{ a lts, } A \in \operatorname{Co}(X), \operatorname{int} A \neq \varnothing, \hat{x} \notin A/ \Rightarrow$$
$$\exists x' \in X': \sup_{x \in A} \langle x', x\rangle \leqslant \langle x', \hat{x}\rangle.$$

Without loss of generality we can suppose that $0 \in \operatorname{int} A$. On the subspace $X_0 = \operatorname{lin}\{\hat{x}\} =: \{\xi: \xi = \alpha\hat{x}, \alpha \in R\}$, we define a linear functional l by putting $\langle l, \alpha\hat{x}\rangle = \alpha\mu A(\hat{x})$. We have $\langle l, \xi\rangle = \mu A(\xi)\ \forall \xi: \xi = \alpha\hat{x}, \alpha > 0$, and for $\alpha \leqslant 0$

$$\langle l, \alpha\hat{x}\rangle = \alpha\mu A(\hat{x}) \leqslant 0 \leqslant \mu A(\alpha\hat{x}).$$

Consequently, by the Hahn-Banach theorem, l can be extended to a linear functional Λ such that $\langle \Lambda, x\rangle \leqslant \mu A(x)\ \forall x$. From the latter relation, Theorem 1 and Corollary 1 of that theorem, it follows that $\Lambda \in X'$. Here, if $x \in A$, then $\langle \Lambda, x\rangle \leqslant \mu A(x) \leqslant 1$, and $\langle \Lambda, \hat{x}\rangle = \mu A(\hat{x}) \geqslant 1$.

Corollary 3 (Second separation theorem). *A non-empty convex closed subset of a locally convex space can be strictly separated by a hyperplane from any point which does not belong to it.*

We give an analytic formulation: $/X$ a lcs, $A \in \operatorname{Cl}\operatorname{Co}(X)$, $A \neq \varnothing$, $\hat{x} \notin A/ \Rightarrow \exists x' \in X': \sup_{x \in A} \langle x', x\rangle < \langle x', \hat{x}\rangle$.

Since A is closed and X is locally convex there is a convex neighbourhood V of $\hat{x}$ which does not intersect A. We seek a hyperplane separating V and A.

We recall that we have already touched on this theme twice—in Section 2.3 and in § 3 of this chapter, where we dealt with finite-dimensional convex analysis. Here we have deduced the separation theorems from the Hahn-Banach theorem;

this is a now a well known method. But the first theorem could be proved directly by induction, essentially as in § 3, but using Zorn's lemma.

4.2. The Krejn-Mil'man Theorem. We recall, that a point x belonging to a convex set A is called *extreme* if there is no segment $\Delta = [x_1, x_2]$ such that $x_1 \in A$, $x_2 \in A$ and $x \in \operatorname{int} \Delta$, that is, $x = \alpha x_1 + (1 - \alpha)x_2$, $0 < \alpha < 1$. The set of extreme points of A is denoted $\operatorname{ext} A$.

Theorem 7 (M.G. Krejn, D.P. Mil'man). *A nonempty compact convex set in a locally convex space is the convex closure of its extreme points.*

We note that in an infinite-dimensional space $\operatorname{ext} A$ may turn out to be not closed.

This theorem is an infinite-dimensional analogue of Minkowski's theorem, which we discussed in the introduction.

4.3. The Banach-Alaoglu-Bourbaki Theorem

Theorem 8. *The polar of a neighbourhood of zero, in any topology compatible with duality, is a weakly compact set in the dual space.*

This result is a corollary of Tikhonov's theorem on the compactness of the Tikhonov product of compact sets. For the proof see Kutateladze [1983], Holmes [1975].

4.4. Appendix. Here we state two results which are often used in concrete examples.

Theorem 9 (Mazur). *Let X be a Banach space and let a point x belong to the weak closure of a set $A \subset X$. Then there is a sequence of convex combinations of elements of A converging to x in norm.*

Theorem 10 (on strict separation). *Two disjoint convex sets of a locally convex space, one compact and the other closed can be strictly separated from each other.*

Chapter 2
Convex Calculus

In this chapter are concentrated the basic results of classical convex analysis: duality (§ 1), Legendre-Young-Fenchel transform (§ 2), the calculus of convex functions and subdifferential calculus (§ 3) and the calculus of convex sets and homogeneous functions (§ 4).

§ 1. Theorems on Involutiveness

In this section we prove the duality theorems, discussed in subsection 0.2 of the introduction.

1.1. Statement of the Theorems. Let X and Y be two spaces in duality (see Chap. 1, § 2), L a subspace, A a subset, K a cone in X, p a homogeneous function, f a proper function on X.

Theorem 1 (on involutiveness).

a) $L^{\perp\perp} = L$ *if and only if* L *is closed*:

$$L \in \mathrm{Lin}(X) \Rightarrow L^{\perp\perp} = L \Leftrightarrow L \in \mathrm{Lin}(X) \cap \mathrm{Cl}(X).$$

b) $\pi^2 A = A$ *if and only if* A *is convex, closed, and contains zero*:

$$\pi^2 A = A \Leftrightarrow A \in \mathrm{Co}(X) \cap \mathrm{Cl}(X) \quad \text{and} \quad 0 \in A.$$

c) $\sigma^2 K = K$ *if and only if* K *is convex and closed*:

$$\sigma^2 K = K \Leftrightarrow K \in \mathrm{Cone}(X) \cap \mathrm{Cl}(X).$$

d) $l^2 f = f$ *if and only if* $f \not\equiv +\infty$ *is convex and closed*:

$$l^2 f = f \Leftrightarrow f \in \mathrm{Cl}\,\mathrm{Co}(X, \mathbf{R}^\circ).$$

Theorem 2 (on the operators inverse to s and ∂).

a) $s\partial p = p$, *for* $\partial p \neq \varnothing$, *if and only if* p *is convex and closed*:

$$s\partial p = p \Leftrightarrow p \in \mathrm{Cl}\,\mathrm{S}\mathscr{L}(X, \bar{\mathbf{R}}).$$

b) $\partial s A = A$ *if and only if* A *is convex and closed*:

$$\partial s A = A \Leftrightarrow A \in \mathrm{Cl}\,\mathrm{Co}(X).$$

In view of the discussion in subsection 4.1, closure in Theorems 1 and 2 can be relative to any topology compatible with duality.

The result b) in Theorem 1 is usually called the *theorem on bipolars*, d) is called the *Fenchel-Moreau theorem.*

All of these results are direct corollaries of the separation theorems. We will prove the Fenchel-Moreau theorem first (since it is the most adaptable in the applications to the theory of extremal problems), and then deduce from it all the remaining results.

1.2. The Proof of the Fenchel-Moreau Theorem

A) **Lemma 1.** a) *The functions* lf *and* $l^2 f$ *are convex and closed.*

b) *Young's inequality holds*:

$$\langle x, y \rangle \leqslant f(x) + (lf)(y) \; \forall x \in X, y \in Y.$$

c) $l^2 f(x) \leqslant f(x) \; \forall x.$

Proof of the lemma. It is clear from the definition of the Young transform that lf is the least upper bound of the family of affine functions $y \to \langle x, y \rangle - f(x)$, and therefore its epigraph is an intersection of closed half-spaces (since $\operatorname{epi}(\bigvee_\alpha f_\alpha) = \bigcap_\alpha \operatorname{epi} f_\alpha$) and thus will be a closed set, that is, $lf \in \operatorname{Cl}\operatorname{Co}(Y, \bar{\mathbf{R}})$. It is proved in a similar way that $l^2 f \in \operatorname{Cl}\operatorname{Co}(X, \bar{\mathbf{R}})$. b) follows immediately from the definition.

c) From Young's inequality $\langle x, y \rangle - lf(y) \leqslant f(x)$, hence $f(x) \geqslant \sup_y(\langle x, y \rangle - lf(y)) =: l^2 f(x)$.

B) Necessity follows directly from part a) of Lemma 1.

C) **Lemma 2.** *The conjugate function of a closed proper function is proper.*

Proof of the lemma. Let g be a proper function $\Rightarrow$ (def) $g(x) > -\infty\ \forall x$ and $\exists \xi: g(\xi) < +\infty \Rightarrow lg(y) =: \sup_x(\langle x, y \rangle - g(x)) \geqslant \langle \xi, y \rangle - g(\xi) > -\infty$. It remains to show that $\operatorname{dom} lg \neq \varnothing$. Consider the point $(g(\xi) - 1, \xi)$. It does not belong to $\operatorname{epi} g$ and, consequently, by the second separation theorem, can be strictly separated from $\operatorname{epi} g$ by a non-zero linear functional, that is, $\exists \gamma \in \mathbf{R}, y \in Y$: $\gamma(g(\xi) - 1) + \langle \xi, y \rangle > \sup\{\gamma\alpha + \langle x, y \rangle: (\alpha, x) \in \operatorname{epi} g\}$. Clearly the inequality $\gamma > 0$ is impossible (otherwise the right hand side becomes $+\infty$), $\gamma = 0$ is likewise impossible since we would then have $\langle \xi, y \rangle > \langle \xi, y \rangle$. Thus $\gamma < 0$ and, dividing both sides of the inequality by $|\gamma|$ and putting $\eta = y/|\gamma|$, we obtain that $lg(\eta) < +\infty$.

D) *Sufficiency.* If $f \equiv \infty$, then the equality $f = l^2 f$ is trivial. The inequality $l^2 f \leqslant f$ follows from Lemma 1. Let f be convex, closed and $\exists x_0 \in \operatorname{dom} l^2 f$, where $l^2 f(x_0) < lf(x_0)$. By the second separation theorem we can separate the point $(l^2 f(x_0), x_0)$ from the convex closed set $\operatorname{epi} f$. This means that there is a number β and a $y \in Y$, not simultaneously equal to zero, such that

$$\beta l^2 f(x_0) + \langle x_0, y \rangle > \sup\{\beta\alpha + \langle x, y \rangle: (\alpha, x) \in \operatorname{epi} f\}. \tag{$*$}$$

In $(*)$ β cannot be a positive number since then the supremum on the right hand side would be equal to $+\infty$. The case $\beta = 0$ is also impossible since then, choosing $y_1 \in \operatorname{dom} lf$ (by Lemma 2), we obtain:

$$\begin{aligned} lf(y_1 + ty) &=: \sup\{\langle x, y_1 + ty \rangle - f(x): x \in \operatorname{dom} f\} \\ &\leqslant \sup\{\langle x, y_1 \rangle - f(x): x \in \operatorname{dom} f\} + t \cdot \sup\{\langle x, y \rangle: x \in \operatorname{dom} f\} \\ &= lf(y_1) + t \cdot \sup\{\langle x, y \rangle: x \in \operatorname{dom} f\}. \end{aligned}$$

Thus

$$\begin{aligned} l^2 f(x_0) &\geqslant \langle x_0, y_1 + ty \rangle - lf(y_1 + ty) \\ &\geqslant \langle x_0, y_1 \rangle - lf(y_1) + t(\langle x_0, y \rangle - \sup\{\langle x, y \rangle: x \in \operatorname{dom} f\}) \to \infty \end{aligned}$$

as $t \to \infty$, that is, $x_0 \notin \operatorname{dom} l^2 f$, contrary to assumption. Thus $\beta < 0$. Dividing both sides of $(*)$ by $|\beta|$ and putting $y_0 = |\beta^{-1}|y$, we obtain

$$\langle x_0, y_0 \rangle - l^2 f(x_0) > \sup\{\langle x, y \rangle - \alpha: (\alpha, x) \in \operatorname{epi} f\} =: lf(y_0),$$

that is,

$$\langle x_0, y_0 \rangle > l^2 f(x_0) + lf(y_0)$$

in contradiction to Young's inequality.

We will denote the collection of closed proper convex functions by $\operatorname{Cl}\operatorname{Co}(X, \mathbf{R})$.

1.3. Some Relations Between the Basic Operators. Before proceeding to the remaining proofs we quote some of the simpler formulae relating the operators introduced earlier. Subsequently A is a convex set, p is a convex homogeneous function and K is a convex cone. Then the following equalities are true:

1. $l\delta A =: sA$.
2. $l\mu A = \delta\pi A$.

$$(l\mu A)(y) =: \sup_x (\langle x, y \rangle - \inf\{\alpha > 0: x/\alpha \in A\})$$

$$= \sup_x \sup_{x \in \alpha A} (\langle x, y \rangle - \alpha) =: \begin{cases} 0, & y \in \pi A, \\ \infty, & y \notin \pi A \end{cases} =: \delta\pi A$$

3. $0 \in A \Rightarrow \mu\pi A = sA$.

In fact, $0 \in A \Rightarrow sA(y) \geqslant 0 \ \forall y$. If $sA(y) = 0$, then this means that the whole of the ray $\alpha y \in \pi A \Rightarrow \mu\pi A(y) = 0$. Let $sA(y) > 0 \Rightarrow \forall \alpha > 0$ we obtain: $sA(y/\alpha) = \alpha^{-1} sA(y) \Rightarrow sA(y) = \inf\{\alpha: \langle y/\alpha, x \rangle \leqslant 1 \ \forall x \in \pi A\} =: \mu\pi A(y)$.

4. $lp = \delta\partial p$.

In fact

$$lp(y) =: \sup_x (\langle x, y \rangle - p(x)) = \begin{cases} 0, & \langle x, y \rangle \leqslant p(x) \ \forall x \\ \infty, & \text{otherwise} \end{cases} =: \delta\partial p.$$

Relations 5.–7. follow at once from the definitions:

5. $\sigma K = -\pi K$.
6. $sK = \delta\sigma K$.
7. $\delta\pi K = sK$.

Let A_1 and A_2 be convex sets in X. Relations 8.–10. follow from the definitions

8. $\delta(A_1 + A_2) = \delta A_1 \oplus \delta A_2$.
9. $\delta(A_1 \cap A_2) = \delta A_1 \vee \delta A_2 = \delta A_1 + \delta A_2$.
10. $\delta(A_1 \operatorname{co}\cup A_2) = \delta A_1 \operatorname{co}\wedge \delta A_2$.
11. $0 \in A_1 \cap A_2 \Rightarrow \mu A_1 + \mu A_2 = \mu(A_1 \boxplus A_2)$.

Let $x \in A_1 \boxplus A_2$. This means that $\exists \hat{\alpha}$, $0 \leqslant \hat{\alpha} \leqslant 1$: $x \in \hat{\alpha} A_1 \cap (1 - \hat{\alpha}) A_2 \Rightarrow \mu A_1(x) + \mu A_2(x) \leqslant 1$. If $\mu A_1(x) = \alpha$, $\mu A_2(x) = \beta$, $\alpha + \beta = 1$, then $\forall a$, $0 \leqslant a \leqslant 1 \Rightarrow ax \in \alpha A_1$, $ax \in \beta A_2 \Rightarrow \mu(A_1 \boxplus A_2) \leqslant 1$.

The next property follows immediately from the definition:

12. $\mu(A_1 \cap A_2) = \mu A_1 \vee \mu A_2$.

The next formula is proved by analogy with 11.

13. $\mu(A_1 \operatorname{co}\cup A_2) = \mu A_1 \operatorname{co}\wedge \mu A_2$.

1.4. The Proofs of Theorems 1 and 2. The necessity in a)–b) is obvious. We will prove the sufficiency of b). Let A be convex, closed and contain zero. Then

$$\delta\pi^2 A \overset{2.}{=}: l\mu\pi A \overset{3.}{=} lsA \overset{1.}{=} l^2\delta A = \delta A.$$

The last equality is the Fenchel-Moreau theorem. As a result $\pi^2 A = A$.

Sufficiency in c). Let K be a convex and closed cone. Then

$$\sigma^2 K \overset{5.}{=} (-\pi)^2 K = \pi^2 K = K.$$

The last equality is from b).

Sufficiency in a). Let L be closed. Then by b)

$$L^{\perp\perp} =: \pi^2 L = L.$$

Let us prove Theorem 2.

Necessity in a) is obvious since sA is a closed function. Now let $p \in \mathrm{Cl}\, S\mathscr{L}(X, \bar{\mathbf{R}})$. Then by the Fenchel-Moreau theorem,

$$p = l^2 p =: l(lp) \overset{4.}{=} l\delta\partial p \overset{1.}{=} s\partial p.$$

Necessity in b) is also obvious since the subdifferential is always closed (as we recall, it is even compact). Now let $A = \mathrm{Cl}\,\mathrm{Co}(X)$. Then, by the Fenchel-Moreau theorem,

$$\delta A = l^2\delta A =: l(l\delta A) \overset{1.}{=} lsA \overset{4.}{=} \delta\partial sA \Rightarrow A = \partial sA.$$

§ 2. The Legendre-Young-Fenchel Transform

In this section we will proves some formulae for the Legendre transform of a function which are obtained using the operations described in Chap. 1, § 1.4.

2.1. Statements of the Theorems. In Chap. 1 § 1.4 six operations were introduced: $+$, $\oplus$, $\vee$, $\mathrm{co}\wedge$, the inverse image and the image under a linear mapping. Corresponding to each of these operations there is a formula for the Legendre-Young-Fenchel transform. In three ("non-local", see § 1.4) cases the formulae are true without restrictions; the remaining three cases require certain conditions. There is a set of sufficient conditions guaranteeing the validity of the formulae. Most of these can be amalgamated into a single scheme of "general position". There is a more detailed discussion of this in the final chapter of this part. Here we choose very simple sufficient conditions in order to clarify the structure of the proofs.

In what follows, if a formula is true without any restrictions (although we will always assume convexity), we will write $=$, if certain "general position" conditions are needed we will write $\cong$.

Theorem 3 (on the calculus of the Legendre-Young-Fenchel transform). *There are the following formulae:*

1. $l(f_1 \oplus f_2) = lf_1 + lf_2$.
2. $l(f_1 + f_2) \cong lf_1 \oplus lf_2$.
3. $l(f_1 \operatorname{co}\bigwedge f_2) = lf_1 \bigvee lf_2$.
4. $l(f_1 \bigvee f_2) \cong lf_1 \operatorname{co}\bigwedge lf_2$.
5. $l(\Lambda f) = lf\Lambda^*$.
6. $l(f\Lambda) \cong \Lambda^* f$.

In 1–4 X and Y are spaces in duality, $f_i \in \operatorname{Co}(X, \bar{\mathbf{R}})$; in 5 and 6 Λ is a continuous linear mapping from X to a linear topological space Z; in 5 $f \in \operatorname{Co}(X, \mathbf{R})$ and in 6 $f \in \operatorname{Co}(Z, \mathbf{R})$.

In formulae 2, 4 and 6 f_1, f_2 and f are convex and proper.

For equality in 2, it is sufficient to assume that there is a point $x_0 \in \operatorname{dom} f_2$, at which f_1 is continuous; for equality in 4. it is sufficient that f_1 and f_2 have compact support in X and f_1 is continuous; for equality in 6. it is sufficient that there is a point $z_0 \in \operatorname{Im} \Lambda$, at which f is continuous.

2.2. Proofs.

1. $$l(f_1 \oplus f_2)(y) =: \sup_x (\langle x, y\rangle - \inf_\xi (f_1(x - \xi) + f_2(\xi)))$$
$$\overset{(x - \xi =: \eta)}{=} \sup_{\xi, \eta} (\langle \xi, y\rangle - f_1(\xi) + \langle \eta, y\rangle - f_2(\eta))$$
$$=: lf_1(y) + lf_2(y).$$

2. A) It follows from Young's inequality that

$$lf_1(y) + lf_2(y_2) \geqslant \langle x, y_1 + y_2\rangle - f_1(x) - f_2(x) \; \forall y_1, y_2$$
$$\overset{\text{def}}{\Rightarrow} lf_1(y_1) + lf_2(y_2) \geqslant l(f_1 + f_2)(y_1 + y_2)$$
$$\Rightarrow (lf_1 \oplus f_2)(y) =: \inf(lf_1(y_1) + lf_2(y_2) \geqslant l(f_1 + f_2)(y).$$

Hence it follows that 2 is true if $\operatorname{dom} l(f_1 + f_2) = \varnothing$.

B) Let $\operatorname{dom} l(f_1 + f_2) \neq \varnothing$ and $\hat{y} \in \operatorname{dom}(f_1 + f_2) \Leftrightarrow l(f_1 + f_2)(\hat{y}) = \hat{\alpha} < \infty$. Since by assumption $\operatorname{dom}(f_1 + f_2) \neq \varnothing$, then $l(f_1 + f_2)$ is proper. In fact,

$$l(f_1 + f_2)(y) =: \sup_x (\langle x, y\rangle - (f_1 + f_2)(x)$$
$$\geqslant \langle \hat{x}, y\rangle - (f_1 + f_2)(\hat{x}) \; (\hat{x} \in \operatorname{dom}(f_1 + f_2)).$$

In particular $|\hat{\alpha}| < \infty$. Consider the set

$$A = \{(\alpha, x) \in \mathbf{R} \times X : \alpha \leqslant \langle x, \hat{y}\rangle - f_2(x) - \hat{\alpha}\}.$$

It is obviously convex. The set $\operatorname{int} \operatorname{epi} f \neq \varnothing$ by Theorem 1 of Chap. 1. Let us prove that $A \cap \operatorname{int} \operatorname{epi} f_1 = \varnothing$. In fact, by Theorem 1 of Chap. 1, $(\alpha, x) \in A \cap \operatorname{int} \operatorname{epi} f_1 \Rightarrow f_1(x) < \alpha \leqslant \langle x, \hat{y}\rangle - f_2(x) - \hat{\alpha}$, which immediately implies the contradiction:

$$\hat{\alpha} < \langle x, \hat{y}\rangle - f_1(x) - f_2(x) \leqslant l(f_1 + f_2)(\hat{y}) = \hat{\alpha}.$$

C) Applying the first separation theorem: there is $(\beta, y_1) \in \mathbf{R} \times Y$:

$$\sup\{\beta\alpha + \langle x, y_1\rangle : (\alpha, x) \in \operatorname{epi} f_1\} \leqslant \inf\{\beta\alpha + \langle x, y_1\rangle : (\alpha, x) \in A\}.$$

It is clear that $\beta \leqslant 0$. If $\beta = 0$, then it follows from the inequality that $\operatorname{dom} f_1$ and $\operatorname{dom} f_2$ are separated, which is impossible since $\operatorname{int} \operatorname{dom} f_1 \cap \operatorname{dom} f_2 \neq \varnothing$ by assumption. Thus $\beta < 0$. Dividing both sides of the inequality by $|\beta|$ and putting $y = |\beta|^{-1} y_1$, we obtain:

$$\begin{aligned} lf_1(y) &=: \sup_x (\langle x, y\rangle - f(x)) \\ &=: \sup_{(\alpha, x) \in \operatorname{epi} f_1} (\langle x, y\rangle - \alpha) \leqslant \inf_{(\alpha, x) \in A} (\langle x, y\rangle - \alpha) \\ &=: \inf_{x \in \operatorname{dom} f} (\langle x, y - \hat{y}\rangle + f_2(x)) + \hat{\alpha} =: -lf_2(\hat{y} - y) + \hat{\alpha} \end{aligned}$$

Hence it follows that

$$\begin{aligned} l(f_1 \oplus f_2)(y) &\overset{\text{def}}{\leqslant} lf_1(y) + lf_2(\hat{y} - y) \\ &\leqslant \hat{\alpha} = l(f_1 + f_2)(y). \end{aligned}$$

The proofs of the remaining formulae proceed in exactly the same way.

2.3. Some Properties of the Legendre-Young-Fenchel Transform. Firstly we direct the attention of the reader to the geometric meaning of the Legendre-Young-Fenchel transform (see Fig. 12). And, in addition, we will write down several properties of this transformation (some of which have been mentioned), which continually turn out to be necessary in working with conjugate functions.

1. $l^2 f \leqslant f$ and $l^2 f$ is the convex closure of f, that is, it is the largest closed convex function not exceeding the given function.
2. $f \leqslant g \Rightarrow lf \geqslant lg$, $l^2 f \leqslant l^2 g$.
3. $f \in \operatorname{Co}(\mathbf{R}^n, \bar{\mathbf{R}}), \Rightarrow l^2 f = \operatorname{cl} f$.
4. $f \in \operatorname{Co}(\mathbf{R}^n, \bar{\mathbf{R}})$, $A \in \mathscr{L}(\mathbf{R}^n, \mathbf{R}^n)$, $A\mathbf{R}^n = \mathbf{R}^n$,

$$\begin{aligned} &f(x) = h(A(x - a)) + \langle x, a^*\rangle + \alpha \\ &\quad\Rightarrow lf(x^*) = lh(A^{*-1}(x^* - a^*)) + \langle x^*, a\rangle + \alpha^*, \\ &\alpha^* = -\alpha - \langle a, a^*\rangle. \end{aligned}$$

2.4. The Conjugate Function of a Convolution Integral. Let us say a few words on the continuous analogue of the formula $l(f_1 \oplus f_2) = lf_1 + lf_2$.

Let (T, Σ, μ) be a space with a positive measure, $f\colon T \times \mathbf{R}^n \to \mathbf{R}$ an integrand. The *convolution integral* of f with μ is the function on $\mathbf{R}^n$ defined by the equality

$$\left(\oint_T f_1 \, d\mu\right)(x) = \inf\left\{\alpha \in R \colon (\alpha, x) \in \int_T \operatorname{epi} f_t \, d\mu\right\}.$$

Here we are integrating a many-valued mapping $t \to \operatorname{epi} f_t$. If the integrand f is superpositionally measurable (see Ioffe & Tikhomirov [1974]), then the value of the convolution integral coincides with the value of the problem:

$$\int_T f(t, x(t))\, d\mu \to \inf, \qquad \int_T x(t)\, d\mu = x,$$

to which can be reduced, in particular, linear problems in optimal control. The following formula holds (if f is a measurable integrand and $\operatorname{dom} \oint_T f_t\, d\mu \neq \varnothing$):

$$l\left(\oint_T f_t\, d\mu\right) = \int_T lf_t\, d\mu.$$

If, in addition, μ is continuous, then $\oint_T f_t\, d\mu$ is a convex function,

$$\operatorname{cl} \oint_T f_t\, d\mu = \operatorname{cl} \oint_T l^2 f_t\, d\mu.$$

For more details on this theme see Ioffe & Tikhomirov [1974].

§ 3. The Calculus of Convex Sets and Sublinear Functions

Here we concentrate formulae about the operators of convex analysis (subdifferential, support, polar, annihilator, conjugate for cones, functions and sets), obtained by means of the operators described in subsection 1.4 of Chap. 1.

3.1. Tables of Basic Formulae. In this subsection we will do the same work as was done in the last section, only with respect to the operators ∂, π, s, $\perp$, σ and the operations $+$, $\boxplus$, $\oplus$, $\operatorname{co}\bigwedge$, $\bigvee$, $\operatorname{co}\bigcup$ and $\bigcap$. In a number of cases the equalities always hold; we will then write $=$. In other cases the formulae hold only in the presence of convexity and the imposition of certain conditions of general position type. In those cases we will write $\cong$.

I. The formulae of subdifferential calculus.

I_1. $\partial(p_1 + p_2) \cong \partial p_1 + \partial p_2$.

I_2. $\partial(p_1 \nabla p_2) = \partial p_1 \boxplus \partial p_2$.

I_3. $\partial(p_1 \bigvee p_2) \cong \partial p_1 \operatorname{co} \bigcup \partial p_2$.

I_4. $\partial(p_1 \operatorname{co}\bigvee p_2) = \partial p_1 \bigcap \partial p_2$.

II. Formulae for the polar. Here, in all cases, $0 \in A_i$, $i = 1, 2$.

II_1. $\pi(A_1 + A_2) = \pi A_1 \boxplus \pi A_2$.

II_2. $\pi(A_1 \boxplus A_2) \cong \pi A_1 + \pi A_2$.

II_3. $\pi(A_1 \bigcap A_2) \cong \pi A_1 \operatorname{co}\bigcup \pi A_2$.

II_4. $\pi(A_1 \operatorname{co}\bigcup A_2) = \pi A_1 \bigcap \pi A_2$.

III. Formulae for support functions.

III_1. $s(A_1 + A_2) = sA_1 + sA_2$.

III_2. $s(A_1 \boxplus A_2) \cong sA_1 \nabla sA_2$.

III_3. $s(A_1 \bigcap A_2) \cong sA_1 \operatorname{co}\bigwedge sA_2 \cong sA_1 \oplus sA_2$.

III_4. $s(A_1 \operatorname{co}\bigcup A_2) = sA_1 \bigvee sA_2$.

IV. Formulae for annihilators.

IV_1. $(L_1 + L_2)^\perp = L_1^\perp \cap L_2^\perp$.

IV_2. $(L_1 \cap L_2)^\perp \cong L_1^\perp + L_2^\perp$.

V. Formulae for conjugate cones.

V_1. $\sigma(K_1 + K_2) = \sigma K_1 \cap \sigma K_2 \Leftrightarrow (K_1 + K_2)^* = K_1^* \cap K_2^*$.

V_2. $\sigma(K_1 \cap K_2) \cong \sigma K_1 + \sigma K_2 \Leftrightarrow (K_1 \cap K_2)^* \cong K_1^* + K_2^*$.

3.2. Proofs. All the relations I_1–VI_4 are obtained below by purely formal application of the relations already obtained in §§ 1 and 2 of this chapter. It is a question of relations 1–10 (§ 1) and 1–6 (§ 2). They turn out to be true under hypotheses, which guarantee the formulae of § 2. Below we will write 1.5, 2.3, meaning relation 5 (§ 1) and 3 (§ 2) respectively, etc.

I_1. $\delta(\partial p_1 + \partial p_2) \overset{1.8}{=} (\delta\partial p_1) \oplus (\delta\partial p_2) \overset{1.4}{=} lp_1 \oplus lp_2 \overset{2.2}{\cong} l(p_1 + p_2) \overset{1.4}{=} \delta\partial(p_1 + p_2)$.

I_2. $s(\partial p_1 \boxplus \partial p_2) \overset{\mathrm{def}}{=} s\partial p_1 \triangledown s\partial p_2 \overset{\mathrm{Th.2a)}}{=} p_1 \triangledown p_2 \overset{\mathrm{Th.2b)}}{\Rightarrow} \partial p_1 \boxplus \partial p_2 = \partial(p_1 \triangledown p_2)$.

I_3. $\delta\partial(p_1 \vee p_2) \overset{1.4}{=} l(p_1 \vee p_2) \overset{2.4}{=} lp_1 \operatorname{co}\wedge lp_2 \overset{1.4}{=} \delta\partial p_1 \operatorname{co}\wedge \delta\partial p_2)$

$\overset{1.10}{=} \delta(\partial p_1 \operatorname{co}\cup \partial p_2)$.

I_4. $\delta(\partial p_1 \cap \partial p_2) \overset{1.9}{=} \delta\partial p_1 \vee \delta\partial p_2 \overset{1.4}{=} lp_1 \vee lp_2 \overset{2.4}{=} l(p_1 \operatorname{co}\wedge p_2) \overset{1.4}{=} \delta\partial(p_1 \operatorname{co}\wedge p_2)$.

II_1. $\mu\pi(A_1 + A_2) \overset{1.3,1.1}{=} l\delta(A_1 + A_2) \overset{1.8}{=} l(\delta(A_1 \oplus \delta A_2) \overset{2.1}{=} l\delta A_1 + l\delta A_2$

$\overset{1.3,1.1}{=} \mu\pi A_1 + \mu\pi A_2 \overset{1.11}{=} \mu(\pi A_1 \boxplus \pi A_2)$.

II_2. $\delta\pi(A_1 \boxplus A_2) \overset{1.2}{=} l\mu(A_1 \boxplus A_2) \overset{1.11}{=} l(\mu A_1 + \mu A_2) \overset{2.2}{\cong} l\mu A_1 + l\mu A_2$

$\overset{1.2}{=} \delta\pi A_1 \oplus \delta\pi A_2 \overset{1.8}{=} \delta(\pi A_1 + \pi A_2)$.

II_3. $\delta\pi(A_1 \cap A_2) \overset{1.2}{=} l\mu(A_1 \cap A_2) \overset{1.12}{=} l(\mu A_1 \vee \mu A_2) \overset{2.4}{\cong} l\mu A_1 \operatorname{co}\wedge l\mu A_2$

$\overset{1.2}{=} \delta\pi A_1 \operatorname{co}\wedge \delta\pi A_2 \overset{1.10}{=} \delta(\pi A_1 \operatorname{co}\cup \pi A_2)$.

II_4. $\delta\pi(A_1 \operatorname{co}\cup A_2) \overset{1.2}{=} l\mu(A_1 \operatorname{co}\cup A_2) \overset{1.12}{=} l(\mu A_1 \operatorname{co}\wedge \mu A_2) \overset{2.3}{=} l\mu A_1 \vee l\mu A_2$

$\overset{1.2}{=} \delta\pi A_1 \vee \delta\pi A_2 = \delta(\pi A_1 \cap \pi A_2)$.

III_1. follows from the definitions.

III_2. $s(A_1 \boxplus A_2) \overset{1.3}{=} \mu\pi(A_1 \boxplus A_2) \overset{\mathrm{II}_2}{=} \mu(\pi A_1 + \pi A_2) \overset{\mathrm{def}}{=} \mu\pi A_1 \triangledown \mu\pi A_2$

$\overset{1.3}{=} sA_1 \triangledown sA_2$.

III_3. $s(A_1 \cap A_2) \overset{1.1}{=} l\delta(A_1 \cap A_2) \overset{1.8}{=} l(\delta A_1 \vee \delta A_2) \overset{2.4}{\cong} l\delta A_1 \operatorname{co}\wedge l\delta A_2$

$\overset{1.1}{=} sA_1 \operatorname{co}\wedge sA_2$.

III_4. $s(A_1 \operatorname{co}\cup A_2) \overset{1.1}{=} l\delta(A_1 \operatorname{co}\cup A_2) \overset{1.8}{=} l(\delta A_1 \operatorname{co}\wedge \delta A_2) \overset{2.3}{=} l\delta A_1 \vee l\delta A_2$

$\overset{1.1}{=} sA_1 \vee sA_2$.

Formulae IV_1, IV_2, V_1 and V_2 follow from II_1–II_4, if we take account of $L^\perp = \pi L$, $\sigma K = -\pi K$.

§4. Subdifferential Calculus

In this section we discuss the fundamental theorems on subdifferentials.

4.1. The Simplest Properties of Subdifferentials. In subsection 2.3 of Chapter 1 the subdifferential $\partial f(\hat{x})$ of a function f at a point $\hat{x}$ was defined as the set $\{y \in Y: f(x) - f(\hat{x}) \geqslant \langle x - \hat{x}, y\rangle\ \forall x \in X\}$. The definition of the subdifferential of a convex homogeneous function p was given there also. From these definitions it follows that $\partial p = \partial p(0)$, that is, the subdifferential of a convex homogeneous function coincides with its subdifferential at the point zero. We can give another definition of the subdifferential at a point. Let $f: X \to \bar{\mathbf{R}}$ have derivatives in all directions at $\hat{x}$. Then $\partial f(\hat{x}) = \partial f'(\hat{x}, 0)$ (subdifferentiating with respect to the second argument). The elements of the subdifferential are sometimes called *subgradients*. f is said to be *subdifferentiable at* $\hat{x}$, if $\partial f(\hat{x}) \neq \varnothing$.

Below X is a locally convex space.

Examples. 1. An affine function $f(x) = \langle x^*, x\rangle + \alpha$ is subdifferentiable at any point and $\partial f(x) = \{x^*\}$.

2. The subdifferential of the indicator function $\delta A(\cdot)$ at any point $\hat{x}$ coincides with the normal cone to A at the point $\hat{x}$:

$$\partial \delta A(\hat{x}) =: NA(\hat{x}) =: \{x^* \in X^*: \langle x^*, x - \hat{x}\rangle \leqslant 0\ \forall x \in A\}.$$

3. The subdifferential of the norm at the origin in a normed space X, as is easy to see, is the unit sphere in X^*.

Lemma 3. *Let f be a convex function on X. The following conditions are equivalent*:

$$\text{a)}\quad x^* \in \partial f(x),$$

$$\text{b)}\quad f(x) + f^*(x^*) = \langle x^*, x\rangle.$$

A convex proper function f is subdifferentiable at $\hat{x} \in \operatorname{dom} f$ if and only if the function $x \to f'(\hat{x}, x)$ is closed at zero.

Theorem 4 (on the compactness of the subdifferential). *Let f be a convex function continuous at a point $\hat{x}$. Then its subdifferential $\partial f(x)$ is a non-empty weak* compact set in X^*.*

This theorem is a corollary of the Banach-Alaoglu-Bourbaki theorem.

We recall that even in the finite-dimensional case the subdifferential may be empty (Example 6d, §2.4 of Chap. 1). In this case ($f \in \operatorname{Co}(\mathbf{R}^n, \bar{\mathbf{R}})$, $|f(\hat{x})| \leqslant \infty$, $\partial f(\hat{x}) = \varnothing$) there must be $x \in \mathbf{R}^n$ such that $f'(\hat{x}, x) = -\infty$. From the above-mentioned it follows that if f is continuous at $\hat{x}$, then $f'(\hat{x}, x) = s(\partial f(\hat{x}))(x)$. From this it can be easily deduced that, under the conditions of Theorem 4, f is Gateau differentiable if and only if $\partial f(\hat{x})$ has just one element.

4.2. The Fundamental Formulae of Subdifferential Calculus. The theorems below are similar to the corresponding results from differential calculus.

Theorem 5 (composition). a) *Let X and Y be locally convex spaces, $\Lambda \in \mathscr{L}(X, Y)$, g a convex function on Y, $f(x) = g(\Lambda x)$. If g is continuous at some point $\bar{y} = \Lambda\bar{x}$, then for any $x \in X$, $\partial f(x) = \Lambda^* \partial g(\Lambda x) \Leftrightarrow \partial(g \circ \Lambda)(x) = \partial(g(\Lambda x)) \circ \Lambda$, where Λ^* is the adjoint operator to Λ.*

b) *Let $f_1, \ldots, f_n$ be convex functions on X, $F = (f_1, \ldots, f_n)$, φ a convex monotone function on $\mathbf{R}^n$ (that is, $x \leqslant y \Rightarrow \varphi(x) \leqslant \varphi(y)$), $f(x) = \varphi(F(x))$. Then, if F is continuous at $\hat{x}$ and φ is continuous at $F(\hat{x})$,*

$$\partial f(\hat{x}) = \left\{x^* \in X^*: x^* = \sum_{i=1}^{n} \alpha_i x_i^* x_i^* \in \partial f_i(\hat{x}), (\alpha_1, \ldots, \alpha_n) \in \partial\varphi(F(\hat{x}))\right\}$$

$$\Leftrightarrow \partial(\varphi \circ F)(\hat{x}) = \partial\varphi(F(\hat{x})) \circ \partial F(\hat{x}).$$

Theorem 6 (the subdifferential of a sum). *Let $f_1, \ldots, f_n$ be convex functions, and moreover at some point $\bar{x} \in \bigcap_{i=1}^{n} \operatorname{dom} f_i$ let all the functions, except possibly one, be continuous. Then for any $x \in X$*

$$\partial\left(\sum_{i=1}^{n} f_i\right)(x) = \sum_{i=1}^{n} \partial f_i(x).$$

Theorem 7 (the subdifferential of the maximum). *Let $f_1, \ldots, f_n$ be convex functions continuous at a point $\hat{x}$. Then*

$$\partial(f_1 \vee \cdots \vee f_n)(\hat{x}) = \left\{\sum_{i \in T(\hat{x})} \lambda_i x_i' : \lambda_i \in \mathbf{R}_+,\ \sum_{i \in T(\hat{x})} \lambda_i = 1, x' \in \partial f_i(x)\right\},$$

where

$$T(\hat{x}) = \{i: f_i(\hat{x}) = (f_1 \vee \cdots \vee f_n)(\hat{x})\}.$$

The theorems stated here are fundamental to the subdifferential calculus. They are all proved by similar means and, as a matter of fact, all are equivalent to the separation theorem. Using the theorems and formulae of the previous sections these results are automatic.

Theorem 5a) was first proved by Rockafellar, 5b) and 7 by A. Ya Dubovitskij and A.A. Milyutin (the latter is sometimes called the *Dubovitskij-Milyutin theorem*), Theorem 3 by Moreau and Rockafellar (and it is usually called the *Moreau-Rockafellar theorem*).

These theorems have a natural completion in the composition theorem for a convex and a monotone mapping in an ordered space, Theorem 6—to the case of integrals depending on parameters and to operators in ordered spaces. This topic is considered in § 5 of Chap. 2 and Chap. 4. But the question of how to generalize Theorem 7, will be presented in the next subsection.

4.3. The Refinement Theorem

Theorem 8. *Let T be a compact topological space, $f: T \times \mathbf{R}^n \to \mathbf{R}$ ($f = f(t, x)$) a function which is convex relative to x for each $t \in T$ and upper semicontinuous* *with respect to t for each* $x \in \mathbf{R}^n$, *where for each* $t \in T$ *the function* $x \to f(t, x)$ *is continuous at* $\hat{x}$.

Put $\varphi(x) = \max_{t \in T} f(t, x)$. *Then* $\operatorname{dom} \varphi = \operatorname{dom} f(t, \cdot)$ $\forall t \in T$ *and for any point* $x \in \operatorname{dom} \varphi$ *each element* $y \in \partial\varphi(x)$ *may be represented in the form* $y = \sum_{i=1}^{r} \alpha_i y_i$, *where* $r \leqslant n + 1$, $\sum_{i=1}^{r} \alpha_i = 1$ $\alpha_i > 0$, $y_i \in \partial f(\tau_i, \cdot)(x)$, $\tau_i \in T_0(x) =: \{t \in T: f(t, x) = \varphi(x)\}$.

This theorem was proved by V.L. Levin 1969. It was preceded by a long history, the beginning of which, in principle, can be found in the classical work of P.L. Chebyshev (Oeuvres, 1947), then in the works of Vallée-Poussin, L.G. Shnirel'man. S.I. Zukhovitskij and S.B. Stechkin, S.Ya. Khavinson and many others.

Subdifferential calculus in a finite-dimensional space has been presented in detail in the monograph by Rockafellar, 1970. The infinite-dimensional case has been presented in detail in the monographs Kantorovich, 1939; Brønsted, 1964 and the survey articles Ioffe & Levin, 1972; Ioffe & Tikhomirov, 1968.

§ 5. Convex Analysis in Spaces of Measurable Vector-Functions

5.1. Introduction. Convex analysis in spaces of measurable vector-functions, which arise in connection with the needs of the theory of extremal problems, has developed into an independent division of convex analysis, forming one of its perspective directions. A significant part of the recently published monograph Levin [1985], is devoted to it. The author of this monograph, V.L. Levin, has kindly provided material for this section.

In this section (T, Σ, μ) is a space with positive measure, with the so-called direct sum property (for the definition see Levin [1985]). We note that the most important classes of measures, σ-finite and Radon measures on locally compact spaces, have the direct sum property. Put $I_f(x(\cdot)) = \int_T f(t, x(t))\, d\mu$. An important rôle in this area is played by theorems on the duality and the subdifferentiation of such integral functionals. Regarding f, we will assume that it is a normal convex integrand (see Ioffe & Tikhomirov (1974)).

Let X be a Banach space and L a Banach space of μ-measurable functions $x(\cdot): T \to X$. Put $K_1(L) = \{A \in \Sigma: \mu(A) > 0, L_\infty(A, X) \subseteq L$ and this embedding is continuous$\}$. Suppose that L has the two properties: (L_1) for any $A \in \Sigma$, $0 < \mu(A) < \infty$ there is a $B \in K_1(L)$, $B \subset A$, (L_2) if $A \in K_1(L)$, $\mu(A) < \infty$, $x(\cdot) \in L$. $\chi_A(\cdot)x(\cdot) \in L_\infty(A, X)$, then $\|\chi_A(\cdot)x(\cdot)\|_{L_1} \leqslant \|x(\cdot)\|_{L_1}$. As examples we quote the spaces $L_p(T, X)$, $1 \leqslant p \leqslant \infty$, and also the Orlicz spaces generated by convex integrands. The most important tool of convex analysis in spaces of measurable vector-functions is the Lebesgue decomposition of the space L^* of linear functionals on L, according to which $x^* \in L^*$ is represented as a sum of absolutely continuous and singular functionals, where the subspaces of each are closed in L^*. Another important tool is the so-called theory of lifting. We give the definition. Denote by $\mathscr{L}_\infty = \mathscr{L}_\infty(T, \Sigma, \mu)$ the space of bounded μ-measurable functions $\varphi\colon T \to \mathbf{R}$ with the sup-norm and with no identification of μ-equivalent functions. $\mathscr{L}_\infty$ is a real commutative Banach algebra relative to pointwise multiplication. Here the mapping $\pi\colon \mathscr{L}_\infty \to L_\infty$ taking a function $\varphi \in \mathscr{L}_\infty$ to its class of μ-equivalent functions, is a homeomorphism of Banach algebras. A *lifting* of $\mathscr{L}_\infty$ is a homomorphism of Banach algebras $\rho\colon \mathscr{L}_\infty \to \mathscr{L}_\infty$ such that 1) $\pi\rho = \pi$, 2) $\pi(\varphi) = 0 \Rightarrow \rho(\varphi) = 0$, 3) $\rho(1) = 1$.

5.2. Subdifferentials of Convex Integral Functionals and Their Applications

Theorem 9 (the regular integral representation). *If X is separable and f is a normal convex integrand, then the subdifferential $\partial I_f(x_0(\cdot))$ admits a regular integral representation, that is,*

$$x^* \in \partial I_f(x_0(\cdot)) \Rightarrow \langle x^*, x(\cdot)\rangle = \int_T \langle x(t), x'(t)\rangle \, d\mu + \langle x_0^*, x(\cdot)\rangle,$$

where $x'(\cdot)\colon T \to X^*$ is weak μ-summable,*

$$x'(t) \in \partial f(t, \cdot)(x_0(t))\mu \text{ a.e.,} \quad x^* \in N(\operatorname{dom} I_f)(x_0(\cdot)).$$

This theorem implies the following *formula* for the *subdifferential of an integral*:

$$\partial\left(\int_T f_t \, d\mu\right)(x) = \int_T \partial f_t(x) \, d\mu + N\left(\operatorname{dom} \int_T f_t \, d\mu\right)(x).$$

Formula of this type in the finite-dimensional case were first obtained in Ioffe & Tikhomirov [1969]. In the finite-dimensional and separable cases the proof uses the existence and properties of measurable selections. In the non-separable case formulae of a similar kind are proved under certain conditions, expressed in terms of a lifting. With the help of the theory of lifting deep progress has been made in the theorems about the subdifferential of a maximum (see Levin [1985].

5.3. The Calculation of the Adjoint Functional I_f^*. Let X and Y be Banach spaces and suppose that on $X \times Y$ there is given a nondegenerate bilinear form $\langle x, y\rangle$ which is separately continuous with respect to each argument. Let L and M be vector spaces of μ-measurable functions $x(\cdot)\colon T \to X$, $y(\cdot)\colon T \to Y$. We will assume that $\langle x(\cdot), y(\cdot)\rangle \in L_1(T)\, \forall x(\cdot) \in L$, $y(\cdot) \in M$. Then there is a bilinear form on $L \times M$ defined by

$$\langle x(\cdot), y(\cdot)\rangle = \int_T \langle x(t), y(t)\rangle \, d\mu.$$

We will assume that it is nondegenerate with respect to $y(\cdot)$ ($\Rightarrow M$ can be identified with a subspace in $L^\#$).

A function $f\colon T \times X \to \mathbf{R} \cup \{+\infty\}$ is called a *convex L-integrand* if it is convex and lower semicontinuous with respect to x for μ a.a. $t \in T$, and the composition $t \to f(t, x(t))$ is μ-measurable for any $x(\cdot) \in L$.

Let f be a convex L-integrand and $I_f(x(\cdot)) > -\infty\ \forall x(\cdot) \in L$. Put $F(x(\cdot)) = f(\cdot, x(\cdot))$ for $f(\cdot, x(\cdot)) \in L_1(T)$ and $= e_\infty$ otherwise (where e_∞ is an improper element majorizing all the proper elements of $L_1(T)$, with the natural properties $L_1(T) \cup \{e_\infty\} =: \hat{L}_1$). It is clear that F is convex on L and $\operatorname{dom} F =: \{x(\cdot): F(x(\cdot)) \neq e_\infty\} = \operatorname{dom} I_f$. If $\operatorname{dom} F \neq \varnothing$, then we can consider the adjoint mapping $F^*: M \to \hat{L}_1$:

$$F^*(y(\cdot)) = \sup\{\langle x(\cdot), y(\cdot)\rangle - F(x(\cdot)): x(\cdot) \in \operatorname{dom} F\}.$$

Lemma 4. *If* $\operatorname{dom} I_f \neq \varnothing$ *and* L *satisfies* (L_1), *then*:

$$I_f^*(y(\cdot)) = I_{F^*}(y(\cdot)), \quad \forall y(\cdot) \in M,$$

$$I_{F^*}(y(\cdot)) =: \int_T F^*(y(\cdot))(t)\, d\mu.$$

This result, due to V.L. Levin, was one of the starting points for the abstract investigation of duality of convex mappings in ordered spaces (see § 2 of Chap. 4).

Theorem 10. *Let* X *be separable,* f *a normal convex integrand,* $\operatorname{dom} I_f \neq \varnothing$, $I_f(x(\cdot)) > -\infty\ \forall x(\cdot) \in L$ *and let* L *satisfy* (L_1). *Then*

$$I_f^*(y(\cdot)) = I_{f^*}(y(\cdot)), \quad \forall y(\cdot) \in M.$$

Theorems of this kind were first obtained by Rockafellar, Castaing and Valadier [1977] and others. For the generalization and development of this theorem see Levin [1985].

Chapter 3
Some Applications of Convex Analysis

Earlier it was said that convexity arises in a natural and interesting way in various branches of mathematics, and also in the natural sciences, engineering problems and economics. And, in the majority of cases, in connection with optimization problems. It was this requirement to solve and study extremal problems which, perhaps, to a large extent stimulated the separation of convex analysis into separate scientific directions. Basically this is the topic of this chapter.

We make several remarks on terminology. Every *extremal problem* has following elements: a functional $f: X \to \bar{\mathbf{R}}$, where X is the domain of definition of f, and $\bar{\mathbf{R}}$ is the extended real line, and a *constraint*, that is, a subset $C \subset X$. A constraint has been called the *set of feasible elements*. Here the problem is: "find the infimum (supremum) of $f(x)$ under the constraint $x \in C$" written as follows:

$$f(x) \to \inf(\sup), \qquad x \in C \tag{z}$$

The set of solutions of problem (z), that is, $\{\xi \in C: f(x) \geqslant f(\xi)(f(x) \leqslant f(\xi))$ $\forall \xi \in C\}$ is denoted $\operatorname{Arg} z$. If $\hat{x} \in \operatorname{Arg} z$ then we write that $\hat{x} \in \operatorname{absmin} z$ ($\operatorname{absmax} z$). The numerical value of the problem, that is, $f(\operatorname{Arg} z)$ is denoted S_z. The assignment of a problem in the form (z) is called *formalization*.

§ 1. Linear and Convex Programming

1.1. Statements of the Problems. Let X be a linear space, $f: X \to \bar{\mathbf{R}}$ a convex function on X and let the constraint C be given by a system of inequalities, equalities and inclusions of the following form:

$$C = \{x \in X: f_i(x) \leqslant 0, i = 1,\dots,m', f_i(x) = 0, i = m' + 1,\dots,m, x \in A\},$$

where the f_i are convex functions if $1 \leqslant i \leqslant m'$, affine functions if $i \geqslant m' + 1$, and A is a convex set in X. The problem (z) in this case is called a *convex programming problem*. If all the functions f_i are affine, and A is a polyhedron, then the problem

$$f_0(x) \to \inf, \qquad f_i(x) \leqslant 0, \qquad 1 \leqslant i \leqslant m', \qquad f_i(x) = 0,$$
$$i \geqslant m' + 1, \qquad x \in A \qquad (z_0)$$

is a *linear programming problem*. The expression $\sum_{i=1}^{m} \lambda_i f_i(x) =: \mathscr{L}(x, \lambda, \lambda_0)$, $\lambda = (\lambda_1,\dots,\lambda_m)$, $\lambda_i \in \mathbf{R}$, $0 \leqslant i \leqslant m$, is called the *Lagrange function* of (z_0), λ and λ_0 are called *Lagrange multipliers*.

Very often linear programming problems are discussed in the form ($X = \mathbf{R}^n$, $A = \mathbf{R}^n_+$):

$$\langle c, x\rangle \to \inf, \qquad Ax \geqslant b, \qquad x \geqslant 0, \qquad (z_1)$$

where $c \in \mathbf{R}^n$, $A = (a_{ij})$, $1 \leqslant i \leqslant m$, $i \leqslant j \leqslant n$. The symbol $z' \geqslant z$ for finite-dimensional vectors means that each component of z does not exceed the corresponding component of z'.

A particular case of a linear programming problem is the *transportation problem* (here given in the form due to L.V. Kantorovich and M.K. Gavurin):

$$\sum_{i,j=1}^{m} c_{ij}x_{ij} \to \inf, \qquad \sum_{i=1}^{m} x_{ik} - \sum_{j=1}^{m} x_{kj} = \rho_k,$$
$$k = 1,\dots,m, \qquad x_{ij} \geqslant 0 \qquad (z_2)$$

The values $|\rho_k|$ characterise the volumes of production (for $\rho_k > 0$) and requirements ($\rho_k < 0$) of a homogeneous product. It is assumed that the sum of the volumes of production is equal to the volume of requirements (that is, $\sum_{k=1}^{m} \rho_k = 0$). The quantity c_{ij} is the cost of transport from the i-th point to point j. It is required to transport with the least possible total cost.

The transportation problem, as with other problems of linear programming, finds its applications in economics. Here the basic results concerning duality have an interesting economic interpretation (we will not pause on this but refer the reader to the books Eremin & Astaf'ev [1976] and Makarov & Rubinov [1973]). We will return to this problem in § 3 of this chapter.

1.2. The Method of Duality in Convex Programming. It has been stressed repeatedly that convex objects have a dual description. This applies, in particular, to convex programming.

The basic principle for the construction of the dual problem is this. Consider the problem

$$f(x) \to \inf, \qquad x \in X, \tag{$\tilde{z}$}$$

where X is a linear (topological) space ((z) reduces to ($\tilde{z}$), if we put $f(x) = \infty$ for $x \notin C$). Let us put it into a class of similar problems depending on a parameter y, taking values in another linear (topological) space Y:

$$F(x, y) \to \inf, \qquad x \in X, \tag{$\tilde{z}(y)$}$$

where $F\colon X \times Y \to \bar{\mathbf{R}}$, $F(x, 0) = f(x)$.

F is called a *perturbation* of f, and the series ($\tilde{z}(y)$) a *perturbation* of ($\tilde{z}$). We will denote the numerical value of the problem $\tilde{z}(y)$ by $S(y)$, and call the function $y \to S(y)$ the *S-function*. If we assume that $F \in \mathrm{Co}(X \times Y, \bar{\mathbf{R}})$, then it is easy to show that $S(y) = \inf_x F(x, y) \in \mathrm{Co}(Y, \mathbf{R})$.

Theorem 1. *If S is lower semicontinuous at zero then*

$$S(0) = S^{**}(0) =: \sup_{y^*}(-F^*(0, y^*)). \tag{1}$$

This result follows immediately from the definitions and the Fenchel-Moreau theorem (according to which $S(0) = S^{**}(0)$). Thus, $S_{\tilde{z}} = S_{\tilde{z}^*}$, where

$$\varphi(y^*) \to \sup, \qquad y^* \in Y^*, \qquad \varphi(y^*) = -F^*(0, y^*) \tag{$\tilde{z}^*$}$$

This problem is called the *dual problem* to (z). We have arrived at a scheme

$$f(x) \to \inf, \quad x \in X, \quad (\tilde{z})$$

$$\varphi(y^*) \to \sup, \quad y^* \in Y^*, \quad (\tilde{z}^*)$$

$$F(x, y) \to \inf, \quad (\text{w.r.t. } x) \quad (\tilde{z}(y))$$

$$\Phi(x^*, y^*) =: -F^*(x^*, y^*) \to \sup(\text{w.r.t. } y^*), \quad (\tilde{z}^*(y^*))$$

$$f(x) = F(x, 0), \quad \varphi(y^*) = \Phi(0, y^*),$$

$$S_{\tilde{z}} = S_{\tilde{z}^*}$$

It is reasonable to apply the principle of duality when the dual problem is simpler or more visible than the primal. In addition, it very often happens that the dual problem has a solution, at precisely the same time as the primal does not. And, finally, it is usually useful to investigate both problems simultaneously.

These reasons explain the fruitfulness of duality in convex programming. For more details on this principle see Ekeland & Temam [1976].

1.3. The Fundamental Theorems of Linear Programming

Theorem 2 (existence). *If the set of feasible elements is non-empty and their values are finite then* (z_1) *has a solution.*

This fact follows immediately from the closure of finite-dimensional polyhedral sets. Applying the principle of duality, described in the previous subsection (putting $F(x, \eta) = \langle c, x\rangle$, if $Ax \geqslant b + \eta$, $x \geqslant 0$ and $= +\infty$ otherwise), we arrive at a pair of dual problems

$$\langle c, x\rangle \to \inf, \quad Ax \geqslant b, \quad x \geqslant 0, \quad (z_1)$$

$$\langle b, y\rangle \to \sup, \quad A^*y \leqslant c, \quad y \geqslant 0, \quad (z_1^*)$$

Theorem 3 (duality in linear programming). *For the pair of dual problems* (z_1) *and* (z_1^*) *there are the following alternatives:* a) *either the values of the problems are finite and equal and both have solutions, or* b) *in one of the problems the set of feasible elements is empty.*

Theorem 4 (criteria for solutions). *In case* a) *of Theorem* 3 *elements* $\hat{x} \in \mathbf{R}^n$, $\hat{y} \in \mathbf{R}^m$ *are solutions of* $(z)_1$ *and* (z_1^*) *if and only if they are feasible and satisfy one of the two relations:* α) $\langle \hat{x}, c\rangle = \langle \hat{y}, b\rangle$ *or* β) $\langle A\hat{x} - b, \hat{y}\rangle = \langle A^*y - c, x\rangle = 0$ (*the conditions in* β) *are called complementary slackness conditions*).

The conditions α) and β) in Theorem 4 may also be interpreted as follows.

Theorem 5 (on saddle points). $\hat{x} \in \operatorname{Arg} z_1 \Leftrightarrow \exists \hat{y} \in \mathbf{R}^m_+$:

$$\mathscr{L}(\hat{x}, y, 1) \leqslant \mathscr{L}(\hat{x}, \hat{y}, 1) \leqslant \mathscr{L}(x, \hat{y}, 1) \ \forall x \in \mathbf{R}^n, \qquad y \in \mathbf{R}^m,_+$$

where $\mathscr{L}(x, y, 1) = \langle c, x\rangle + \langle b - Ax, y\rangle$ *is the Lagrange function of the problem* (z_1).

1.4. Convex Programming. Convex problems have a number of significant properties, determining for them a special position in the theory of extremal problems. One of the most important is that necessary and sufficient conditions for them, as a rule, coincide and hence the Lagrange principle (see Alekseev, Galeev & Tikhomirov [1984], Alekseev, Tikhomirov & Fomin [1979]) is true for convex problems in the strongest possible form. Namely,

Theorem 6 (Kuhn-Tucker) a) $/f \in \operatorname{Co}(X, \mathbf{R})/ \Rightarrow \hat{x} \in \operatorname{absmin} f \Leftrightarrow 0 \in \partial f(\hat{x})$,
b) $/\hat{x} \in \operatorname{Arg} z_0/ \Rightarrow \exists \lambda = (\lambda_1, \ldots, \lambda_m), \lambda_0 (\lambda_i \in R, 0 \leqslant i \leqslant m)$:

$$\alpha) \ \min_{x \in A} \mathscr{L}(x, \lambda, \lambda_0) = \mathscr{L}(\hat{x}, \lambda, \lambda_0),$$

$$\beta) \ \lambda_i \geqslant 0, \quad 0 \leqslant i \leqslant m',$$

$$\gamma) \ \lambda_i f_i(\hat{x}) = 0, \quad 1 \leqslant i \leqslant m'.$$

If $\lambda_0 \neq 0$, then it follows from $\alpha)$–$\gamma)$ that $x \in \operatorname{absmin} z_0$. A sufficient condition for $\lambda_0 \neq 0$ is (in the absence of equality constraints) that the *Slater condition* hold, that is, there is an $\bar{x} \in A$ such that $f_i(\bar{x}) < 0$, $i \geqslant 1$.

In condition α) of Theorem 6b we find the perfect expression of the *Lagrange principle "of removal of constraints"*: at $\hat{x}$ the Lagrange function takes a minimum with respect to those constraints which do not enter into the Lagrange function.

Theorem 6 (under the Slater condition) can give the appearance of an assertion on the existence of a saddle point:

$$\min_{x \in A} \mathscr{L}(x, \hat{\lambda}, 1) = \mathscr{L}(\hat{x}, \hat{\lambda}, 1) = \max_{\lambda \in R^m_+} \mathscr{L}(\hat{x}, \lambda, 1).$$

All the theorems of this section are corollaries of the basic facts of convex calculus. For more detail on linear and convex programming see Dantzig [1963]; Eremin & Astaf'ev [1976]; Gass [1958]; Gavurin & Malozemov [1984]; Gol'shtein [1971]; Karmanov [1980]. On duality in problems with operator constraints see Kusraev, [1985].

§ 2. Convexity in Geometry

Geometry throughout its development has been an inexhaustible source, from which problems connected with convexity have been extracted.

2.1. Extremal Geometric Inequalities. The most ancient extremal problem is the isoperimetric problem of the greatest area enclosed by a curve of given perimeter. This problem had been mentioned at the time of Aristotle (V B.C.). The problem itself trivially reduces to the case of convex curves.

Theorem 7 (on the isoperimetric property of the circle). *Let A be a convex planar figure, $p(A)$ its perimeter, $S(A)$ its area. Then the following* (*isoperimetric*) *inequality holds*: $p^2(A) \geqslant 4\pi S(A)$, *in which equality is possible only if A is a circle $\rho B + a$ ($B =: B_2^2$ is the unit circle with centre at the origin).*

This result has been proved in many different ways. One of the simplest is as follows. Making use of the Steiner formula (i) (Theorem 22 of Chap. 1), the trivial relations of homogeneity for p and S and the Brunn-Minkowski inequality (Theorem 21 of Chap. 1) (ii), we obtain

$$\begin{aligned} S^{1/2}(\alpha A + (1-\alpha)B) &\overset{\text{i}}{=} (S(\alpha A) + p(\alpha A)(1-\alpha) + \pi(1-\alpha)^2)^{1/2} \\ &= (\alpha^2 S(A) + \alpha(1-\alpha)p(A) + \pi(1-\alpha)^2)^{1/2} \\ &\overset{\text{ii}}{\geqslant} \alpha S^{1/2}(A) + (1-\alpha)\pi^{1/2}. \end{aligned}$$

Squaring and cancelling like terms we arrive at the isoperimetric inequality. Let us give an n-dimensional generalization of it.

Theorem 8 (on the isoperimetric property of the sphere). *Let a closed $(n-1)$-dimensional surface ∂A be the boundary of an n-dimensional set A. Then ($V(C)$ is the volume of the set C)*

$$V^{1/(n-1)}(\partial A) \leqslant V^{1/n}(A)\frac{V^{1/(n-1)}(S^{n-1})}{V^{1/n}(B_2^n)},$$

where equality is attained only if $A = \rho B_2^n + a$.

The proof may be obtained in a similar way to that of Theorem 7. This result can be proved using the method of symmetrization (see Yaglom & Boltyanskii, [1951]). The following result is also proved using symmetrization.

Theorem 9 (Bieberbach inequality).

$$A \subset \mathbf{R}^n \Rightarrow V(A) \leqslant 2^{-n}V(B_2^n)(\operatorname{Diam} A)^n.$$

We quote several more facts from convex geometry. Denote by β_n the infimum over all $A \in \operatorname{Co}(\mathbf{R}^n)$ of the ratio $r(A)/h(A)$, where $r(A)$ is the radius of the inscribed sphere, and $h(A)$ is the width of the set A.

Theorem 10 (Blaschke). *For n even β_n is equal to $(n+2)^{1/2}(n+1)^{-1}$, and for n odd $\beta_n = n^{-1/2}$.*

Further, let j_n be the supremum over all $A \in \operatorname{Co}(\mathbf{R}^n)$ of the ratio $R(A)/\operatorname{Diam} A$, where $R(A)$ is the radius of the escribed sphere.

Theorem 11 (Young). $j_n = (2n/(n+1))^{1/2}$.

Theorems 10 and 11 are dual to each other. Young's theorem, using the refinement theorem (or Helly's theorem) reduces to the case when A is an n-dimensional simplex, and further it is easily proved that amongst simplexes the regular ones are optimal (see Ioffe & Tikhomirov [1974]).

By essentially the same means the following formula may be proved. Let X be an n-dimensional Banach space, $I(X) = \sup\{R(A, X)/\operatorname{Diam}(A, X) \colon A \in \operatorname{Co}(X)\}$.

Theorem 12 (Bonenblast). *There is the sharp inequality $I(X) \leqslant n/(n+1)$.*

The dual is the following result.

Theorem 13 (Leichtweiss). *Let $B(X) = \inf r(A, X)/h(A, X)$. Then there is the sharp inequality $B(X) \leqslant (n+1)/2$.*

We give two more theorems which have numerous applications. Let $A \in \operatorname{Co}(\mathbf{R}^n)$ and let $f_A(H, x)$ be the ratio, no greater than one, of the volumes of the two parts into which A is divided by a hyperplane H, passing through a point $x \in A$. Put

$$F_n(A, x) = \inf_{x \in H} f_A(H, x)$$

Theorem 14. *Let $g = g(A)$ be the centre of gravity of A. Then there is the sharp inequality $F_n(A, g) \geqslant n^n((n+1)^n - n^n)^{-1}$. In particular, $F_2(A, g) \geqslant 4/5$ and equality is attained only on triangles.*

Theorem 15. *For each convex body there is a unique ellipsoid of minimal volume, which contains it and a unique ellipsoid of maximal volume contained in it.*

In the proof of this result John [1948] demonstrated the application of the then just created methods of mathematical programming.

At this point we interrupt the discussion of this very interesting theme, referring the reader to the books, Berger, [1977]; Danzer, Grünbaum & Klee, [1963]; Eggleston, [1958]; Encyclopaedia of Elementary Mathematics, [1966]; Leichtweiss, [1980]; Lyusternik, [1956], etc. We stress once again that in many cases results in convex geometry can be fairly simply deduced from the basic facts of convex analysis which were presented in Chaps. 1 and 2.

2.2. Convex Geometry and the Approximation Theory. The theory of approximations is closely related to geometry. Already the original problem itself—on Chebyshev polynomials (see Chap. 1 of Part II)—admits, as we recall, a natural and fruitful geometric interpretation, being the problem of shortest distance

$$\|x - y\| \to \inf, \qquad x \in A. \tag{2}$$

The numerical value of this problem is denoted by $\rho A(x)$ or $d(x, A, X)$. If A is convex, then (2) is related to our topic. As with any extremal problem, there are the problems of existence, uniqueness, criteria for optimal elements, continuous dependencies, etc. We discuss some of these themes.

2.2.1. On Existence, Uniqueness and Continuous Dependence

Theorem 16. a) *In a normed space any finite-dimensional subspace is a space of existence.* b) *For any closed subset of a Banach space to be a subset of existence, it is necessary and sufficient that the space be reflexive.*

a) follows trivially from the local compactness of a finite-dimensional subspace. Here it is useful to recall that property a) does not hold in Fréchet spaces (Al'binus, 1966). The brilliant result b) is due to James [1963]. We note further that there is an incomplete space, every closed subspace of which is a space of existence.

There will be an account on the uniqueness of extremal elements in Chap. 1 of the article "Theory of approximations" in this volume.

The continuity of the metric projection, basic in considerations of compactness, was investigated in the finite-dimensional case a long time ago (Nikolesku, 1938; S.M. Nikol'skij, 1940; Klee, 1953, 1961, etc.)

Theorem 17 (Ky Fan and Glicksberg (1958), Singer (1964)). *Approximative compactness of a convex closed set is sufficient for the upper semi-continuity of the metric projection.* (Regarding approximative compactness see §4 of Chap. 4 of the next part of this volume).

The same authors pointed out the geometric properties of spaces, for which any closed convex set is approximatively compact.

In an independent direction we can choose papers, in which Banach spaces with a continuous (in some sense) metric projection are characterised (Brown, L.M. Vlasov, E.V. Oshman and others). We will quote a typical example.

Theorem 18 (E.V. Oshman, 1983). *In a Banach space each convex closed set is a set of existence with an upper semi-continuous metric projection if and only if it is an Efimov-Stechkin space* (On Efimov-Stechkin spaces see also § 4 of Chap. 4 of the next part of this volume).

2.2.2. Extremal Problems in the Approximation Theory and Convex Analysis. It has already been mentioned that if A is a convex set then problem (2) is related to convex programming and, thus, has a dual.

Theorem 19 (The duality formula for least distance problems). *X a normed space, $A \in \mathrm{Co}(X) \Rightarrow$*

$$\rho A(x) = \sup\{\langle x^*, x\rangle - sA(x^*) \colon \|x^*\| \leqslant 1\}$$

This result follows immediately from the standard formulae of convex calculus. For the space L the duality formula

$$\rho L(x) = \sup\{\langle x^*, x\rangle \colon x^* \in L^{\perp}, \|x^*\| \leqslant 1\}$$

was obtained by S.M. Nikol'skij. It plays a major role in the theory of approximations. In particular, the reduction of quadratures to monosplines, the relation between Kolmogoroy and Gel'fand widths the inequalities of Bernstein-Nikol'skij and Bohr-Favard and many others can in fact be described by duality formulae.

We give yet another form of the problem of least distance, which is used in approximation theory. Below X is a normed space $x^* \in X^*$, $i = 1, \ldots, n$.

Theorem 20 (Duality in the problem of moments). *The value of the problem $\|x\|_X \to \inf$, $\langle x_i^*, x\rangle = \alpha_i$ is equal to*

$$\sup\left\{\sum_{i=1}^{n} \alpha_i \beta_i \colon \left\|\sum_{i=1}^{n} \beta_i x_i^*\right\|_{X^*} \leqslant 1\right\}.$$

This fact is also proved by an application of the principle of duality, mentioned in § 1.

Other connections between the theory of approximations and convex programming are shown in the next section, the monographs Krejn, Nudel'man [1973], Tikhomirov [1976] and in the following part of this volume.

2.2.3. The Problem of Convexity of Chebyshev Sets. A set in a normed space is called a *Chebyshev* set if it is a set of existence and uniqueness (in the sense of least distance) for any element of the space. The question of convexity for Chebyshev sets has been long and actively discussed in the scientific literature.

Theorem 21 (Bunt-Motskin, 1934). *In a strictly convex finite-dimensional space with second order modulus of smoothness (in particular, in a Euclidean space) any Chebyshev set is convex.*

Bunt in his work had noted that the condition of strict convexity can, in general, be weakened. Motskin (1935) (independently of Bunt) proved, that for a set in a finite dimensional Euclidean space to be Chebyshev it is necessary and sufficient that it be convex and closed.

We give a sketch of the proof (due to L.P. Vlasov) of this beautiful theorem. A central place in this proof is the necessity of convexity of the set. For brevity we denote the projection of x onto A by $P(x)$. Let $x \notin A$. For positive α we consider the mapping $F_\alpha\colon B_2^n \to B_2^n$ (the Euclidean sphere $B_2^n \subset \mathbf{R}^n$), given by the formula:

$$F_\alpha(y) = (x - P(x + \alpha y)/|x - P(x + \alpha y)|.$$

This mapping is continuous. By the Brouwer fixed point theorem there is a y_α such that $F_\alpha(y_\alpha) = y_\alpha$. Since $|F_\alpha(y)| = 1, \forall y \in B_2^n$, we obtain $|y_\alpha| = 1$. In addition, y_α is parallel to $x - P(x + \alpha y_\alpha)$, that is, $x \in (y_\alpha,\ P(x + \alpha y_\alpha)) \Rightarrow P(x) = P(x + \alpha y_\alpha)$, thus, at the same time, $y_\alpha = (x - P(x))/|x - P(x)|$. So $\forall \alpha > 0$ $\operatorname{int} B(x + \alpha y_\alpha, |x - P(x)| + \alpha) \cap A \neq \varnothing$. But the union of these spheres is an open subspace Π, where $\partial \Pi$ passes through $P(x)$ orthogonal to $[x, P(x)]$. We have proved that $\operatorname{int} \Pi \cap A \neq \varnothing$, that is, $x \notin \operatorname{co} A$. Since $x \notin A$ is arbitrary, then $A = \operatorname{co} A$, as required.

Fixed point theorems were first applied, in similar problems, by Klee (1953).

The convexity of Chebyshev sets has brought into being an entire scientific area, which will be discussed in § 4 of Chap. 4 of the next part of this volume.

Theorem 22 (Brunn). *Let any section of a convex body $A \subset \mathbf{R}^3$ be centrally symmetric. Then A is an ellipsoid.*

Theorem 24 (Blaschke). *Let $A \subset \mathbf{R}^3$ be a convex centrally symmetric body with the property that for any plane Γ passing through the origin there is a line l which lies interior to the cylinder $(\partial A \cap \Gamma) + l$, then A is an ellipsoid.*

Hence it can be deduced that any Hilbert space has the property that on each subspace there is a projection of norm one. In connection with this there is the interesting question of how to characterize those spaces with the property that in any space containing them there is a projection of norm one onto them.

Theorem 24 (Kelley). *A Banach space X has the property described if and only if it is isometric to a space $C(T)$, where T is a completely disconnected compactum.*

Here there lies hidden an intrinsic description of all parallelopipeds.

We have not touched at all on the theme: "convex geometry in functional analysis"—it will be reflected in other volumes of this series. But it is difficult to refrain from mentioning the universality of $C(I)$ (any convex set from $\mathbf{R}^n$ can be obtained as a section of the ball of $C(I)$ by some n-dimensional surface—*Banach's theorem*) and the uniqueness of Hilbert spaces (in any infinite-dimensional space there are "almost Hilbert" finite-dimensional subspaces—*Dvoretskij's theorem*).

§ 3. Convex Analysis and Variational Problems

It has already been mentioned that the separation of convex analysis into independent divisions of mathematics was to a large extent initiated by the requirements of the theory of extremal problems. In the first section we spoke of the new divisions of optimization theory—linear and convex programming. However, convexity is inherent, in general, in all the problems of variational calculus and optimal control. It can even be said: the theory of variational problems comprises the synthesis of differential calculus and convex analysis, We declared this thesis in the introduction. Here it will be expanded with concrete examples.

3.1. Lyapunov's Theorem and its Generalisations

Theorem 25 (A.A. Lyapunov [1940]). *Let (T, Σ) be a measurable space (that is, a set with a σ-algebra of subsets Σ), in which there is given a continuous ($\Leftrightarrow$ without atoms) vector-measure (that is, a set m of continuous finite measures $m = (\mu_1, \ldots, \mu_n)$). Then the set $\{x\colon x = m(A),\ A \in \Sigma\}$ is convex and compact.*

This is one of the most beautiful and fundamental theorems of convex analysis. Its proof is based on the basic theorems of Krejn-Milman and Alaoglu, mentioned in Chap. 1.

Lyapunov's theorem links two ideas: integral mappings and convexity. Using it the following result, which joins both ideas even more expressively, can be proved.

Theorem 26 (Aumann). *Let (T, Σ, μ) be a space with a continuous measure and let F be a measurable and closed-valued many-valued mapping of T to $\mathbf{R}^n$ such that for some symmetric function $r(\cdot)\colon T \to \mathbf{R}_+$ $x \in F(t)$ implies $|x| \leqslant r(t)$. Then*

$$\int_T F(t)\, d\mu = \int_T (\operatorname{co} F)(t)\, d\mu$$

It is clear from Theorem 26 that the integral over a continuous measure "convexifies" a many-valued mapping. To realize this phenomenon geometrically is not too difficult. The heart of the matter is in the possibility of realizing a sliding regime, where part of the time is spent in a neighbourhood of one point of $F(\cdot)$, part of the time in a neighbourhood of another, and the integral "averages", and we find ourselves near a linear combination of points. This idea of "sliding" permeates variational calculus and optimal control, being reflected in the Weierstrass condition, the maximum principle, Bogolyubov's theorem and many other results. The same idea explains the appearance of convexity in economics, the presence of a huge number of not very significant subsystems which, added together, present a global economic effect, reduces to the fact that sums of large numbers of small terms are well approximated by convex sets.

Later on we will mention the applications of convex analysis in the problems of existence, duality and conditions for extrema in variational problems.

3.2. The Existence of Solutions and Extensions. Consider *the simplest problem of classical variational calculus*:

$$J(x(\cdot)) = \int_{t_0}^{t_1} L(t, x, \dot{x})\, dt \to \inf,$$

$$x(t_0) = x_0, \qquad x(t_1) = x_1, \qquad (L \in C(\mathbf{R} \times \mathbf{R}^n \times \mathbf{R}^n) \tag{3}$$

We will consider the problem in the space

$$W_1^1([t_0, t_1], \mathbf{R}^n) = \left\{x(\cdot)\colon x(\cdot) \in AC([t_0, t_1], \mathbf{R}^n), \int_{t_0}^{t_1} |\dot{x}(t)|\, dt < \infty\right\}$$

We can give three reasons why this problem cannot have a solution: the non-convexity of the integrand, its insufficient growth relative to x and the presence of conjugate points. We give the corresponding examples (in these $t_0 = 0, t_1 = 1$).

Example 1. $L = (1 - \dot{x}^2)^2 + x^2$, $x_0 = x_1 = 0$. Here the integrand is non-convex and sliding, mentioned above, is possible.

Consider the sequence $x_n(0) = 0$, $\dot{x}_n(t) = \operatorname{sign} \sin 2\pi n t$, $n = 1, 2, \ldots$

Here the derivative "slides" between one and minus one.

It is clear that $\mathcal{T}(x_n(\cdot)) \to 0$ and that there is no $\dot{x} \in W_1^1([0, 1])$ for which $\mathcal{T}(\dot{x}(\cdot)) = 0$.

Example 2. $L = t^2\dot{x}^2$, $x_0 = 0$, $x_1 = 1$ (*Weierstrass example*). Here the infimum is also equal to zero and is achieved by the sequence $x_n(t) = \arctan nt/\arctan n$, $n = 1, 2, \ldots$. The cause is the absence of growth of the integrand at $t = 0$.

Conjugate points have no direct connection with our theme and so we will omit the corresponding example (for more details see Ioffe & Tikhomirov [1974]).

In the presence of growth of L there are no variable conjugate points. Convexity and growth give the existence of a solution.

Convexity relative to $\dot{x}$ and growth of the integrand guarantee the existence of a solution. The first theorems of this type were obtained by Tonelli in 1915. An essential element turned out to be the fact that, although the functionals of classical variational calculus are not continuous in the natural topologies, when the integrand is convex with respect to $\dot{x}$ they turn out to be lower semicontinuous. (The first remarks on this were circulated by Lebesgue.) We give one recent result concerning existence and regularity of a solution for problem (z).

Theorem 27. a) *Suppose there is a neighborhood U of $(\tau, \xi) \in \mathbf{R} \times \mathbf{R}^n$ such that the function $(t, x, v) \to L(t, x, v)$ is Lipschitz on compact subsets of $U \times \mathbf{R}^n$ and for any $(t, x) \in U$ the function $v \to L(t, x, v)$ is strictly convex on $\mathbf{R}^n$, Then for any $C > 0$ there are $\varepsilon > 0$ and $\gamma > 0$ such that for any (t_i, x_i), $i = 1, 2$, with $|t_i - \tau| < \varepsilon$, $|x_i - \xi| \leqslant C/|t_1 - t_0|$, problem (3), with the extra condition $|x(t) - \xi| < \gamma$, $t \in [t_0, t_1]$ has at least one continuously differentiable solution (theorem on existence in the small).*

b) *Let the function $(t, x, v) \to L(t, x, v)$ be locally bounded on $[t_0, t_1] \times \mathbf{R}^n \times \mathbf{R}^n$, measurable with respect to t, convex in v, locally Lipschitz uniformly in t and satisfy the growth condition: $L(t, x, v) \geqslant -\alpha|x| + \varphi(|v|)$ for all (t, x, v) in $[t_0, t_1] \times \mathbf{R}^n \times \mathbf{R}^n$, where $\alpha \geqslant 0$, $\varphi(\cdot) \in \mathrm{Co}(\mathbf{R}, \mathbf{R})$ and $\varphi(e)\rho^{-1} \to \infty$ as $\rho \to \infty$. Then there exists a solution $\hat{x}(\cdot)$ of problem (3). If we assume that $L \in C^r([t_0, t_1] \times \mathbf{R}^n \times \mathbf{R}^n)$ in a neighbourhood of $(t, \hat{x}(t), \dot{x}(t))$ and $\hat{L}_{\dot{x}\dot{x}}(t) > 0$, then $\dot{x}(\cdot) \in C^r$ on $[t_0, t_1]$.*

Closely related to this theme is the *extension problem.* Hilbert proposed that "each regular variational problem has a solution, although it is necessary to add an extended interpretation to the notion itself of a solution". Corresponding to the above two examples there are two types of extensions: an extension of the problem with integrands which are non-convex in $\dot{x}$ and the extension of the class of solutions to problems integrands which are convex in $\dot{x}$ integrands with insufficient growth.

Theorem 28 (N.N. Bogolyubov). *For any vector-function $x(\cdot) \in W^1_\infty([t_0, t_1], \mathbf{R}^n)$ there is a sequence of functions $\{x_n(\cdot)\}_{n \in N}$, $x_i(\cdot) \in C^1([t_0, t_1], \mathbf{R}^n)$ such that $x_n(t_j) = x_j$, $j = 0, 1$, $x_n(\cdot) \to x(\cdot)$ uniformly on $[t_0, t_1]$ and, in addition,*

$$\liminf_{n \to \infty} I(x_n(\cdot)) = \tilde{I}(x(\cdot)),$$

where

$$\tilde{I}(x(\cdot)) = \int_{t_0}^{t_1} \tilde{L}(t, x, \dot{x})\, dt,$$

and $\tilde{L}(t, x, \cdot)$ is the "convexification" of $L(t, x, \cdot)$ relative to $\dot{x}$ (that is, $\tilde{L} = L^{**}(t, x, \cdot)$, and the adjoint is taken with respect to $\dot{x}$).

The problem

$$\tilde{I}(x(\cdot)) = \int_{t_0}^{t_1} \tilde{L}(t, x, \dot{x})\, dt \to \inf, \qquad x(t_j) = x_j, \qquad j = 0, 1,$$

is called the *lower semi-continuous extension of* (3) (since $\tilde{I}(x(\cdot)) = \liminf_{n \to \infty} I_n(x_n(\cdot))$, $x_n(\cdot) \to x(\cdot)$ uniformly, and here, in general, I is lower semi-continuous). It follows from Theorem 28 that the second problem has the same value as the first. From the same result it also follows that from the theoretical point of view one can always regard the quasi-regularity condition to be satisfied.

3.3. Duality of Variational Problems. Lyapunov's theorem allows the reduction of the following class of problems with integral functionals virtually to convex problems:

$$\begin{aligned} &F_0(u(\cdot)) + g_0(x) \to \inf, \\ &F_i(u(\cdot)) + g_i(x) \leqslant 0, \qquad 1 \leqslant i \leqslant m', \\ &F_i(u(\cdot)) + g_i(x) = 0, \qquad i = m' + 1, \ldots, m, \\ &u(\cdot) \in \mathscr{U}, \qquad x \in A, \end{aligned} \tag{4}$$

where

$$F_i(u(\cdot)) = \int_T f_i(t, u(t))\, d\mu(t), \qquad 0 \leqslant i \leqslant m,$$

(T, Σ, μ) is a space with a continuous measure, U is a topological space, $\mathscr{U}$ is a collection of mappings $u(\cdot)\colon T \to U$, $f\colon T \times U \to R$ is a family of superpositionally measurable functions, X is a linear space, $g_i\colon X \to \mathbf{R}$ is convex for $0 \leqslant i \leqslant m'$ and affine for $i \geqslant m' + 1$, and A is a convex set. We note that for $f \equiv 0$ a convex programming problem, the topic of § 1, is obtained. But the whole point is this, that problem (4), even though the functions f may not have any relation with convexity (for example U may be a discrete space), is in fact related to convex programming. A corollary of Lyapunov's theorem is that the image of $F(u(\cdot)) = (F_0(u(\cdot)), \ldots, F_n(u(\cdot)))$ is convex in $\mathbf{R}^{n+1}$.

Problems of type (4) are called *Lyapunov problems*. The simplest example of a Lyapunov problem is this:

$$\int_T f_0(t, u(t))\, d\mu \to \inf, \qquad u(t) \in U \text{ a.e.} \tag{4'}$$

This is called the *simplest optimal control problem*. Under very general assumptions it can be proved that

$$\hat{u}(\cdot) \in \operatorname{absmin}(4') \Leftrightarrow \min_{u \in U} f_0(t, u) = f(t, \hat{u}(t)) \quad \text{a.e.}$$

(that is, the *minimum principle* holds). The so-called measurable selection theorems play a fundamental role in the statement and proof of this result (see Arkin & Levin [1972]; Levin [1985]).

This allows one also to prove the Lagrange principle in its strongest form and to apply the concept of duality to problem (4).

Theorem 29 (The Lagrange principle for Lyapunov problems). *If a pair $(\hat{u}(\cdot), \hat{x})$ attains an absolute minimum for problem* (4), *then there is a vector $\lambda \in \mathbf{R}^m$, $\lambda = (\lambda_1, \ldots, \lambda_m)$ and a number λ_0 such that*

a) *the minimum principle holds, according to which*

$$\min_{u \in U} \sum_{i=0}^{m} \lambda_i f_i(t, u) = \sum_{i=0}^{m} \lambda_i f_i(t, \hat{u}(t)) \quad \text{a.e.},$$

$$\min_{x\in A}\sum_{i=0}^{m}\lambda_i g_i(x)=\sum_{i=0}^{m}\lambda_i g_i(\hat{x}),$$

b) $\lambda_i \geqslant 0, 0 \leqslant i \leqslant m'$ (*non-negativity condition*)

c) $\lambda_i(F_i(\hat{u}(\cdot)) + g_i(\hat{x})) = 0, 1 \leqslant i \leqslant m'$ (*complementary slackness condition*).

If $\lambda_0 \neq 0$ then a)–c) are sufficient for $(\hat{u}(\cdot), \hat{x}) \in \operatorname{absmin}(4)$. (For the proof see Alekseev, Tikhomirov & Fomin [1979].

The dual to problem (4), where $g_i \equiv 0$ and there are no equalities, is the problem

$$-\Phi(\lambda)=\int_T L(t,\lambda,1)\,dt\to\sup,\qquad \lambda\geqslant 0,$$

where $L(t, \lambda, 1) = \inf_{u\in U}(f_0(t, u) + \sum_{i=1}^{m}\lambda_i f_i(t, u))$.

We will see that the dual problem is of another character—it is an ordinary extremum problem for a function of several variables.

This phenomenon is repeatedly used in classical analysis. Here is one of the more significant examples. Consider an optimal control problem, linear in the phase variables:

$$I_0(x(\cdot),u(\cdot))\to\inf,\qquad \dot{x}=A(t)x+F(t,u(t)),$$

$$I_i(x(\cdot),u(\cdot)\begin{cases}\leqslant c_i, & 1\leqslant i\leqslant m',\\ = c_i, & m'+1\leqslant i\leqslant m,\quad u(\cdot)\in\mathscr{U},\end{cases}$$

$$I_i(x(\cdot),u(\cdot))=\int_{t_0}^{t_1}(\langle a_i(t),u(t)\rangle+f_i(t,u(t))\,dt$$

$$+\langle\gamma_{0i},x(t_0)\rangle+\langle\gamma_{1i},x(t_1)\rangle,\quad 0\leqslant i\leqslant m \tag{5}$$

It reduces to the following Lyapunov problem

$$I_0(x(\cdot),u(\cdot))=-\int_\Delta G_0(t,u(t))\,dt+\langle\beta_0,x(t_0)\rangle\to\inf,$$

$$I_i(x(\cdot),u(\cdot))=-\int_\Delta G_i(t,u(t))\,dt+\langle\beta_i,x(t_0)\rangle\begin{cases}\leqslant c_i, & 1\leqslant i\leqslant m',\\ = c_i, & m'+1\leqslant i\leqslant m,\end{cases} \tag{6}$$

where $\Delta = [t_0, t_1]$, $G(\tau, u) = F^*(\tau, u)p_i(\tau) - f_i(\tau, u)$, $\beta_i = \gamma_{0i} - p(t_0)$, $p_i(\tau) = -\Omega(\tau_i, t_1)\gamma_{1i} - \int_{t_0}^{t_1}\Omega(\tau, t)a_i(\tau)\,dt$, $\Omega(\cdot, \cdot)$ is a fundamental matrix of the homogeneous linear system, $\dot{x} = Ax$.

Theorem 30 (The maximum principle for problems linear in the phase variables). *Let $a_i\colon \Delta \to \mathbf{R}^n$, $i = 0, 1, \ldots, m$ and $A\colon \Delta \to \mathscr{L}(\mathbf{R}^n, \mathbf{R}^n)$ be integrable over $\Delta = [t_0, t_1]$, U a topological space, $f_i\colon \Delta \times U \to \mathbf{R}$, $i = 0, 1, \ldots, m$, $F\colon \Delta \times U \to \mathbf{R}^n$ continuous functions. Then, if $(\hat{x}(\cdot), \hat{x}(\cdot))$ attains the absolute minimum of problem (5), there are a number $\lambda_0 \geqslant 0$ and a vector $\lambda = (\lambda_1, \ldots, \lambda_n)$, not simultaneously zero, and an absolutely continuous function $p\colon \Delta \to \mathbf{R}^n$, such that the following hold:*

a) *the Euler-Lagrange equation* $-\dot{p} = A^*(t)p + \sum_{i=0}^{m} \lambda_i a_i(t)$,
b) *the transversality condition* $p(t_j) = (-1)^j \sum_{i=0}^{m} \lambda_i \gamma_{ji}$, $j = 0, 1$,
c) *the maximum principle* $\max\{G(t, u): u \in U\} = G(t, \hat{u}(t))$ *almost everywhere* $(G(\tau, u) = F^*(\tau, u)p(\tau) - \sum_{i=0}^{m} \lambda_i f_i(\tau, u))$,
d) *the non-negativity condition* $\lambda_i \geqslant 0$, $1 \leqslant i \leqslant m'$,
e) *the complementary slackness condition*: $\lambda_i(I_i(\hat{x}(\cdot), \hat{u}(\cdot)) - c_i) = 0$, $1 \leqslant i \leqslant m'$.

If $\lambda_0 \neq 0$, then conditions a)–e) are sufficient for $(\hat{x}(\cdot), \hat{u}(\cdot)) \in \operatorname{absmin}(5)$.

For the proof see Alekseev, Tikhomirov & Fomin [1979]; Arkin & Levin [1972]. If $I_i = 0$, $i = 1, \ldots, m$, and the stronger conditions $x(t_j) = x_j$, $j = 0, 1$, are satisfied, then (5) reduces to the problem.

$$I(x(\cdot)) = \int_{\Delta} f(t, \dot{x}(t))\, dt \to \inf,$$

$$x(t_0) = 0, \quad x(t_1) = x, \tag{7}$$

where $\Delta = [t_0, t_1]$, $x = y_1 - \Omega(t_1, t_0)y_0$, $f(t, v) = \langle a(t), \Omega(t, t_0)x_0\rangle + \inf\{\langle \int_{\tau}^{t_1} \Omega(\tau, t)a(\tau)\, d\tau, F(t, u)\rangle + f_0(t, u): u \in U, \Omega(t_1, t)F(t, u) = v\}$ (see Ioffe & Tikhomirov [1974]). We have obtained a problem of classical variational calculus, independent of the phase variables. Convex analysis allows us to practically carry through the study of problems of the type (7). We denote by $S(x)$ the value of the problem (7).

Theorem 31 (duality). *If* $x \in \operatorname{ri} \operatorname{dom} S$, *then*

$$S(x) = \max_p \left(\langle p, x\rangle - \int_{\Delta} f^*(t, p)\, dt \right)$$

$$(f^*(t, p) = \sup_z (\langle z, p\rangle - f(t, z))) \tag{8}$$

Theorem 32. *In problem* (7) *let* f *be a normal integrand on* $\Delta \times R^n$ *(for the definition see Ioffe & Tikhomirov* [1974]) *and in* (8) *let the maximum be attained at* p_0. *Then* $\hat{x}(\cdot) \in \operatorname{absmin}(7) \Leftrightarrow f(t, \dot{\hat{x}}(t)) + f^*(t, p_0) = \langle p_0, \dot{\hat{x}}(t)\rangle$ a.e., *where, if* $p_0 \in \operatorname{int}(\operatorname{dom} S^*)$, *then a solution to the problem exists.*

If p_0 lies on the boundary of $\operatorname{dom} S^*$, then there is a generalized solution for which the regular solution, defined in b), gives a solution everywhere except at the singular points, where there are discontinuities. Discontinuities arise because of insufficient growth of the integrand—we recall that this was mentioned in subsection 3.2.

3.4. Duality in Many-dimensional Problems. In the previous subsection we twice came across the situation of a problem of classical variational calculus or optimal control reducing to a problem of simpler structure—a problem in finite-dimensional convex programming. In many-dimensional variational calculus the dual problem, as a rule, has the same structure as the primal. Here the fruitfulness of the method of duality reveals itself in the fact that in the dual problem a usual solution exists (whereas in the original that may not be the case).

The monograph Ekeland & Temam [1976] is devoted to duality in variational calculus. We quote two typical results from this book.

Let Ω be an open bounded domain in $\mathbf{R}^n$, and suppose it is required to find a solution of the problem

$$-\Delta x = f \quad \text{on } \Omega, \qquad x = 0 \quad \text{on } F =: \partial\Omega \tag{9}$$

The variational form of this problem is:

$$\begin{gathered}\int_\Omega (\langle x', x'\rangle - xf)\, dt \to \inf,\\ x(\cdot) \in \mathring{W}_2^1(\Omega) \quad (=: H_0^1(\Omega)),\\ x'(\cdot) = (\partial x/\partial t_1, \ldots, \partial x/\partial t_n).\end{gathered} \tag{10}$$

Theorem 33. *Problem* (9), *described in the form* (10), *has as dual problem*

$$\begin{gathered}-2^{-1}\int_\Omega \sum_{i=1}^n p_i^2(t)\, dt \to \sup,\\ \sum_{i=1}^n \frac{\partial p_i}{\partial t_i} = f, \qquad p_i(\cdot) \in L_2(\Omega)\end{gathered} \tag{11}$$

Both problems have a unique solution $\hat{x}(\cdot)$ and $\hat{p}(\cdot) = (\hat{p}_1(\cdot), \ldots, \hat{p}_n(\cdot))$ with the same value, and the solutions are connected by the relation $\hat{p}(\cdot) = -\hat{x}'(\cdot)$. From this, properly speaking, it also follows that $\hat{x}(\cdot)$ is a solution of (9).

Now let Ω be a domain in $\mathbf{R}^n$, with a boundary having certain regularity properties and let $\xi(\cdot)$ be a function from $W(\Omega)$. Consider the problem

$$\int_\Omega (1 + \langle x', x'\rangle)^{1/2}\, dt \to \inf, \qquad x(\cdot) = \xi(\cdot) + \mathring{W}_1^1(\Omega) \tag{12}$$

This is the variational form of the Dirichlet problem for the equation of minimal surfaces.

Theorem 34. *Problem* (12) *has as dual problem*

$$\begin{gathered}-\left(\int_\Omega \langle p(t), \xi'(t)\rangle\, dt + \int_\Omega (1 - |p(t)|^2)^{1/2}\, dt \to \sup,\right.\\ p(\cdot) \in L_\infty(\Omega, R^n), \qquad |p(t)| \leqslant 1 \text{ a.e.}, \qquad \operatorname{div} p = 0\end{gathered} \tag{12'}$$

Here problems (12) *and* (12′) *have the same value.* (12′) *has a unique solution $\hat{p}(\cdot)$, whereas* (12) *(because of the insufficient growth of the integrand) may have no solutions. But if* (12) *has a solution $\hat{x}(\cdot)$, then $|\hat{p}(\cdot)| < 1$ a.e. and $\hat{x}'(t) = \hat{p}(t)/(1 - |\hat{p}(t)|^2)$ a.e.*

The latter is analogous to the equality in b) of Theorem 31. Here if there is no solution it is possible to describe a generalized solution in a similar way to that mentioned after Theorem 31.

Let us say a few words on the *Monge-Kantorovich problem*. Consider an infinite-dimensional analogue of the transportation problem (see § 1). It was first discussed by L.V. Kantorovich (see Kantorovich & Rubinshtein [1957]). The finite set of points is replaced here by an arbitrary compact set X with the σ-algebra of Borel sets $\mathscr{B}$, and the matrix c_{ij} by a function $c(\cdot,\cdot)\colon X \times X \to \mathbf{R} \cup \{+\infty\}$. The analogue of the vector ρ will be a Radon measure $\rho(\cdot)$ with the condition $\rho(X) = 0$, the analogue of the matrix x_{ij} will be a measure $\mu \geqslant 0$ on $X \times X$, subject to the condition $\mu(X, A) - \mu(A, X) = \rho(A)$.

Thus we obtain the problem:

$$\int_{X \times X} c(x_1, x_2)\, d\mu \to \inf,$$

$$\mu(X, A) - \mu(A, X) = \rho(A), \qquad \forall A \in \mathscr{B}, \qquad \mu \geqslant 0. \tag{13}$$

This problem is a generalization of an old problem, which had been discussed by Monge (on the rational displacement of earth in the construction of ditches and embankments). Many works are devoted to the Monge-Kantorovich problem (L.V. Kantorovich, L.V. Kantorovich & G. Sh. Rubinshteĭn, V.L. Levin, V.L. Levin & A.A. Milyutin etc.). Significant problems in the theory of plasticity (Mosolov & Myasnikov [1971] [1981]), the theory of probability (Rachev [1984]), mathematical economics (Levin [1985]), can be reduced to it. We give one of the definitive results.

The problem, dual to (13), is posed in the class of continuous functions $u(\cdot)$:

$$\int_X u(x)\, d\rho \to \sup, \qquad u(x_1) - u(x_2) \leqslant c(x_1, x_2) \qquad \forall x_1, x_2 \subset X.$$

Denote by $A(c, \rho)$ and $B(c, \rho)$ the values of the original problem (that is, (13)) and the dual problem and suppose that

$$A(c, \rho) = \lim_{N \to \infty} A(\min(c, N), \rho) \qquad \forall \rho, \rho(X) = 0 \tag{i}$$

Theorem 35 (B.L. Levin & A.A. Milyutin [1979]). *Let X be a compact metric space and let $c\colon X \times X \to \mathbf{R} \cup \{+\infty\}$ be bounded below and have an analytic Lebesgue set $\{(x_1, x_2)\colon c(x_1, x_2) \leqslant \alpha\}$, $\alpha \in \mathbf{R}$. Then for the duality relation $A(c, \rho) = B(c, \rho)$ $\forall \rho$, $\rho(X) = 0$, it is necessary and sufficient that condition* (i) *is satisfied and the function $\bar{c}(x_1, x_2)$, equal to $c_*(x_1, x_2)$ for $x_1 \neq x_2$ and* 0 *for $x_1 = x_2$, is bounded below and lower semi-continuous on $X \times X$ (in this case $A(c, \rho) = B(c, \rho) > -\infty$ $\forall \rho$, $\rho(X) = 0$). Here $c_*(x, x) = \inf c_n(x_1, x_2)$, $c_n(x_1, x_2) = \min\{c(x_1, x_2), \inf\{c(x_1, z_1) + c(z_1, x_2)\colon z_1 \in X\}, \ldots, \inf\{c(x_1, z_1) + c(z_1, z_2) + \cdots + c(z_n, x)\colon z_i \in X\}\}$.*

3.5. Appendices

3.5.1. Convexity in Variational Calculus and Optimal Control. Consider the simplest problem of variational calculus, (3), under the assumption that L is a smooth (say, class C^1) function of both variables. This problem, considered as a weak extremum belongs to smooth analysis, since the mapping $x(\cdot) \to I(x(\cdot))$ is smooth. But if it is written as an optimal control problem

$$I(x(\cdot), u(\cdot)) = \int_{\Delta} L(t, x, u)\, dt \to \inf,$$

$$\dot{x} = u, \qquad x(t_j) = x_j, \qquad j = 0, 1, \qquad u \in \mathbf{R}^n \tag{3'}$$

and we consider $u(\cdot) \to I(x(\cdot), u(\cdot))$ for fixed $x(\cdot)$, then, as we showed at the beginning of this section, this functional has turned out to be connected with convex analysis.

The variables $x(\cdot)$ and $u(\cdot)$ are unequivalent. The phase variables $x(\cdot)$ are obtained from $u(\cdot)$ be integration. It is possible, by choosing two controls $u_1(\cdot)$ and $u_2(\cdot)$, to construct a "sliding" between them, jumping with one to the other, remaining a time α in $u_1(\cdot)$ and $(1-\alpha)$ in $u_2(\cdot)$, and we then obtain, disintegrating the sliding, a function $\alpha x_1(\cdot) + (1-\alpha)x_2(\cdot)$, $\dot{x}_1(\cdot) = u_1$, $\dot{x}_2(t) = u_2$, $j = 0, 1$ with greater and greater accuracy. Such a sliding is possible if relative to $u(\cdot)$ the conditions for an extremum take the form of minimum (or maximum, depending on the choice of sign) principles. For example, if $\hat{x}(\cdot)$ achieves a strict minimum in problem (3), then the following condition holds:

$$\min_{u \in \mathbf{R}^n} (L(t, \hat{x}(t), u) - \langle \hat{L}_x(t), u \rangle) = \hat{L}(t) - \langle \hat{L}_x(t), \dot{x}(t) \rangle,$$

$$\hat{L}(t) = L(t, \hat{x}(t), \dot{\hat{x}}(t)), \qquad \hat{L}_{\dot{x}}(t) = L(t, \hat{x}(t), \dot{\hat{x}}(t)).$$

This is called the *Weierstrass condition.* A similar condition also holds in optimal control. It is called the *maximum principle.* Jointly with the Euler equation it forms the *Pontryagin maximum condition.* The Euler equation is connected with the smooth structure of the optimal control equation, the maximum principle with its convex structure.

3.5.2. Necessary and Sufficient Conditions for an Extremum in Mathematical Programming Problems. Consider the problem.

$$f_0(x) \to \inf, \qquad f_i(x) \leqslant 0, \qquad i = 1, \dots, m, \qquad F(x) = 0, \tag{14}$$

where $f_i\colon U \to R$, $0 \leqslant i \leqslant m$, $F\colon U \to Y$, X, Y are normed spaces, and U is a neighbourhood in X. This problem is concerned with mathematical programming, a particular case of which is convex programming (see § 1).

In the derivation of necessary and sufficient conditions for an extremum for problems of type (14) convex analysis plays a central roles play. (14) (with smoothness conditions on f_i and F) reduces to the following convex programming problem:

$$\langle f_0'(\hat{x}), x\rangle \to \inf,$$

$$\langle f_1'(\hat{x}), x\rangle \leqslant 0, \qquad i = 1,\ldots,m, \qquad F'(\hat{x})x = 0 \tag{14'}$$

Namely, it can be shown, that $\hat{x} \in \text{locmin}\,(14) \Rightarrow 0 \in \text{absmin}\,(14')$.

(14′) is a "convex approximation to (14). Convex approximations are also possible with fewer smoothness conditions. The problems of convex programming itself have been widely discussed in the literature (see the monographs Boltyanskij [1973]; Pshenichnyj, [1980, 1982]).

An application of the Kuhn-Tucker theorem to problem (14) immediately leads to the following result.

Theorem 36 (the Lagrange multipliers rule for the mathematical programming problem). */X, Y Banach spaces, $f_i \in C^1(U)$, $F \in C^1(U, Y)$, $F'(\hat{x})X$ closed in Y, $\hat{x} \in \text{locmin}\,(14)/ \Rightarrow \exists\lambda = (\alpha, y^*) \in R^{m+1} \times Y^*$, $\lambda = (\alpha_0, \ldots, \alpha_m, y^*)L$, $\lambda \neq 0$*

a) $L_x(\hat{x}, \lambda) = 0$ *(stationarity condition),*

b) $\alpha_i \geqslant 0$ *(non-negativity condition)*

c) $\alpha_i f_i(\hat{x}) = 0$, $i \geqslant 1$ *(complementary slackness condition).*

So convex analysis plays the same major role even in the investigation of sufficient conditions for an extremum for mathematical programming problems (Alekseev, Tikhomirov & Fomin [1979]; Arkin & Levin [1972]).

3.5.3. Convex Analysis and Inequalities. The enormous topic "Inequalities in analysis" is in the closest way linked with convexity. Inequalities for means, between the geometric mean and arithmetic mean, Hölder, Minkowski and many many others are proved by a standard application of the basic results of convex calculus The reader may verify this, by looking through the material in the well-known book Hardy, Littlewood & Pólya [1934].

A classical topic is the theory of linear inequalities. It started at the beginning of this century in the works of Farkas, Minkowski, Voronoj and others. The majority of the classical results in this domain can also be deduced in a standard way from the basic theorems of duality or convex calculus. We quote some theorems on inequalities.

Theorem 37 (Minkowski-Farkas). *For the inequality $\langle y_0, x\rangle \leqslant 0$ to follows from the inequalities $\langle y_i, x\rangle \leqslant 0$, $i = 1,\ldots,m$, $y_i, x \in \mathbf{R}^n$, it is necessary and sufficient that $y_0 \in \text{cone}\{y_1,\ldots,y_m\}$.*

Theorem 38 (A.D. Aleksandrov, Ky Fan). *A necessary and sufficient condition for consistency of the system of inequalities $\langle y_i, x\rangle - b_i \leqslant 0$, $i = 1,\ldots,m$, is that*

$$\left\{\sum_{i=1}^{m} \lambda_i y_i = 0, \lambda_i \geqslant 0\right\} \Rightarrow \sum_{i=1}^{m} \lambda_i b_i = 0.$$

Theorem 39. *For the inequality $\langle y_0, x\rangle \leqslant 0$ to follows from the inequalities $\langle y_\alpha, x\rangle \leqslant 0$, $\alpha \in \mathscr{A}$, it is necessary and sufficient that $y_0 \in \text{cl}\,\text{cone}\{y_\alpha\}_{\alpha \in \mathscr{A}}$.*

Theorem 40 (the conjugate cone). *Let X and Y be Banach spaces, $\Lambda: X \to Y$ a linear, continuous, surjective operator $x^* \in X^*$, $i = 1,\ldots,s$, $K = \{x \in X: \langle x_i^*, x\rangle \leqslant 0, i = 1,\ldots,s, \Lambda x = 0\}$. Then $K^* = \operatorname{cone}\{x_1^*,\ldots,x_s^*\} + \operatorname{Im}\Lambda^*$.*

These results may be deduced from the general theorems on convex calculus or by applying the method of duality. On the theory of linear inequalities see Chernikov [1962]; Rockafellar [1970].

3.5.4. Convexity in the Theory of Manifolds of Mappings. We quote two classical theorems.

Theorem 41 (Michael). *Let T be paracompact, X* a Banach space, *F a many-valued lower semi-continuous mapping from T to X such that $F(t) \neq \varnothing$, $F(t) \in \operatorname{Cl}\operatorname{Co}(X)$ $\forall t \in T$. Then there is a continuous function $f \in C(T, X)$ such that $f(t) \in F(t)$ $\forall t$.*

Theorem 42 (Kakutani, Tikhonov, Ky Fan). *Let X be a locally convex space, A a non-empty compact convex set in X, F an upper semi-continuous many-valued mapping from A to a non-empty closed subset of it. There there is a point x_0: $x_0 \in F(x_0)$.*

3.5.5. Newton Polyhedra and Other Convex Polyhedra in Concrete Problems of Analysis. Let $A = (\alpha^1,\ldots,\alpha^m)$ be points of $\mathbf{Z}^n$ or $\mathbf{R}^n$. The convex hull $\operatorname{co}(\alpha^1,\ldots,\alpha^m)$ is sometimes called a *Newton polyhedron*. Recently there have appeared a huge number of papers where the number of solutions of systems of polynomial equations, asymptotic formulae for the roots of exponential sums, etc. are expressed in terms of Newton polyhedra (their volumes, mixed volumes, etc.). Since we cannot comment on this in detail we refer the reader to Khovanskij [1983].

Chapter 4
Extensions of the Sphere of Convex Analysis and Generalizations of Convexity

§ 1. Non-smooth Analysis

1.1. Introduction. The foundations of smooth analysis of functions of one real variable were laid in the works of Newton and Leibniz at the end of the seventeenth century. During the eighteenth and nineteenth centuries differential calculus for a special class of functionals—functionals of variational calculus (Euler, Lagrange)—was created. At the end of the nineteenth century the calculus of functions of several variables (the source of which was contained in the works of Euler) was elaborated in the works of Weierstrass, Dini, Stolz, Young and others. Finally, in the first half of the twentieth century, smooth analysis in infinite-dimensional spaces was created. The basic concept of smooth analysis is the idea of a derivative or differential. A smooth mapping is approximated (in some sense) in a neighbourhood of a given point by an affine mapping; the corresponding linear operator is called the derivative, and the linear mapping itself the differential. Differential calculus is the set of rules, which allow us to calculate the differentials of complicated functions from the differentials of the simple ones, and also a number of general theorems (the mean value theorem, the inverse function theorem, etc.). Convex analysis pursues similar aims, although its objective is narrower. Convex functions do not have to be smooth, although to a certain extent they are endowed with smoothness properties. This manifests itself in two aspects. To one of them—convex calculus, which has lines of similarity with differential calculus—we have given enough attention. We recall: a convex function also admits an approximation; indeed, it is approximated not by a linear function but by a convex homogeneous function $x \to f'(x_0, x) =: \lim_{t \downarrow 0} (f(x_0 + tx) - f(x_0))t^{-1}$. A convex homogeneous function is a supremum of linear functions and this leads to the idea of a subdifferential, which is the union of the differentials, which participate in the approximation $f(x) - f(x_0)$.

Another aspect is the fact that a convex function is in fact almost smooth, for example, a finite-dimensional function is differentiable almost everywhere. We begin the interesting part of this section with this theme of "convexity and differentiability". Basically we pay attention to a new direction, properly non-smooth analysis, to the attempts to find a synthetic calculus, feeding on two sources—smooth and convex analysis.

1.2. Convexity and Differentiability. A convex function does not have to be smooth, but it is differentiable in massive sets. We give the original result on this topic.

Theorem 1 (Mazur, on differentiability, 1933). *A convex function f which is continuous on an open convex set A of a separable Banach space X is Gâteaux differentiable on a dense G_δ subset of A.*

Corollary 1. *In a finite-dimensional space a convex function is differentiable almost everywhere.*

Corollary 2. *In a separable Banach space every closed convex set with a non-empty interior has a G_δ-set of points of smoothness on the boundary.*

Recently the attention of many mathematicians has been attracted to this cycle of questions. This topic, certainly, is fairly narrow but several beautiful results have been obtained, one of which we wish to mention here.

A Banach space X, with the property that every convex function which is continuous on an open subset $A \subset X$ is Fréchet differentiable in a dense G_δ-subset of A, is called an *Asplund space*, in honour of the mathematician, who introduced these spaces in 1968 and suggested the study of their properties.

Theorem 2 (criterion for a space to be Asplund). *X is an Asplund space if and only one of the following properties is satisfied:*

a) *Any equivalent norm is Fréchet differentiable at at least one point (and is then differentiable in a dense G_δ-subset);*

b) *the dual space is separable.*

A significant literature has been devoted to these problems. The interested reader is referred to two monographs Giles [1982]; Phelps [1978].

And now we speak briefly on non-smooth analysis.

1.3. The Origin of Non-smooth Analysis. Clarke's Approach. Very often one meets functions which are neither smooth nor convex—for example, the distance from a non-convex set. There is a natural desire to extend the concepts of differential calculus and convex analysis to a possibly broader class of functions. The study of smoothness properties of non-smooth mappings was initiated a fairly long time ago. Thus, for example, Bouligand (1932) introduced the idea of the *contingency* of an arbitrary set A to a point x_0, as the union of the semitangents to A passing through x_0. A *semitangent* passing through x_0 is a ray l, with origin at x_0, which is a limit of rays passing through x_0 and points x_n, $x_n \in A$. This leads to the definition of a *contingent* or *tangent cone* $TA(x_0) = \limsup_{t \downarrow 0} (A \backslash x_0)t^{-1}$. The study of the contingency derivatives of sets occupied Besicovitch, Kolmogorov and Verchenko and other mathematicians. But this was only an episode. The origin of the modern stage of development of non-smooth analysis has been placed with Clarke in his paper Clarke [1975].

Clarke's original plan was very simple. He suggested extending the idea of differential and subdifferential to the case of Lipschitz functions and mappings. As is well-known.

Theorem 3 (Rademacher). *A Lipschitz (finite-dimensional) function is differentiable almost everywhere.*

Clarke suggested taking the subdifferential of a Lipschitz function $f: \mathbf{R}^n \to \mathbf{R}$ at a point x_0 to be the convex hull of the limits of all gradients of f in a neighbourhood of x_0. More precisely, he introduced the set $\partial_C f(x_0) = \bigcap_{\delta > 0} \mathrm{co}\{f'(x) : x \in B(x_0, \delta)\}$. This set came to be called the *Clarke Subdifferential.*

It is impossible to define the Clarke subdifferential in so simple a way in the infinite-dimensional case, and it has to be defined in another way. Let X be a Banach space, f a Lipschitz function in a neighbourhood of x_0 and x an arbitrary vector from X. We denote by $f^\circ(x_0, x)$ the quantity $\limsup_{y \to x_0, t \downarrow 0} (f(y + tx) - f(y))t^{-1}$. In contrast to the usual directional derivative $f^\circ(x_0, x)$ is always defined. It is easy to verify that the function $x \to f^\circ(x_0, x)$ is finite, convex, homogeneous of the first degree, and also upper semi-continuous, thus the functions $x \to f^\circ(x_0, x)$ and $(x_0, x) \to f^\circ(x_0, x)$ also satisfy the condition $f^\circ(x_0, x) = (-f^\circ)(x_0, x)$. The subdifferential of the sublinear function $x \to f^\circ(x_0, x)$ is denoted by $\partial_C(f(x_0)$.

Theorem 4. *Let X be a Banach space, $f: X \to \mathbf{R}$ Lipschitz in a neighbourhood of x_0. Then $\partial_C f(x_0)$ is a nonempty convex weak* compact set in X^* and $f^\circ(x_0, x) = s\partial_C f(x_0)$.*

The Clarke subdifferential generalizes both the notion of differential of a smooth function and that of subdifferential of a convex function.

The function $f: X \to \mathbf{R}$ is called *strictly differentiable* at x_0 if it is (Fréchet) differentiable at x_0 and for any $\varepsilon > 0$ there is a $\delta > 0$ such that $\|x' - x_0\| < \delta$, $\|x'' - x_0\| < \delta$ implies $|f(x'') - f(x') - \langle f'(x_0), x'' - x' \rangle| \leqslant \varepsilon \|x'' - x'\|$.

Theorem 5. *Let X be a Banach space, f Lipschitz in a neighbourhood of x_0 and $\partial_C f(x_0) = \{x^*\}$. Then f is strictly differentiable at x_0 and $f'(x_0) = x^*$. The converse is also true: if f is strictly differentiable at x, then it is Lipschitz in a neighbourhood of x_0 and $f'(x_0) = \partial_C f(x_0)$.*

Example. $f(x) = x^2 \sin x^{-1}$ shows $f'(0) = 0$, but $\partial_C f(0) = [-1, 1]$, so the condition of strict differentiability is essential.

We trace how Clarke's subdifferential calculus is related to differential calculus and the subdifferential calculus of convex functions. We give the basic formulae in the form of a table.

Differential calculus	*Convex analysis*	*Nonsmooth analysis*
1. $\hat{x} \in \operatorname{loc\,ext} rf$ $\Rightarrow f'(\hat{x}) = 0$	$\hat{x} \in \operatorname{abs\,min} f$ $\Leftrightarrow 0 \in \partial f(x)$	$x \in \operatorname{loc\,ext} rf$ $\Rightarrow 0 \in \partial_C f(\hat{x})$
2. $(af)'(x) = af'(x)$, $a \in \mathbf{R}$	$\partial(af)(x) = a\partial f(x)$, $a \in \mathbf{R}^+$	$\partial_C(af)(x) = a\partial_C f(x)$, $a \in \mathbf{R}$
3. $(f_1 + f_2)'(x)$ $= f_1'(x) + f_2'(x)$	$\partial(f_1 + f_2)(x)$ $\cong \partial f_1(x) + \partial f_2(x)$	$\partial_C(f_1 + f_2)(x)$ $\subset \partial_C f_1(x) + \partial_C f_2(x)$
4. ———	$\partial(f_1 \vee f_2)(x)$ $\cong \operatorname{co}(\partial f_1(x) \cup \partial f_2(x))$	$\partial_C(f_1 \vee f_2)(x)$ $\subset \operatorname{co}(\partial_C f_1(x) \cup \partial_C f_2(x))$
5. $(f \circ g)'(x)$ $= f'(g(x)) \circ g'(x)$	———	$g(\cdot) \in SD'(x)$ $\Rightarrow \partial_C(f \circ g)(x)$ $\subset \partial_c f(g(x)) \circ g'(x)$

The "Rademacher" method of definition of the Clarke subdifferential also allows us to define a generalisation of the Jacobian of finite-dimensional Lipschitz mappings. Consider a Lipschitz mapping $F: \mathbf{R}^n \to \mathbf{R}^m$, $F(x) = (f_1(x), \ldots, f_m(x))$, f_i Lipschitz in a neighbourhood of a point x. The *generalised Jacobian* of F at x, denoted $\partial_C F(x)$, is the convex hull of the limits of the sequences of Jacobians $F'(x_i)$, $x_i \to x$ (and $F'(x_i)$ exist).

Theorem 6 (Composition, finite increments and inverse functions). a) *Let $f = g \circ F$, where $F: \mathbf{R}^n \to \mathbf{R}^m$ is a Lipschitz mapping in a neighbourhood of x and $g: \mathbf{R}^m \to \mathbf{R}$ is Lipschitz in a neighbourhood of $F(x)$. Then f is Lipschitz in a neighbourhood of x and $\partial_C f(c) \subset \mathrm{co}(\partial_C g(F(x))\partial_C F(x)$.*

b) *If X is a normed space and $f: X \to R$ is a Lipschitz function, then $f(x) - f(x') \in \{z: z = \langle x^*, x - x' \rangle, x^* \in \partial_C f(\xi), \xi \in (x, x')\}$*

c) *Let F be a mapping of $\mathbf{R}^n$ to itself which is Lipschitz in a neighbourhood of a point x_0, and such that all the matrices from $\partial_C F(x_0)$ have maximal rank. Then there are neighbourhoods U and V of x_0 and $F(x_0)$ respectively and a Lipschitz mapping $G: V \to \mathbf{R}^n$ such that $G \circ F(u) = u\ \forall u \in U$, $F \circ G(v) = v\ \forall v \in V$.*

We can see some similarities and with them radical differences between the Clarke calculus and the two classical calculi: equalities (say, in the formulae for the subdifferential of sum) are changed into inclusions (and so the very conception of "calculus", as a means of calculating subdifferentials of complex objects in terms of simple ones, is significantly weakened). Even the sense of approximation is transformed—the function $x \to f^\circ(x, x)$ does not approximate f globally, and is somewhat onesided. And, apparently, these defects in principle cannot be removed. However, there has recently been undertaken an effort to extend even more the limits of nonsmooth analysis, by introducing more interesting notions of subdifferential.

1.4. Other Approaches. Varga (1975) proposed another way of uniting all the derivatives close to a given point, but this has not yet been sufficiently developed, and so we will not give it. V.F. Dem'yanov and A.M. Rubinov suggested considering not the whole of the space of Lipschitz functions, but only the part of it consisting of functions which have a directional derivative at each point which is a difference of two convex homogeneous functions. For such a function the pair (A_1, A_2), such that $f'(x_0, x) = sA_1(x) - sA_2(x)$, is called the *quasidifferential* of f at a point x.

The calculus of quasidifferentials is developed in Dem'yanov & Rubinov [1980]. It has not been excluded that the functions studied by the authors, can be recovered from their quasidifferential.

We give one more definition of subdifferential. It was given by B.Sh. Mordukhovich (1976) in the finite-dimensional case and by A.D. Ioffe (1981)—in the form that we give here—in the infinite-dimensional case.

Let X be a Banach space, $f: X \to \bar{\mathbf{R}}$. The *lower Dini derivative in the direction* x at x_0 is the quantity

$$d^- f(x, x) = \liminf_{y \to x\ t \downarrow 0} \frac{f(x_0 + ty) - f(x_0)}{t},$$

The sets

$$\partial^- f(x_0) = \{x^* \in X^*: \langle x^*, x \rangle \leqslant d^- f(x_0, x)\ \forall x\},$$

$$\partial_\varepsilon^- f(x_0) = \{x^* \in X^*: \langle x^*, x \rangle \leqslant d^- f(x_0, x) + \varepsilon \|x\|\ \forall x\}$$

are called, respectively, the *Dini-subdifferential* and the ε-Dini-subdifferential.

The set

$$\partial_A f(X_0) = \bigcap_{L \in L} \bigcap_{\delta > 0} \bigcup_{x \in U(f,z,z)} \partial_s^- f_{x+L}(x),$$

where $U(f, z, \delta) = \{x: |f(x) - f(z)| \leqslant \delta, \|x - z\| \leqslant \delta\}$, and $\mathscr{L}$ is the collection of all finite-dimensional subspaces in X is called the *approximative subdifferential* (or *Ioffe-Mordukhovich subdifferential*) of f at x. This set is contained in $\partial_C f$ if f is Lipschitz.

Every time that a collection of approaches is discovered we wish to understand which of them is preferable and, in general, what we want of them. Here it is natural to try to use an axiomatic approach. What properties are reasonable to keep for the subdifferential $\partial f(x)$ of a nonsmooth function? It is natural to want the following:

a) If f takes a minimum at $\hat{x}$ then $0 \in \partial f(\hat{x})$;

b) if f is convex then ∂f coincides with the subdifferential in the sense of convex analysis;

c) $\partial(f + g)(x) \subset \partial f(x) + \partial g(x)$, if one of the functions is Lipschitz in a neighbourhood of x;

and, finally, one more, a topological requirement:

d) the mapping $x \to \partial f(x)$ must be upper semi-continuous in some neighbourhood (from the norm topology to the weak* topology of X^*) for a function f which is Lipschitz in this neighbourhood.

A.D. Ioffe (1984) proved that a subdifferential ∂f, of a function f Lipschitz in some neighbourhood, which satisfies conditions a)–d), must contain the approximative subdifferential.

We refer the reader, interested in the further development of nonsmooth analysis and its applications, to the monograph Clarke [1983].

§ 2. Convexity and Order

Convexity is closely related to the order relation on the set of real numbers. Namely, it allows us to introduce the notion of a segment or to link numbers by inequalities, and that, strictly, enables us to define the ideas of convex set and convex function. In an attempt to extend convex analysis to operators we naturally turn to spaces in which there is an order relation. Amongst ordered spaces the most suitable are the so-called *K-spaces* or *Kantorovich spaces* (or *order complete lattices*). It is natural to expect that many of the concepts of convex analysis will evolve during an investigation of convex mappings in K-spaces.

2.1. Basic Definitions. A linear space Y which is simultaneously an ordered set with the properties $y_1 \geqslant y_2 \Rightarrow y_1 + \eta \geqslant y_2 + \eta \ \forall \eta \in Y$ and $\lambda y_1 \geqslant \lambda y_2 \ \forall \lambda > 0$, is called a *linearly ordered* or *ordered vector space*. If for any elements $y_1, y_2 \in Y$

there is a supremum $y_1 \vee y_2$ and an infimum $y_1 \wedge y_2$, then Y is called a *vector lattice*. We recall that an element ξ is called the *supremum of a set A* if $\xi \geqslant y$ $\forall y \in A$ and $\xi' \geqslant y$ $\forall y \in A$ implies $\xi' \geqslant \xi$. The infimum is defined similarly. A vector lattice Y is called a *K-space* if any nonempty bounded set has a supremum and an infimum. An example of a K-space is the space $B(T)$ of bounded functions on a set T.

Let Y be an ordered space with a convex cone of positive elements Y, $\bar{Y} =: Y \cup \{+\infty\} \cup \{-\infty\}$ (with the natural extensions to $\bar{Y}$ of the algebraic operations), $Y^\circ =: Y \cup \{+\infty\}$, $F: X \to Y$. As in the classical case, sets $\operatorname{dom} F$ and $\operatorname{epi} F$ (the effective set and epigraph of F) can be defined.

$$\operatorname{dom} F =: \{x \in X: F(x) \in Y \cup \{-\infty\}\},$$

$$\operatorname{epi} F =: \{(x, y) \in X \times Y: F(x) \leqslant y\}$$

If $\operatorname{epi} F$ is a convex set then F is called a convex operator. $P: X \to Y^\circ$ is called sublinear if $P(\alpha x) = \alpha P(x)$ $\forall \alpha > 0$, $x \in X$ and $P(x_1 + x_2) \leqslant P(x_1) + P(x_2)$. The set of sublinear operators is denoted $S\mathscr{L}(X, Y)$. The *subdifferential of a sublinear operator* P is the set $\partial P =: \{\Lambda \in \mathscr{L}(X, Y): \Lambda x \leqslant P(x)\, \forall x \in X\}$. The *subdifferential of an operator* F at a point x_0 is the set $\partial F(x_0) =: \{\Lambda \in \mathscr{L}(X, Y): \Lambda x - \Lambda x_0 \leqslant F(x) - F(x_0)\, \forall x \in X\}$.

Let Y be a K-space and $F: X \to \bar{Y}$. Put $lF(\Lambda) = \sup_{x \in X}(\Lambda x - F(x))$. The operator lF is called the *Legendre-Young-Fenchel transform* (briefly the *Young transform*) of F.

We see these definitions are all analogues of the classical definitions.

2.2. Theorems on Extensions and Their Corollaries. In the introduction the role of separation in classical convex analysis was revealed. But separation, in its turn, is closely connected with the Hahn-Banach theorem. For mappings in K-spaces (and in fact only for them) there is a complete analogue of this theorem. The precise meaning of what has been said is the following:

Theorem 7 (Hahn-Banach-Kantorovich). $/Y$ *a K-space*, $P \in S\mathscr{L}(X, Y^\circ)$, $X_0 \in \operatorname{Lin}(X)$, $\Lambda_0 \in \mathscr{L}(X_0, Y)$, $\Lambda_0 x \leqslant P(x)$ $\forall x \in X_0/ \Rightarrow \exists \Lambda \in \mathscr{L}(X, Y)$: $\Lambda x \leqslant Px$ $\forall x \in X$.

Theorem 8 (Bonnice-Silverman-To). *Let Y be an ordered space (with reproducing cone of positive elements) such that Theorem 7 holds for vector space X, $X_0 \in$* $\operatorname{Lin}(X)$, $P \in S\mathscr{L}(X, Y^\circ)$, *then Y is a Kantorovich space.*

We give a number of results concerning the calculus of sublinear operators, which map a linear space X to a K-space Y (without a topology).

Theorem 9. $/P \in SL(X, Y)/ \Rightarrow \partial P \neq \varnothing$, $Px = \sup\{\Lambda x: \Lambda \in \partial P\}$.

Theorem 10. $/P_i \in SL(X, Y)$ $(i = 1, \ldots, n,)/ \Rightarrow \partial(P_1 + \cdots + P_n) = \partial P_1 + \cdots + \partial P_n$.

The analogue of the Dubovitskij-Milyutin theorem appears more complicated. But here is the *analogue of the theorem on composition with a linear operator.*

Theorem 11. *Let X and Z be linear spaces, Y a K-space, Λ a linear operator, mapping Z to X, $P \in S\mathscr{L}(X, Y)$. Then $\partial(P \circ \Lambda) = \partial P \circ \Lambda$.*

2.3. The Subdifferential Calculus of Convex Operators. We have mentioned that in the theory of convex functions there is essentially only one non-trivial sublinear operator, namely, the operator of taking the supremum. This is because every closed sublinear function is representable as a supremum of linear operators. The case of sublinear operators with values in K-spaces is somewhat similar. Here connected with each cardinal and each K-space there is one important sublinear operator—the so-called *Kutateladze canonical operator.* Let Y be a K-space, $\mathscr{A}$ an arbitrary nonempty set. We denote by $l_\infty(\mathscr{A}, Y)$ the collection of all bounded mappings from $\mathscr{A}$ to Y. The operator $\mathscr{E}_{\mathscr{A}}$: $l_\infty(A, Y) \to Y$, given by the formula $\mathscr{E}_{\mathscr{A}} f = \sup\{f(A)\colon A \in \mathscr{A}\}$, is called *canonical.* Any everywhere defined sublinear operator is a composition of a canonical operator and a linear operator. Thus, in the construction of sublinear calculus it is only necessary to understand the canonical operator. For details and references see Kutateladze [1979].

2.4. The Method of General Position. An important role in convex analysis is played by the conditions, under which basic results of convex calculus turn out to be true. One of the coarsest conditions of general position for cones is the following: say that cones K_1 and K_2 in a linear topological space are in *general position* if there is a complementary subspace $X_0 \subset X$ such that the set $(V \cap K_1 - V \cap K_2) \cap (V \cap K_1 - V \cap K_2)$ is a neighbourhood of zero in X_0 for any neighbourhood V in X. If $X_0 - X$, then it has been said that K_1 and K_2 form a *non-flat pair.* Thus, the concept of general position of cones is an immediate generalization of the idea of non-flatness, originating from M.G. Krejn. Among sufficient conditions of general position we mention two:

$$\text{(i) } \operatorname{int} K_1 \cap K_2 \neq \varnothing, \qquad \text{(ii) } \dim(K_1 - K_2) < \infty, \quad \operatorname{ri} K_1 \cap \operatorname{ri} K_2 \neq \varnothing.$$

Convex sets A_1 and A_2 in an l.t.s. X *are in general position* if the conical hulls of the sets $A\,x\{1\}$ and $A\,x\{1\}$ are in general position in $X \times \mathbf{R}$.

Let us give examples illustrating the applicability of the concept of general position.

Theorem 12. *Let X be an l.t.s., and Y a topological K-space, $P_i \in S\mathscr{L}(X, Y)$, $i = 1, 2$. If* $\operatorname{epi} P_1$ *and* $\operatorname{epi} P_2$ *are in general position, then $\partial^C(P_1 + P_2) = \partial^C P_1 + \partial P_2$ (here $\partial^C P = \partial P \cap \mathscr{L}(X, Y)$).*

For further reference see Levin [1985].

Theorem 13. *Let $P \in S\mathscr{L}(X Y^0)$, $\Lambda \in \mathscr{L}(Z, Y)$ and the sets $Z \times$* epi P *and $\Lambda \times Y$ be in general position. Then $\partial(P \cdot \Lambda) = \partial P \cdot \Lambda$.*

It is not possible for us to say anything here on other aspects of vector duality. We refer the reader to the recent monograph Kusraev [1985].

2.5. The Legendre-Young-Fenchel Transform. For an operator $f\colon X \to Y$ of a linear space X to a K-space Y the following relations, analogous to the vector case, hold:

(α) The operators lf and $l^2 f$ are convex;

(β) for any $x \in X$ and $\Lambda \in \mathbf{L}(X, Y)$ there is the *Young inequality*: $\Lambda x \leqslant f(x) + lf(\Lambda)$;

(γ) $l^2 f \leqslant f$;

(δ) $f \leqslant g \Rightarrow lf \leqslant lg$.

Theorem 14. *If X is a linear topological space, Y a topological K-space, f_1 and f_2 convex operators not identically equal to $+\infty$ and in general position, then* $l(f_1 + f_2) = lf_1 \oplus l_2 f$.

Here $\oplus$ is defined quite analogously with the scalar case, and general position of f_1 and f_2 means general position of their Hörmander transforms.

In concludion we give an analogue of the *Fenchel-Moreau theorem.*

Theorem 15. *Every lower semi-continuous convex operator, acting from a locally convex space to a topological K-space with a solid cone of positive elements, coincides with its second Legendre-Young-Fenchel transform.*

§ 3. Choquet Theory and Generalisations of Convexity

3.1. Introduction. Sometimes (see, for example, Kusraev, Kutateladze [1982] in the scientific literature *local convex analysis*, the study of individual convex objects, is compared with *global convex analysis*, the study of classes of convex objects (the collection of convex sets, functions, etc.) and duality. Choquet theory has an essential place in the second direction.

We have repeatedly touched upon the notion of the "set of extreme points" of a convex subset. The introduction of this notion was moulded by the idea of "refinement", the wish to single out *a*, possibly small, subset carrying all the information about the original object. Here the question of the recovery of the elements of the object arises naturally. The Krejn-Milman theorem can be reformulated as follows: each point of a compact convex subset of a locally convex space is the centre of gravity of a probability measure, concentrated on the closure of the extreme points of the set.

The requirements of classical analysis (probability theory, potential theory, etc.) have dictated the necessity of a more intense study of those minimal sets, from which it is possible to recover all the points of convex objects. In a cycle of papers in the mid fifties (see Choquet [1962]) Choquet initiated the detailed investigation of the problem of existence and uniqueness of the representation of the elements of a convex set via probability measures. As a result a separate chapter of mathematical analysis, called *Choquet theory*, was begun. On this theory see Fuchssteiner & Lusky [1981]; Kutateladze & Rubinov [1976].

3.2. Existence and Uniqueness Theorems

Theorem 16 (Choquet, on existence). *Any point of a metrizable compact convex set, in a locally convex space, has a representation by a regular Borel measure, concentrated on the extreme points of the set.*

In the non-metrizable case the set of extreme points may turn out to be not Borel (Bishop. De Leeuw), however, even here the basic intention of the representation is justified.

Theorem 17 (Bishop de Leeuw, on existence). *Any point of a compact convex set, in a locally convex space, has a representation by a probability measure defined on the σ-algebra of subsets of the compact set generated by the set of extreme points and the Baire sets. In this connection the measure of the set of extreme points is equal to one.*

As we recall, in a finite-dimensional space uniqueness of the representation is possible only for simplexes—the simplest geometric objects of convex analysis. There arises the question: what is the analogue of a simplex in an infinite-dimensional space?

Let X be a locally convex space and A a compact convex subset of X. We embed A in $X \times \mathbf{R}$, putting $\tilde{A} = \{(x, 1): x \in A\}$ and consider the cone $K_A = \alpha\tilde{A}$, $\alpha \geqslant 0$. A is called a *simplex* (or, *Choquet simplex*), if it turns $X \times \mathbf{R}$ into a partially-ordered space, for which $K_A - K_A$ is a lattice.

It is not difficult to see that in the finite-dimensional case A is a normal (full dimensional) simplex. There are several equivalent definitions. One of them is this, for any convex function f on A the function

$$\bar{f}(x) = \inf\{h(x): f \leqslant h, h \in \operatorname{Aff}(A, R)\}$$

is affine (the infimum in the definition of $\bar{f}$ is taken over all affine h such that $h \geqslant f$).

Theorem 18 (Choquet, on uniqueness). *A convex metrizable compact set, in a locally convex space, is a simplex if and only if each of its points has a unique representing measure concentrated on the set of extreme points.*

Related to theorems of Choquet type are such classical results as the Riesz theorem on the general form of a linear functional in a space C, the Bernstein theorem on absolutely monotone functions, etc.

An essential element of Choquet theory is the description of the order in spaces of measures on convex compact sets and the related ordering theorem on decompositions (see, for example, the Cartier-Fell-Meyer theorem) and extensions. For this see Kutateladze & Rubinov [1976], Fuchssteiner & Lusky [1981].

3.3. Other Generalizations of Convexity. A large number of attempts have been made to generalize the idea of convexity in order to extend the most important theorems of convex analysis to other objects. We give some examples.

We recall that if $A \in \mathrm{Co}(\mathbf{R}^n)$, then A is an intersection of a family of half-spaces, and if $f \in \mathrm{Co}(\mathbf{R}^n, \bar{\mathbf{R}})$, then f is a supremum of affine functions. This provides the possibility for the following generalisation. Let X be a set and $\mathscr{H}$ a family of subsets of X. A set A is called $\mathscr{H}$-convex, if it is the intersection of a subfamily of $\mathscr{H}$. For functions a similar method leads to the notion of Φ-convexity of Ky Fan. Here a family of functions Φ is given and a function f is called *Φ-convex* if it is the supremum of a subfamily of Φ.

Convexity generalises to nonlinear geometrical objects in a natural way. An example is the so-called *d-convexity*. Let (X, d) be a metric space. A set $A \subset X$ is called *d-convex* if together with any two points $x, y \in A$ it contains the metric segment $[x, y] = \{z \in X : d(x, z) + d(z, y) = d(x, y)\}$. This notion is due to Menger. Various versions of the axiomatic approach to convexity are possible. In the axiomatic approach the aim is to select conditions, as broad as possible, which allow the formulation of analogues of the most important theorems of convex analysis (say, the Krejn-Milman theorem, the Kuhn-Tucker theorem, etc.). Regarding this see the recently published book of V.P. Soltan, [1984].

3.4. Supremal Generators. We begin with a function of one variable. Let $\Phi = \mathrm{Aff}(R, R)$. Then, as we know, the Φ-convex functions in the sense of §3.3 will all be convex closed functions. Add to $\mathrm{Aff}(R, R)$ the parabolic form $-(t + \alpha)^2$. Then.

Theorem 19. *The Φ-convex functions on a finite segment, for* $\Phi = \mathrm{Aff}(R, R) \cup (\bigcup_{\alpha \in R}(-(t + \alpha)^2)$ *are precisely the lower semicontinuous functions.*

In this way, the collection of all quadratic trinomials "supremally generate" the space C on a finite segment and even more, generate the collection of lower semi-continuous functions.

A system of functions $\{f_1, \ldots, f_m\}$ defined on a compact set is called a *Korovkin system*, if the subspace of $C(T)$ it spans supremally generates $C(T)$.

P.P. Korovkin proved that for three functions $\{f_1, f_2, f_3\}$ to generate a subspace which supremally generates $C(A)$ (A a finite segment of $\mathbf{R}$), it is necessary and sufficient that they generate a Chebyshev system.

Connected with this is a remarkable phenomenon, also discovered by P.P. Korovkin, that if $\{T_n\}_{n \in N}$ is a family of positive linear operators which converges to the identity on a set of three functions, which generates a Chebyshev system, then it converges to the identity operator everywhere. These facts have been the source of a complete direction in approximation theory.

They found their place in Choquet theory united with the notion of H-convexity (on this see Kutateladze & Rubinov, [1976].

At this point it seems reasonable to us to end this survey of the directions to which recent works have been devoted.

Summary of the Literature

Let us mention, first of all, basic works: Minkowski [1910], Moreau [1962], Helly [1923], Fenchel [1949, 1951]. Of the numerous monographs specially devoted to convexity or convex analysis we name the following: Levin [1985], Bonnesen & Fenchel [1934], Eggleston [1958], Leichtweiss [1980], van Tiel [1984] and particularly Rockafellar [1970]. There are interesting and substantial books on convexity: Encyclopaedia of Elementary Mathematics [1966], Danzer, Grünbaum & Klee [1963], Lyusternik [1956], Yaglom & Boltyanskij [1951]. We list some courses on functional analysis and books on the theory of extremal problems, which present the foundations of convex analysis and its applications: Alekseev, Galeev & Tikhomirov [1984], Alekseev, Tikhomirov & Fomin [1979], Ekeland & Temam [1976], Girsanov [1970], Holmes [1975], Ioffe & Tikhomirov, Kantorovich & Akilov [1984], Kolmogorov & Fomin [1981], Kutateladze [1983], Kutateladze & Rubinov [1976], Laurent [1972], Pshenichnyj [1980, 1982].

An introduction to convex analysis is presented in many works. For the definitions of convex objects see Alekseev, Galeev & Tikhomirov [1984], Alekseev, Tikhomirov & Fomin [1979], Kutateladze [1983], Pshenichnyj [1980, 1982, Rockafellar [1970] and others, for duality theory in linear spaces see Bourbaki [1953], Kutateladze [1983], Robertson & Robertson [1954], for the topological properties of convex objects Bourbaki [1953], Rockafellar [1970].

The basic facts about convex calculus are presented in Alekseev, Galeev & Tikhomirov [1984], Alekseev, Tikhomirov & Fomin [1979], Boltyanskij [1973], Ioffe & Tikhomirov [1974], Kutateladze [1983], Kutateladze & Rubinov [1976], Levin [1985], Pshenichnyj [1980, 1982], Rockafellar [1970]. On convex analysis of integral functionals see first of all the monographs Ioffe & Tikhomirov [1974], Levin [1985].

On linear programming see Dantzig [1963], Eremin & Astaf'ev [1976], Gass [1958], Gavurin & Malozemov [1984], Karmanov [1980]. On duality see Gol'shtein [1971], on extensions of the duality theorem to bounded operators see Kusraev [1985]. There is a collection of book devoted to convex geometry, see, for example, Berger [1977], Bonnesen & Fenchel [1934], Danzer, Grünbaum & Klee [1963], Encyclopaedia of Elementary Mathematics [1966], Hadwiger [1957], Lyusternik [1956], Yaglom [1951]. On convexity in the approximation theory see Laurent [1972], Tikhomirov [1976].

Lyapunov's theorem is discussed, for example, in Alekseev, Galeev & Tikhomirov [1984], Arkin & Levin [1972], Ioffe & Tikhomirov [1974].

For more details on theorems of Aumann type [Theorem 28] see Ioffe & Tikhomirov [1974].

The theme "Extension of variational problems" was initiated with the work of N.N. Bogolyubov [1969], it was continued by Young, McShane and many others. It is touched on in the monograph Ioffe & Tikhomirov [1974] and the survey article Ioffe & Tikhomirov [1968]. The extension of the optimal control problem was initiated with the work of R.V. Gamkrelidze [1962]. The survey article Arkin & Levin [1972] is devoted to the Lyapunov problem, see also Alekseev, Tikhomirov & Fomin [1979]. The monograph Ekeland & Temam [1976] is devoted to duality of many-dimensional problems. On the Monge-Kantorovich problem see Kantorovich & Akilov [1984], Kantorovich & Rubinshtein [1957], Levin [1985], Levin & Milyutin [1979]. Many problems reduce to this problem: in elasticity theory Mordukhovich [1976, 1986]; probability theory [see the survey Rachev [1984]]; economics [see Levin [1985]]; etc.

The theory of necessary and sufficient conditions for extrema in mathematical programming problems was constructed in Levitin, Milyutin & Osmolovskij [1978]. For an exposition see also Alekseev, Galeev & Tikhomirov [1984], Alekseev, Tikhomirov & Fomin [1979].

Convexity in variational calculus and optimal control is discussed in the monographs Alekseev, Tikhomirov & Fomin [1979], Boltyanskij [1973], Gamkrelidze [1977], Ioffe & Tikhomirov [1974], Pshenichnyj [1980, 1982].

The idea of convexity has been applied in algebra and analysis [for example, through Newton polyhedra [Khovanskij [1983]]], in the theory of function spaces, Krasnosel'skii & Rutitskii [1958], games theory, Karlin [1959], in the theory of embeddings of Orlicz function spaces, in numerical methods and in many other questions.

On differentiability of convex functions see Giles [1982], Phelps [1978].

Nonsmooth analysis originated with the work of Clarke [1975]. The results of the development of this division are given in Clarke [1983]. Many aspects of nonsmooth analysis are presented in the monograph of B.Sh. Mordukhovich, [1988]. Of the works, devoted to nonsmooth analysis, we mention Aubin [1984], Ioffe [1981, 1984], Mosolov & Myasnikov [1971]. Dem'yanov, Vasil'ev [1987], Dem'yanov, Rubinov [1980].

The theory of ordered spaces is presented in Akilov & Kutateladze [1978], Kantorovich & Rubinshtein [1957]. Convex analysis in ordered spaces is elucidated in Fuchssteiner & Lusky [1981], Krejn & Rutman [1948], Kusraev [1985], Kusraev & Kutateladze [1982, 1983], Kutateladze [1979]. On Choquet theory see Choquet [1962], Fuchssteiner & Lusky [1981], Kutateladze [1975], Kutateladze & Rubinov [1976].

References

For the convenience of the reader, references to reviews in Zentralblatt für Mathematik (Zbl.), compiled using the MATH database, and Jahrbuch über die Fortschritte der Mathematik (Jrb.) have, as far as possible, been included in this bibliography.

Akhiezer, N.I., Krejn, M.G.

[1938] Some questions in the theory of moments. Khar'kov: Gos. Ob"ed. Nauchn.-Tekhn. Izdat., English transl.: Amer. Math. Soc. Transl. of Mathematical Monographs 2, 1962. Zbl.117,327

Akilov, G.P., Kutateladze, S.S.

[1978] Ordered vector spaces. Novosibirsk: Nauka. Zbl.395.46010

Aleksandrov, A.D.

[1950] Convex polyhedra. Moscow-Leningrad: ITTI. Zbl.41,509. German transl.: Math. Mon. VIII. Berlin: Akademie-Verlag, 1958. Zbl.79,163

Alekseev, V.M., Galeev, E.M., Tikhomirov, V.M.

[1984] A collection of problems on optimization. Moscow: Nauka

Alekseev, V.M., Tikhomirov, V.M., Fomin, S.V.

[1979] Optimal control. Moscow: Nauka. Zbl.516.49002

Archimedes

[1963] Werke. Darmstadt: Wissenschaftliche Buchgesellschaft

Arkin, V.I., Levin, V.L.

[1972] Convexity of walues of vector integrals, theorems on measurable choice and variational problems. Uspekhi Mat. Nauk *27* No. 3, 21–77. English transl.: Russ. Math. Surv. *27* No. 3, 21–85, 1973

Asplund, E.

[1968] Fréchet differentiability of convex functions. Acta Math. *121*, 31–47. Zbl.162,175

Aubin, J-P.

[1984] L'analyse non linéaire et ses motivations économiques Paris: Masson. Zbl.551.90001

Aubin, J-P., Ekeland, I.

[1984] Applied nonlinear analysis. New York: John Wiley. Zbl.641.47066

Berger, M.

[1977] Géométrie. Vol. 1, 2. Paris: Cedic. Zbl.382.51011, Zbl.382.51012

Bieberbach, L.

[1915] Über eine Extremaleigenschaft des Kreises. Jber. Deutsch. Math.-Ver. *24*, 247–250

Blaschke, W.

[1914] Über den grössten Kreis in einer konvexen Punktmenge. Jber. Deutsch. Math. Ver. *23*, 369–374

[1956] Kreis und Kugel. Berlin: Walter de Gruyter. Zbl.70,175

Bogolyubov, N.N.

[1969] On some new methods in variational calculus. Izbr. Trudy V. 1. Kiev: Naukova Dumka, 63–86

Boltyanskij, V.G.

[1973] Optimal control of discrete systems. Moscow: Nauka. Zbl.268.49017. English transl.: New York, Toronto: John Wiley & Sons, 1978. Zbl.397.49001

[1975] The method of tents in the theory of extremal problems. Uspekhi Mat. Nauk *30* No. 3–55. English transl.: Russ. Math. Surveys *30* No. 3, 1–54, Zbl.318.49017

Bonnesen, T., Fenchel, W.

[1934] Theorie der konvexen Körper. Berlin: Springer. Zbl.8,77

Bonnice, W.E., Silverman, R.S.

[1967] The Hahn-Banach extension and the least upper bound properties are equivalent. Oroc. Amer. Math. Soc. *18*, 843–849. Zbl.165,468

Bouligand, G.

[1932] Introduction à la géométrie infinitésimale directe. Paris: Vuibert. Zbl.5,375

Bourbaki, N.

[1953] Espaces vectoriels topologiques. Paris: Hermann. Zbl.50,107

Brønsted, A.

[1964] Conjugate convex functions in topological vector spaces. Mat.-Fys. Medd. K. Dan. Vidensk Selsk. *34* No. 2, 27 p. Zbl.119,100

Busemann, H.

[1958] Convex surfaces. New York: Interscience. Zbl.196,551

Carathéodory, C.

[1907] Über den Variabilitätsbereich der Koeffizienten von Potenzreihen, die gegebene Werte nicht annehment. Math. Ann. *64*, 95–115

Castaing, C., Valadier, M.

[1977] Convex analysis and measurable multifunctions. Lect. Notes Math. *580*, Springer-Verlag, Berlin Heidelberg New York. Zbl.346.46038

Cauchy, A.

[1913] Sur les polygones et les polyèdres. Second Memoire J. École Polytechn. *9*. Oeuvres Complètes (2) Vol. 1, Paris: 1896

Chebyshev, P.L.

[1947] Questions on least values, connected with approximate representation of functions Collected Works V. 2 Moscow-Len. Izdat. A. N. 152–237. French transl.: Oeuvres. 1. New York: Chelsea Reprint. Zbl.33,338

Chernikov, S.N.

[1962] Linear inequalities. Moscow: Nauka. Zbl.169,41

Choquet, G.

[1962] Les cônes convexes faiblement complets dans l'analyse. In: Proc. Int. Congr. Math. Stockholm, 317–330 Zbl.121,331

Clarke, F.H.

[1975] Generalized gradients and its applications. Trans. Amer. Math. Soc. *205*, 247–262. Zbl.307.26012

[1983] Optimization and nonsmooth analysis. New York: John Wiley. Zbl.582.49001

Clarke, F.H., Vinter, R.B.

[1983] Local optimality conditions and Lipshitzian solutions to Hamilton-Jacobi equation. SIAM J. Control Optimization, *21*, 856–870. Zbl.528.49019

[1985] a) Existence and regularity in the small in the calculus of variations. J. Differ. Equations *59*, 336–354

b) Regularity properties of solutions to the basic problems in the calculus of variations. Trans. Amer. Math. Soc. *289*, 73–98. Zbl.563.49009

Courant, R., Hilbert, D.

[1962] Methods of mathematical physics. Vol. 2. New York: Interscience. Publishers, Zbl.99,295

Dantzig, G.B.
[1963] Linear programming and extensions. Princeton: Princeton Univ. Press. Zbl.108,331
Danzer, L., Grünbaum, B., Klee, V.
[1963] Helly's theorem and its relatives. In: Convexity, Proc. Symp. Pure Math. VII, 101–180. Providence: Amer. Math. Soc. Zbl.132,174
Dem'yanov, V.F., Vasil'ev, L.V.
[1981] Non-differentiable optimization. Moscow: Nauka. Zbl.559.49001
Dem'yanov, V.F., Rubinov, A.M.
[1980] Quasi-differnentiable functionals. Dokl. Akad. Nauk SSSR *250*: 1, 21–25. English transl.: Sov. Math., Dokl. *21*, 14–17. Zbl.456.49016
Dubovitskij, A.Ya., Milyutin, A.A.
[1965] Extremum problems in the presence of restrictions. zh. Bychisl. Mat. Mat. Fiz *5*: 3, 395–453. Zbl.158,1335. English transl.: USSR Comput. Math. Math. Phys. 5, No. 3, 1–80 1965
Eggleston, H.
[1958] Convexity. Cambridge: Cambridge Univ. Press. Zbl.86,153
Ekeland, I., Temam, R.
[1976] Convex analysis and variational problems. New York: North-Holland. Zbl.322.90046
Encyclopaedia of Elementary Mathematics:
[1966] Book on Mathematics. Nauka.
Eremin, I.I., Astaf'ev, N.N.
[1976] Introduction to the theory of linear and convex programming. Moscow: Nauka. Zbl.463.90059
Fedorov, E.S.
[1949] Symmetry and the structure of crystals. Major papers. Moscow-Leningrad: Izdat, Akad. Nauk SSSR
Fenchel, W.
[1949] On conjugate convex functions. Can. J. Math. *1*, 73–77. Zbl.38,209
[1951] Convex cones, sets and functions. Mimeographed lecture notes. Princeton Univ.
Fuchssteiner, B., Lusky, W.
[1981] Convex cones. Amsterdam: North Holland. Zbl.478.46002
Gamkrelidze, R.V.
[1962] Optimal sliding states. Dokl. Akad. Nauk SSSR *143*, 1243–1245. English transl.: Sov. Math. Dokl. *3*, 559–562. Zbl.131,324
[1977] Principles of optimal control. Tbilisi: Izdat. Tbilisi Inst. English transl.: Plenum Press, New York-London 1978. Zbl.401.49001
Gamkrelidze, R.V., Kharatishvili, G.L.
[1969] Extremal problems in linear topological spaces. Izv. Akad. Nauk SSSR, Ser. Mat. *33*, 781–839. Zbl.182,184 English transl.: Math. USSR, Izv. 3, 737–794 (1969)
Gass, S.I.
[1958] Linear programming. Methods and applications. New York: McGraw Hill Book Co. Zbl.81,367
[1961] Recent developments in linear programming. Advances in computers Vol 2. New York-London: Academic Press, 295–377. Zbl.136,139
Gavurin, M.K., Malozemov, V.N.
[1984] Extremal problems with linear constraints. Leningrad: Izdat. Leningrad State Univ. Zbl.593.90046
Giles, J.R.
[1982] Convex analysis with applications in differentiation of convex functions. London: Pitman. Zbl.486.46001
Girsanov, I.V.
[1970] Lectures on the mathematical theory of extremal problems. Moscow: Izdat. Moscow State Univ. English transl.: Lect. Notes Econ. Math. Syst. *67*. Berlin: Springer-Verlag, 1972. Zbl.214,145

Gol'shtein, E.G.
[1971] Duality theory in mathematical programming and its applications. Moscow: Nauka
Hadwiger, H.
[1957] Vorlesungen über Inhalt, Oberfläche und Isoperimetrie. Berlin: Springer-Verlag. Grandlehren der math. Wissenschaften, Bd. 93. Zbl.78,357
Hadwiger, H., Debrunner, H., Klee, V.
[1964] Combinatorial geometry in the plane. New York: Holt. Zbl.89,173
Hardy, G.H., Littlewood, J.E., Pólya, G.
[1934] Inequalities. Cambridge: Cambridge Univ. Press. Zbl.10,102
Helly, E.
[1923] Über Mengen konvexer Körper mit gemeinschaftlichen Punkten. Jber. Deutsch. Math. Ver. *32*, 175–176. Jrb.49,534
Holmes, R.B.
[1975] Geometric functional analysis and its applications. New York: Springer-Verlag. Zbl.336.46001
Hörmander, L.
[1955] Sur la fonction d'appui des ensembles convexes dans un espace localement convexe. Ark. Mat. *3*, 181–186. Zbl.64,105
Ioffe, A.D.
[1981] Nonsmooth analysis: differential calculus of nondifferentiable mappings. Trans. Amer. Math. Soc. *266*, 1–56
[1984] Approximate subdifferentials and applications I. The finite-dimensional theory. Trans. Amer. Math. Soc. *281*, 389–416. Zbl.531.49014
Ioffe, A.D., Levin, V.L.
[1972] Subdifferentials of convex functions. Tr. Mosk. Mat. O.-va. *26*, 3–73. English transl.: Trans. Mosc. Math. Soc. 26, 1–72. Zbl.257.46042
Ioffe, A.D., Tikhomirov, V.M.
[1968] a) Duality of convex functions and extremal problems. Usp. Mat. Nauk *23* No. 6 51–116. English transl.: Russ. Math. Surv. *23*: 6 53–124. Zbl.167,422
b) Extensions of variational problems. Tr. Mosk. Mat. O-va *18*, 187–246. English transl.: Trans. Mosc. Math. Soc. 18, 207–273. Zbl.169,139
[1969] On minimization of integral functionals. Funkts. Anal. Prilozh. *3* No. 3, 61–70. English transl.: Funct. Anal. Appl. 3, 218–227, 1970. Zbl.193,76
[1974] The theory of extremal problems. Moscow: Nauka
James, R.S.
[1963] Characterisations of reflexivity. Stud. Math. *23*, 205–216. Zbl.113,93
Jensen, J.
[1906] Sur les fonctions convex et les inégalités entre les valeurs moyennes. Acta Math. *30*, 175–193. Jrb.37,422
John, F.
[1948] Extremum problems with inequalities as subsidiary conditions. Studies and Essays, presented to R. Courant on his 60th Birthday. Courant Ann. Vol., 187–204. New York: Interscience. Zbl.34,105
Kakutani, S.
[1939] Some characterizations of Euclidean space. Jap. J. Math. *16*, 93–97. Zbl.22,150
Kantorovich, L.V.
[1935] On semi-ordered linear spaces and their applications in the theory of linear operations. Dokl. Akad. Nauk SSSR *4*: 1, 13–16. Zbl.13,168
[1939] Mathematical methods for the organization and planning of production. Leningrad: Izdat. Leningrad State Univ.
Kantorovich, L.V., Akilov, G.P.
[1984] Functional analysis. Moscow: Nauka. English transl. (of 2nd Ed.): Oxford: Pergamon Press, 1982. Zbl.555.46001, Zbl.484.46003

Kantorovich, L.V., Rubinshtein, G.Sh.
[1957] On a functional space and certain extremal problems. Dokl. Akad. Nauk SSSR *115*, 1058–1061. Zbl.81,115
Karlin, S.
[1959] Mathematical methods and theory in games, programming and economics, I, II. London: Pergamon Press. Zbl.139,127
Karmanov, V.G.
[1980] Mathematical programming Moscow: Nauka. Zbl.496.90052
Khovanskij, A.G.
[1983] Newton polyhedra (resolution of singularities). Itogi Nauk Tekh., Ser. Sovrem. Probl. Mat. *22*, 207–239. English transl.: J. Sov. Math. *27*, 2811–2830 (1984) Zbl.544.14006
Klee, V.L.
[1958] Extremal structure of convex sets. I, II. Arch. Maths. *8*, 234–240, 1957. Math. Z. *69*, 90–104. Zbl.79,125
Kolmogorov, A.N., Fomin, S.V.
[1981] Elements of the theory of functions and functional analysis. Moscow: Nauka. English transl.: (of an earlier edition) Rochester: Graylock Press, 1957. Zbl.501.46002
Krasnosel'skij, M.A., Rutitskij, Ya.B.
[1958] Convex functions and Orlicz spaces. Moscow: Fizmatgiz. Zbl.84,101
Krejn, M.G., Milman, D.P.
[1940] On extreme points of regular convex sets. Stud. Math. *9*, 133–138
Krejn, M.G., Nudel'man, A.A.
[1973] The Markov moment problem and extremal problems. Moscow: Nauka. English transl.: Amer. Math. Soc. Transl. of Math. Monographs *50*, 1977. Zbl.265.42006
Krejn, M.G., Rutman, M.A.
[1948] Linear operators, leaving invariant a cone in a Banach space. Usp. Mat. Nauk *3*, 3–95. Zbl.301,129. English transl.: Amer. Math. Soc. Transl. *26*, 128 p., 1950. Zbl.30,129
Kuhn, H., Tucker, A.
[1951] Nonlinear programming. Proc. Second Berkeley Symposium. Berkeley: Univ. of Calif. Press, 481–492. Zbl.44,59
Kusraev, A.G.
[1985] Vector duality and its applications. Novosibirsk: Nauka. Zbl.616.49010
Kusraev, A.G., Kutateladze, S.S.
[1982] Local convex analysis. Itogi Nauki Tekh, Ser. Jovrem. Probl. Mat. 19 155–206. Zbl.516.46026
[1983] Subdifferential calculus. Novosibirsk: Gos. Univ. Novosihivski Nauka 1987
Kutateladze, S.S.
[1975] Choquet boundaries in *K*-spaces. Usp. Mat. Nauk. *30* No. 4, 107–146. English transl.: Russ. Math. Surv. *30* No. 4, 115–155, 1975. Zbl.308.46010
[1979] Convex Operators. Usp. Mat. Nauk *34* No. 1, 167–196. English transl.: Russ. Math. Surv. *34* No. 1, 181–214, 1979. Zbl.415.47034
[1983] Foundations of functional analysis. Novosibirsk: Nauka. Zbl.531.46001
Kutateladze, S.S., Rubinov, A.M.
[1976] Minkowski duality and its applications. Novosibirsk: Nauka
Laurent, P.-J.
[1972] Approximation et optimization. Collection Enseignement des Sciences, No. 13. Paris: Hermann. Zbl.238.90058
Leichtweiss, K.
[1980] Konvexe Mengen. Berlin: Springer-Verlag. Hochschulbücher für Mathematik 81. Zbl.427.52001
Levin, V.L.
[1969] An application of Helly's theorem in convex programming, problems of best approximation and related questions. Mat. Sb., Nov. Ser. *79*(121), 250–263. English transl.: Math. USSR, Sb. 8, 235–247. Zbl.187,176

[1972] Subdifferentials of convex mappings and compositions of functions. Sib. Mat. Zh. *13*, 1295–1303. English transl.: Sib. Math. J. *13*, 903–909, 1973. Zbl.261.28007

[1975] Convex integral functionals and the theory of lifting. Usp. Mat. Nauk *30*: No. 2, 115–178. English transl.: Russ. Math. Surv. *30t* No. 2, 119–184, 1975. Zbl.332.46031

[1985] Convex analysis in spaces of measurable functions and its application to mathematics and economics. Moscow: Nauka. Zbl.617.46035

Levin, V.L., Milyutin, A.A.

[1979] The problem of mass transfer with a discontinuous cost functions and mass statement of the duality problem for convex extremal problems. Usp. Mat. Nauk *34* No. 3, 3–68. English transl.: Russ. Math. Surv. *34* No. 3, 1–78, 1979. Zbl.422.46060

Levitin, E.S., Milyutin, A.A., Osmolovskii, N.P.

[1978] Conditions of high orders for a local minimum in problems with constraints. Usp. Mat. Nauk *33* No. 6, 85–148. English transl.: Russ. Math. Surv. *33*: No. 6, 97–168, 1978. Zbl.456.49015

Lyapunov, A.A.

[1940] On completely additive vector functions. Izv. Akd. Nauk SSSR, Ser. Mat. *4*, 465–478. Zbl.24,385

Lyusternik, L.A.

[1956] Convex figures and polyhedra, Moscow: Goztekhizdat. Zbl.71,378

Makarov, V.L., Rubinov, A.M.

[1973] Mathematical theory of economic dynamics and equilibrium. Moscow: Nauka. English transl.: Springer-Verlag, Berlin-Heidelberg-New York. 1977. Zbl.323.90012

Minkowski, H.

[1910] Geometrie der Zahlen. Leipzig: Teubner.

Mordukhovich, B.Sh.

[1976] The existence of optimal controls. Itogi Nauki Takh., Ser. Sovrem. Probl. Mat. *6*, 207–261. English transl.: J. Sov. Math. 7, 850–886 (1977) Zbl.403.49004

[1988] Approximation methods in problems of optimization and control. Moscow: Nauka

Moreau, J.-J.

[1962] Fonctions convexes en dualité. Inf. convolution. Multigraph. Sem. Math. Fac. Sci. Univ. Montpellier

Mosolov, P.P., Myasnikov, P.P.

[1971] Variational methods in the theory of flows of rigid visco-plastic media. Moscow: Izdat Moscow State Univ. Zbl.258.73006

[1981] Mechanics of rigid plastic media. Moscow: Nauka. Zbl.551.73002

Nikaido, H.

[1968] Convex structures and economic theory. New York: Academic Press. Zbl.172,445

Phelps, R.R.

[1966] Lectures on Choquet's theorem. Princeton: Van Nostrand. Zbl.135,362

[1978] Differentiability of convex functions on Banach spaces. London: Saunders

Pogorelov, A.V.

[1952] Unique determination of general convex surfaces. Kiev: Naukova Dumka. Zbl.44,508

[1969] Extrinsic geometry of convex surfaces. Moscow: Nauka. English transl.: Amer. Math. Soc. Transl. of Math. Monographs *35*, 1973. Zbl.311.53067

Pontryagin, L.S., Boltyanskij, V.G., Gamkrelidze, R.V., Mishchenko, E.F.

[1983] Mathematical theory of optimal processes. Moscow: Nauka. English transl. (of an earlier edition): Oxford: Pergamon Press, 1964. Zbl.516.49001

Pshenichnyj, B.N.

[1980] Convex analysis and extremal problems. Moscow: Nauka. Zbl.477.90034

[1982] Necessary conditions for an extremum. Moscow: Nauka. English transl. (of an earlier edition): New York-Basel: Marcel Dekker, 1971. Zbl.522.90055

Rachev, S.T.

[1984] The Monge-Kantorovich problem on mass transition and its applications in stochastics, Teor. Veroyatn. Primen. *29*: No. 4, 625–653. Zbl.565.60010

Radon, I.
[1921] Mengen konvexer Körper, die einen gemeinsamen Punkt enthalten. Math. Ann. *83*, 113–115. Jrb.48,834
Riesz, F.
[1928] Sur la decomposition de operations fonctionelles. Atti Congresso Bologna *3*, 143–148
Robertson, A.P., Robertson, W.
[1964] Topological vector spaces. Cambridge: Cambridge University Press. Zbl.123,302
Rockafellar, R.T.
[1970] Convex analysis. Princeton: Princeton Univ. Press. Zbl.193,184
Rubinov, A.M.
[1977] Sublinear operators and their applications. Usp. Mat. Nauk *32* No. 4, 113–174. English transl.: Russ. Math. Surv. *32* No. 4, 115–175, 1977. Zbl.355.47002
[1980] Sublinear multivalued mappings and their applications in economical mathematical problems. Leningrad: Nauka. Zbl.466.47041
[1982] Economic dynamics. Itogi Nauki Tekh., Ser. Sovrem. Probl. Mat. 19, 59–110. Zbl.511.90046
Soltan, V.P.
[1984] Introduction to the axiomatic theory of convexity. Kishinev: Shtiinitsa. Zbl.559.52001
Stolz, O.
[1893] Grundzüge der Differential- und Integralrechnung. I. Leipzig: Teubner
Tiel, J. van
[1984] Convex analysis. New York: John Wiley. Zbl.565.49001
Tikhomirov, V.M.
[1976] Some questions in approximation theory. Moscow: Izdat. Moscov. Univ.
Tonelli, L.
[1921, 1923] Fondomenti di calcolo delle variazioni I, II. Bologna: Zanichelli. Jrb.48,581, Jrb.49,349
Yaglom, I.M., Boltyanskij, V.G.
[1951] Convex figures. Moscow: Gostekhizdat. English transl.: New York: Holt, Rinehart & Winston, 1961. Zbl.44,378
Young, H.W.
[1901] Über die kleinste Kugel, die eine räumliche Figur einschliesst. J. leine Aangew. Math. *123*, 241–257
Young, L.C.
[1969] Lectures on the calculus of variations and optimal control theory. London: Saunders. Zbl.177,378

II. Approximation Theory

V.M. Tikhomirov

Translated from the Russian
by D. Newton

Contents

Introduction 96
1. What is Approximation Theory? 96
2. The Fundamental Stages in Approximation Theory 98
3. Notations 100
Chapter 1. Classical Approximation Theory 104
§ 1. Prehistory 104
1.1. The Calculation of π 104
1.2. Approximation of Functions 105
1.3. Quadratures 106
1.4. Interpolation 106
1.5. Approximation of Functionals and Operators 106
§ 2. Orthogonal Polynomials and Polynomials Deviating Least from Zero 106
2.1. Introduction 106
2.2. Legendre Polynomials and Orthogonal Polynomials 108
2.3. Chebyshev Polynomials of the First Kind 112
2.4. Chebyshev Polynomials of the Second Kind 113
2.5. Gaussian Quadratures 114
§ 3. The Circle of Ideas Due to P.L. Chebyshev 115
3.1. Introduction 115
3.2. Existence, Uniqueness and Alternation in the Case of Polynomial Approximations 115
3.3. Exact Solutions 117
3.4. Extremal Properties of Polynomials 118
3.5. Appendix 120
§ 4. Weierstrass' Theorem and its Generalizations 121
4.1. Introduction 121
4.2. Proofs of Weierstrass' Theorem 121

4.3. The Development and Generalization of Weierstrass' Theorem . 123
4.4. The Stone-Weierstrass Theorem 124
4.5. Appendix........................ 125
§5. The Constructive Theory of Functions........................ 126
5.1. Introduction........................ 126
5.2. Direct Theorems. Approximation of the Function $|t|$ 128
5.3. Inverse Theorems........................ 130
5.4. The Statement of the Problem on Comparison of Different Methods of Approximation. Approximation in Classes 132
5.5. The Latest Developments in the Constructive Theory of Functions 136
Chapter 2. Classical Methods of Approximation........................ 140
§1. Methods of Approximation 141
1.1. Algebraic Polynomials 141
1.2. Rational Functions 142
1.3. Trigonometric Polynomials and Some Polynomial Operators... 143
1.4. Splines........................ 146
1.5. Interpolation 148
§2. Spaces of Smooth Functions 149
2.1. Smoothness and Classes of Smooth Functions........................ 149
2.2. Other Approaches to Smoothness. Statement of the Approximation Problem for Smooth Functions 151
2.3. Basic Theorems of Harmonic Analysis in T^n and R^n and Representation Theorems........................ 153
2.4. Embedding Theorems for the Intersection of Spaces........................ 154
§3. Harmonic Analysis and the Approximation of Classes of Smooth Functions........................ 156
3.1. Introduction........................ 156
3.2. Best Approximation of Periodic Functions........................ 156
3.3. Best Approximation of Functions on R^n........................ 159
3.4. The Bernstein-Nikol'skij Inequality for Sobolev Classes........................ 160
3.5. Appendix........................ 161
§4. Linear Methods of Summation of Fourier Series........................ 161
4.1. Introduction........................ 161
4.2. Conditions for Convergence........................ 162
4.3. Asymptotic Results 163
4.4. The Precise Asymptotic Form of the Kolmogorov-Nikol'skij Constant for Fourier Sums........................ 164
4.5. Appendix........................ 166
§5. Approximation by Rational Functions........................ 167
5.1. The Speed of Rational Approximation and Structural Properties of Functions........................ 168
5.2. Comparison of Rational and Polynomial Approximations........................ 169
5.3. Rational Approximations and Singularities of Functions........................ 170

5.4. Rational Approximation of Analytic Functions ... 171
5.5. Appendix ... 171
§6. Splines in Approximation Theory ... 172
6.1. Some General Properties of Splines ... 173
6.2. Extremal Properties of Splines ... 175
6.3. Exact Solutions in Problems of Approximation of Classes of Smooth Functions by Splines ... 177
§7. Appendix ... 179
7.1. Approximation by Positive Linear Operators ... 179
7.2. Approximation of Functions by Polynomials in the Complex Domain ... 180
7.3. Approximation of Functions of Many Variables ... 182
7.4. Polynomial Interpolation and Quadratures ... 183
7.5. Appendix ... 185
Chapter 3. Best Methods of Approximation and Recovery of Functions . 187
§1. Preliminary Information ... 187
1.1. Definitions ... 187
1.2. Historical Information and Commentary ... 188
1.3. Relations Between Widths ... 189
1.4. Some Calculations and Estimates of Widths of Finite-Dimensional Sets ... 190
§2. Widths and Entropies of Classes of Smooth Functions ... 193
2.1. Widths of Sobolev Classes of Functions of One Variable ... 193
2.2. Trigonometric Widths ... 195
2.3. Widths of Intersections of Sobolev and Nikol'skij Classes in the Many-Dimensional Case ... 196
2.4. Some Discussion ... 198
§3. Widths of Classes of Analytic Functions ... 198
3.1. ε-Entropy of the Pair $(BA^G, C(T))$ ... 198
3.2. Rational Approximations of the Classes BA^G in the Space $C(T)$ ... 200
3.3. The Construction of Special Bases for the Pair $(BA^G, C(T))$ ... 201
3.4. Other Classes of Functions ... 202
§4. Exact Solutions of Approximation Problems ... 203
4.1. Best Bernstein Constants for Trigonometric Polynomials and Splines and Bernstein Widths ... 204
4.2. Best Linear Methods of Summation of Fourier Series ad Linear Widths ... 205
4.3. Best Favard Constants for Trigonometric Polynomials and Splines and Gel'fand Widths ... 206
4.4. Widths of Classes of Smooth Functions ... 206
4.5. Appendix ... 208
§5. Exact Solutions of the Problem of Recovery ... 209
5.1. Recovery of Functionals. Best Quadratures ... 209

5.2. Best Recovery of Operators ... 212
5.3. Optimal Recovery of Smooth and Analytic Functions ... 214
Chapter 4. Approximation Theory and its Connections with Neighbouring Domains of Mathematics ... 216
§ 1. Approximation Theory and the Theory of Extremal Problems ... 216
§ 2. Approximation Theory and Harmonic Analysis ... 218
2.1. Harmonic Analysis on Homogeneous Spaces and Approximation Theory ... 218
2.2. Approximation of Functions on the Whole Line ... 220
§ 3. Approximation Theory and Functional Analysis ... 222
§ 4. Approximation Theory and Geometry ... 223
§ 5. Some Results and Thoughts About the Future ... 226
Summary of the Literature ... 228
References ... 231

Introduction

1. What is Approximation Theory? What is the subject of its research? How do we sketch the shape of a mathematical discipline? To respond to these questions is not easy since very different opinions on this score are possible.

Let us take a very broad view of the aims of approximation theory. According to this, the theory was created to quantitatively and qualitatively evaluate the methods of "finitization of the infinite", in other words, methods of transforming finite information into infinite information, founded on a model of the mathematical world given by our representation of number, set, function and mapping. From this point of view even the problem of quadrature of the circle must belong to approximation theory.

A narrower view is possible, in which the aim of approximation theory is to provide theoretical support for computational mathematics. Then works on the calculation of the number of operations for multiplication, division, inversion of matrices, evaluation of polynomials, solution of equations etc. (see for example Knuth [1969]) should belong to approximation theory. But, nevertheless, this is not usually done.

In reality, when the words "approximation theory" are used, in asking the question of whether or not a given mathematician or a given paper should be in this domain, then we have in mind that division of mathematical analysis, which was formed under the influence of the work of P.L. Chebyshev, Weierstrass, Vallée-Poussin, Lebesgue, S.N. Bernstein, Jackson, A.N. Kolmogorov, S.M. Nikol'skij and many of their followers.

Included in this direction are works on the approximation of individual functions by algebraic and trigonometric polynomomials, by rational functions and by splines, estimates of approximation by these methods of various classes of smooth and analytic functions, the solutions of the problems of recovery of the functons, widths and similar problems.

A sketch of the division embraces the material of such monographs and foundational articles as: P.L. Chebyshev [1947], Weierstrass [1885], Borel [1905], Jackson [1911, 1930], S.N. Bernstein [1952] No 3, [1937], Vallée Poussin [1911, 1919], Walsh [1926], A.N. Kolmogorov [1985 No 27, No 28], S.M. Nikol'skij [1945, 1969, 1974], I.P. Natanson [1949], V.L. Goncharov [1954], P.P. Korovkin [1959], A.N. Kolmogorov and V.M. Tikhomirov [1959], A.G. Vitushkin [1959], A.F. Timan [1960], V.I. Smirnov, N.A. Lebedev [1964], Rice [1964], N.I. Akhiezer [1965], Cheney [1966], Lorentz [1966], S.B. Stechkin [1967], Singer [1970], Butzer, Nessel [1971], Collatz, Krabbs [1973], V.M. Tikhomirov [1976], N.P. Kornejchuk [1976, 1984], R.S. Ismagilov [1968, 1974, 1977], Rivlin [1974], B.S. Kashin [1977, 1981], S.B. Stechkin, Yu.N. Subbotin [1976], Karlin, Studden [1972], de Boor [1978], Sendov [1979], A.I. Stepanets [1981], V.K. Dzyadyk [1953], Traub, Wozniacowscki [1980], Pinkus [1985], etc.

Approximation theory was the subject of the all union conferences in Leningrad (1959), Baku (1962), Dnepropetrovsk (1985), a series of international conferences in Budapest (1969, 1978, 1980), Varna (1970, 1981, 1984), Poznan (1972), Cluj (1973), Kaluga (1975), Blagoevgrad (1977), Gdansk (1979), Kiev (1983), Texas (1980, 1983), Oberwolfach, (1983, 1989) and many international conferences in the USA, the Federal Republic of Germany, Canada and other countries. There is a special journal devoted to approximation theory, Journal of Approximation Theory[1], and also many survey and reference publications, whose titles contain the words "approximation theory", "approximation" and such like.

Certainly our definition is purely descriptive, it may be assumed that the field of activity of approximation theory will eventually be broadened, but doubtless the collective of mathematicians within approximation theory, will occupy a special place in the mathematical world and its members will quickly and naturally find a common language. The size of this collective is very significant, and the number of results, worthy of a mention in survey articles can be counted in thousands. To mention them all is impossible, and therefore we should, in a very general way, say how the author came to his choice of material.

It was the author's wish to show the evolution of the ideas, directions and results from the beginning up to the present day. Naturally one of the author's wishes was to attempt to look into the future and consider the present from this vantage point. Prognosis is dangerous, but we cannot manage without it, and the author has devoted a part of this work to the discussion of those problems which the believes must be discussed in the future. It must not be forgotten that

[1] Recently a journal on approximation theory has been started in China.

after many years it will not be concrete facts or theorems that are remembered, but more general directions of thought, particular approaches to problems, the ideological spirit, which has developed during the history of approximation theory. And to this also it was desired to give some attention.

In conclusion the author must express thanks to all those who helped him in the writing of this work and with whom he has descussed its content. To a great extent this refers to A.P. Buslaev, E.M. Galeev, V.A. Ivanov, A.I. Kamzolov, B.S. Kashin, V.N. Konovalov, S.V. Konyagin, N.P. Kornejchuk, G.G. Magaril-Il'yaev, V.E. Majorov, K.Yu. Osipenko, S.B. Stechkin, Yu.A. Farkov.

2. The Fundamental Stages in Approximation Theory. Of course, until the seventeenth century, which saw the origin of mathematical analysis, there was a unique object which could be "approximated"; this was the number. In particular, the number π. The era of the formation of mathematical analysis (the seventeenth and eighteenth centuries) was, as it were, the "prehistoric" period of approximation theory. Then, in the work of Kepler, Mercator, Gregory, Wallis, Newton, I. Bernoulli, J. Bernoulli, Euler, Lagrange and many other mathematicians, methods for the approximate calculation of numbers, functions and operators, solutions of equations, etc., were developed.

Among the prophets of approximation theory, one must first place the name of Gauss. He not only developed methods for the optimal calculation of integrals, solutions of equations etc. but essentially, by his work on the method of least squares, developed approximation theory in the quadratic metric.

The proper history of approximation theory is usually calculated from the moment when, in 1854. Chebyshev, in his memoir (Chebyshev [1947]) attempted to calculate how effective, from the applied viewpoint, was the computation of analytic functions using Taylor's formula. He came to the conclusion that the formula was insufficiently effective when the variable "was subject only to the condition to remain within certain broad limits,—then the finding of approximate expressions for functions requires essentially different methods".

Chebyshev gave methods for optimal approximations of many concrete functions. Now, in the age of electronic computers, many of his methods and arrangements have only historical interest, but one must not discount the possibility that in the near future, in a totally new historical stage, it may be suitable to return to the sources and again try to find optimal calculations by methods which take account of available techniques. But it is important to note that already the significance of Chebyshev's work has turned out to be much broader than narrow pragmatic aims. His ideas brought into being a new direction in analysis, where the subjects of the research were ideas such as analyticity, smoothness, approximability, etc.

In particular, even the very idea of the uniform or Chebyshev metric crystallized under the influence of Chebyshev's work.

P.L. Chebyshev, as it were, opened up a new world to research. Here there were gripping problems, frequently far removed from the original problems with which it began. The approximation of individual functions by means of polynomials and rational fractions; this is the core of the first, Chebyshev, stage of approximation theory. For more details on Chebyshev's work and the first stage of the theory see Akhiezer [1945], Goncharov [1945].

Gradually, the practical orientation slipped into the background, the attention of mathematicians became more and more drawn to the purely theoretical problem of describing the properties of functions by their approximative characteristics (the Taylor and Fourier coefficients, the Padé approximation, best approximations etc.). This led to the second stage in approximation theory.

Here is how this stage was described in a paper by S.M. Nikol'skij at the Amsterdam mathematical congress, Nikolskij [1954]. "Let there be given a class of functions (..., analytic, differentiable, satisfying a Lipscitz condition, etc.). Also let there be given a method of approximation of this class by polynomials (Fourier sums, interpolation, best polynomial, etc.). It is required to say everything possible about the size or character of the variation of the approximation of a function of this class."

In the fifties, under the influence of the works of Kolmogorov, a new viewpoint on approximation theory was promoted. Classical approximation theory at this time had accumulated a huge amount of material touching on the approximative possibilities of algebraic and trigonometric polynomials and rational fractions. But at the same time, "non-classical" tools for approximations, for example splines, had begun to permeate computational practice. It was natural to raise the question of comparison of different methods of approximation and to search for optimal methods. This led to the notion of widths, ε-entropy and so on. Then the best (in the sense of the accuracy of the approximation) methods of approximation of functions, and the best methods of recovery of functions given certain information were studied. These papers form the third stage in approximation theory.

We would like to believe that we stand on the threshold of the next, fourth stage, when the efforts of analysts, logicians and specialists in discrete mathematics are to be recombined. In this connection it is inevitable that the concepts of accuracy and complexity must be merged. Until now the first of these has dominated in approximation theory, the second was even devised outside it. Here it is natural that an approximation theory of random objects must be developed.

We have seen that approximation theory has its own circle of problems. But it is important that its speciality is a tendency to "expand". It has the power to stimulate interest in its problems among its "neighbours" and to produce "bordering" directions. This phenomena cannot be avoided in a survey article.

In accordance with what has been said we have formed the plan of this work. In the next subsection we introduce notations for the basic quantities, classes of functions and tools of approximation. (Approximation theory is in the class of "formal" mathematical quantities. The aim of the investigator usually consists of finding relations, such as equalities or inequalities, between the properties of a function, or class of functions, and its approximations. Formal exposition allows us to include in this survey a very large number of results (somewhat more than two hundred). But even this number represents only a small portion of those that, in principle, are worthy of mention.)

Next comes Chap. 1, devoted to the sources of classical approximation theory. Here there is a reflection of the Chebyshev period and the first phase of the second stage. Chapter 2 is devoted to the development of the second stage, Chap. 3 to the third stage. In the last chapter we briefly mention the interaction of approximation theory with neighbouring areas of mathematics and in conclusion try to look into the future.

3. Notations. In the majority of cases the standard notations are used for the logical connectives and set theory ($\forall$, $\exists$, $\Rightarrow$, $\Leftrightarrow$, $\in$, $\notin$, $\cup$, $\cap$, $\setminus$, $\varnothing$, $\subset$, etc.), numbers and vector spaces: $\mathbf{N}$, $\mathbf{R}$, $\mathbf{C}$, $\mathbf{Z}$, $\mathbf{N}^n$, $\mathbf{R}^n$, $\mathbf{C}^n$, $\mathbf{Z}^n$, etc. $\mathbf{Z}_+ = \mathbf{N} \cup \{0\} = \{0, 1, 2, \ldots\}$, $\mathbf{Z}_+^n = (\mathbf{Z}_+)^n$, $\mathbf{R}_+ = \{x \in \mathbf{R}: x \geqslant 0\}$, $\mathbf{R}_+^n = (\mathbf{R}_+)^n$ (see the list of notations in the first part), $\mathbf{T}^1$ is the circle or one-dimensional torus, realised as the interval $[-\pi, \pi]$ with $-\pi$ and π identified, $\mathbf{T}^n = (\mathbf{T}^1)^n$ is the n-dimensional torus. $I = [0, 1]$, $\hat{I} = [-1, 1]$. The conditions of a theorem (formulated without words) are contained in diagonal lines, followed by an implication sign ($\Rightarrow$) and then the conclusions of the theorem. $=:$ or $\stackrel{\text{def}}{=}$ means "equal by definition", ! means "unique", $\bigvee$ means "or". A function $t \to x(t)$, $t \in T$ is denoted $x(\cdot)$; $\langle x^*, x\rangle$ is the action of a linear functional x^* on an element x of a normed linear space; X^* is the space conjugate to X; $\mathscr{L}(X, Y)$ is the set of all continuous linear mappings of X to Y; Diam A is the *diameter of a set* A; $\omega_k(\delta, x(\cdot), X)$ is the k-th order *modulus of continuity* of a function $x(\cdot)$ in the space X (if $X = C(\Delta)$, then $\omega_k(\delta, x(\cdot), C(\Delta)) =: \omega_k(\delta, x(\cdot))$, if $k = 1$, then $\omega_1(\delta, x(\cdot)) =: \omega(\delta, x(\cdot)) = \sup\{|x(t+h) - x(t)|: |h| \leqslant \delta\}$; $BC(\mathbf{R})$ is the space of bounded continuous functions on $\mathbf{R}$, $D_\rho = \{z \in \mathbf{C}: |z| \leqslant \rho\}$; $K_\rho =: \partial D_\rho =: \{z \in \mathbf{C}: |z| = \rho\}$; $x_+ = \max(0, x)$; $\mathbf{1} = (1, 1, \ldots, 1)$, $\mathbf{0} = (0, 0, \ldots, 0)$, $\boldsymbol{\infty} = (\infty, \infty, \ldots, \infty)$. $\varphi_r(t) = \pi^{-1} \sum_{k \in N} \cos(kt - \pi r/2) k^{-r}$ is the *Bernoulli monospline*; $\tilde{x}_{nr}(t) = 4\pi^{-1} \sum_{k \in N} \sin((2k-1)nt - \pi r/2)(2k-1)^{-(r+1)}$ is the *Euler ideal spline* or *Favard function*; $K_r = 4\pi^{-1} \sum_{k \in N} (-1)^{(k+1)(r+1)} (2k-1)^{-(r+1)}$ is the *Favard constant*.

We give the basic notations for the objects of approximation theory.

a) *Methods of approximation* (see Chap. 2, § 1).

$\mathscr{P}_N$ is the set of algebraic polynomials of degree $\leqslant N$;

$\mathscr{T}_N$ is the set of trigonometric poynomials of degree $\leqslant N$;

$\mathscr{R}_{MN}$ is the set of rational fractions of order (M, N), that is,

$$\{r(\cdot) = p(\cdot)/q(\cdot) | p(\cdot) \in \mathscr{P}_M, q(\cdot) \in \mathscr{P}_N, q(\cdot) \not\equiv 0\};$$

$S_m^k(\Delta, R_n)$ is the set of splines of order m, defect k, with nodes $\{\tau_i\}$, which form a partition R_n. $\hat{S}_{2n,m}(\mathbf{T})$ is the set of splines of order m, defect 1 relative to the uniform partition of $\mathbf{T}^1$ into $2n$ parts.

b) *Characteristics of approximation.* Let X be a normed space with unit sphere B, A an (approximating) subset of X. In the first stage of approximation theory the quantity

$$d(x, A, X) =: \inf\{\|x - \xi\| \,|\, \xi \in A\},$$

none other than the *distance of x* from A, was investigated. Here we try to describe the *set of elements* ξ *of best approximation*, that is, elements such that

$$\|x - \hat{\xi}\| = d(x, A, X),$$

and the properties of the (*metric projection*) operator πA, associating with x all elements of best approximation πAx.

Next the approximation of classes of functions was investigated. Let C be a subset to be approximated. We denote the *deviation* of C from A by

$$d(C, A, X) =: \sup\{d(x, A, X) | x \in C\}$$

In connection with inequalities for polynomials (of Bernstein-Nikol'skij and Bohr-Favard type) it is natural to consider quantities of the following type (the *Bernstein* and *Favard constants*) (L is a subspace of X, ∂C is the boundary of C):

$$b(C, L, X) =: \sup\{\rho \geqslant 0 | \rho B \cap L \subset C\} = \inf\{\|\xi\| \,|\, \xi \in \partial(C \cap L)\},$$

$$f(C, L, X) =: \sup\{\|x\| \,|\, x \in C \cap L^{\perp}\},$$

where $L^{\perp}$ is the *anihilator* of L ($L^{\perp} =: \{x^* \in X^* \colon \langle x^*, x\rangle = 0, \forall x \in L\}$.

In the introduction of the quantities $b(C, L, X)$ and $f(C, L, X)$ we have concentrated attention on the geometric side of the question. In fact in approximation theory the classes of functions are usually the spheres of certain spaces of functions. Therefore an "embedding" approach to these characteristics is possible, G.G. Magaril-Il'yaev [1985].

Let X and Z be normed spaces, $Z \subset X$ and L a (closed) subspace in X. We will say that a *Bernstein-Nikol'skij inequality* is satisfied for the ordered triple $\theta = (Z, L, X)$, if

$$\|x\|_Z \leqslant B(\theta) \|x\|_X, \qquad x \in L.$$

The least constant in this inequality is denoted $b^{-1}(Z, L, X)$.

Associated with the triple θ is a conjugate triple $\theta = \{Z^*, L^{\perp}, X^*\}$, where Z^* and X^* are the dual spaces and $L^{\perp}$ is the annihilator of L. We say that a *Bohr-Favard inequality* is satisfied for the triple θ if θ^* satisfies a Bernstein-Nikol'skij inequality.

If some (in general, nonlinear) *apparatus of approximation* $A \colon C \to A$ is given, then there arise the quantities

$$\lambda(x, (A, A), X) =: \|x - Ax\|, \qquad x \in C,$$

$$\lambda(C, (A, A), X) =: \sup\{\lambda(x, (A, A), X) | x \in C\},$$

$$\lambda(C, L, X) =: \inf\{\lambda(C, (A, L), X) | A \in \mathscr{L}(X, L)\}.$$

Important examples of linear approximating apparati are

$(S_N x)(\cdot)$—*the Fourier operator* (*Fourier sum*);

$(\mathscr{F}_N x)(\cdot)$—*the Fejér operator* (*Fejér sum*);

$(\mathscr{J}_N x)(\cdot)$—*the Jackson operator* (*Jackson sum*)

and many others.

Let C be a class of periodic functions. The quantity $\lambda(C, (S_N, \mathscr{T}_N), X)$ is given the special notation

$$\varphi(C, \mathscr{T}_N, X) =: \sup\{ \|x(\cdot) - (S_N x)(\cdot)\| \,|\, x(\cdot) \in C\}$$

and is called the *Fourier-deviation* of the class C from $\mathscr{T}_N$.

Now we introduce the ideas of various widths. Concentrating our attention in what follows on the description of best Fourier methods, best linear methods, best from the point of view of approximation of subspaces, of best nonlinear methods of approximation and best methods of approximation by a finite number of points, we first give the definitions of Fourier widths, linear, Kolmogorov, Aleksandrov and entropy widths, and widths, connected with the Bohr-Favard and Bernstein-Nikol'skij inequalities.

The fundamental widths are defined by the following scheme.

Let X be a normed space, $\mathrm{Lin}_N(X)$ the collecion of subspaces of X of dimension $\leqslant N$, C a subset of X, $\mathscr{A} = \{A\}$ the collection of approximating subsets of X, $\mathscr{F} = \mathscr{F}(C, A) = \{F\}$ a family of mappings $F\colon C \to A$. The *width* generated by the family $\mathscr{F}$ is defined as follows:

$$p_{\mathscr{F}}(C, X) = \inf_{F \in \mathscr{F}} \sup_{x \in C} \|x - F(x)\| =: \inf_{F \in \mathscr{F}(C, A)} p(C, (F, A), X).$$

In this way we have defined:

the Fourier width

$$\varphi_N(C, X) = \inf\{\varphi(C, L_N, X) \,|\, L_N \in \mathrm{Lin}_N(H)\},$$

characterizing the best Fourier operator; here it is assumed that there is a preHilbert space H everywhere densely situated in the space X, $C \subset H$, and $\varphi(C, L_N, X)$ for $L_N \in \mathrm{Lin}_N(H)$ is $\sup\{ \|x - \mathrm{Pr}\, L_N x\| \,|\, x \in C\}$, where $\mathrm{Pr}\, L_N x$ is the orthogonal projection of x onto L_N;

the linear width

$$\lambda_N(C, X) = \inf\{\lambda(C, (\Lambda, L_N), X) \,|\, L_N \in \mathrm{Lin}_N(X), \Lambda \in \mathscr{L}(X, L_N)\},$$

characterizing the approximative possibilities of N-dimensional linear operators:

the Kolmogorov width

$$d_N(C, X) = \inf\{d(C, L_N, X) \,|\, L_N \in \mathrm{Lin}_N(X)\}$$

$$= \inf_{F \in \mathscr{F}(C, \mathrm{Lin}_N(X))} \sup_{x \in C} \|x - F(x)\|,$$

where $\mathscr{F}(C, \mathrm{Lin}_N(X))$ is the set of all mappings F from C to the set of linear subspaces $L_N \in \mathrm{Lin}_N(X)$; the width $d_N(C, X)$ characterises the approximative possibilities of N-dimensional subspaces,

the Aleksandrov width

$$a_N(C, X) = \inf_{\{K_N\}, F \in \mathscr{F}(C, \{K_N\})} \sup_{x \in C} \|x - F(x)\|,$$

where the infimum is taken over all possible compact sets K_N of dimension N from X and continuous mappings $F: C \to K_N$ ($\Leftrightarrow \mathscr{F}(C, \{K_N\})$), characterising the degree of "N-dimensionality" of the given set, that is, the possibility of approximating it using N parameters;

entropy width

$$\varepsilon_N(C, X) = \inf \lambda(C, (f, \Sigma_N), X)$$

where the infimum is taken over all mappings f to the set of N-point sets $\Sigma_N \subset X$, characterising the approximative possibilities of approximation by N-point sets, which is of interest in tabulation, information storage, etc.

In addition, we have met already the *Bernstein and Favard widths*

$$b_N(C, X) = \sup\{b(C, L_{N+1}, X) | L_{N+1} \in \mathrm{Lin}_{N+1}(X)\},$$

$$f_N(C, X) = \inf\{f(C, L_N, X) | L_N \in \mathrm{Lin}_N(X)\}$$

(they characterise the best subspaces from the point of view of the Bernstein-Nikol'skij and Bohr-Favard inequalities).

In problems of recovery quantities arise, which it is natural to call cowidths. They can be defined by a common scheme. Let C be a metric space, $\mathscr{S} = \{S\}$ a family of coding sets, $\Phi = \Phi(C, S)$ a set of mappings $\varphi: C \to S$. The *cowidth* generated by the family Φ, is defined by the formula

$$p^{\Phi}(C) = \inf_{\varphi \in \Phi} \sup_{x \in C} \mathrm{Diam}(\varphi^{-1}(\varphi(x))),$$

where $\varphi^{-1}(z)$ is the inverse image of z under φ.

By considering successively (for X normed, $C \subset X$) as S and Φ: $\mathbf{R}^N$ and $\mathscr{L}(X, \mathbf{R}^N)$, we get the *linear cowidth* $\lambda^N(C)$; the set of all N-dimensional compact sets and continuous mappings on them, we get the *Aleksandrov cowidth* $a^N(C)$.

We note one more important quantity

$$d^N(C) = \inf_{L_N} \sup_{x \in C \cap L_N} \|x\| \ (\textit{the Gel'fand width}).$$

Here the infimum is taken over all subspaces of codimension N. This width plays an important role in interpolation questions.

c) *Classes and spaces of functions.* $\mathscr{S}(T)$ denotes the spaces of fundamental functions on $T = \mathbf{R}^n \bigvee \mathbf{T}^n$, $\mathscr{S}'(T)$ denotes the space of generalized functions $\mathscr{S}(\mathbf{T}^n) = C^\infty(\mathbf{T}^n)$, $\mathscr{S}(\mathbf{R}^n)$ is the Schwartz space. We will, primarily, investigate the approximation of balls in Sobolev spaces and Nikol'skij classes. We give their definitions in the one-dimensional case for integer values of smoothness. $W_p^r(\mathbf{T}^1)$, $r \in \mathbf{N}$, denotes the (Sobolev) space of functions on $\mathbf{T}^1$ for which the $(r-1)$-st derivative is absolutely continuous and $\int_{T^1} |x^{(r)}(t)|^p \, dt < \infty$,

$$BW_p^r(\mathbf{T}^1) =: \left\{ x(\cdot) \in W_p^r(\mathbf{T}^1) \,\middle|\, \int_{\mathbf{T}^1} |x^{(r)}(t)|^p \, dt \leqslant 1 \right\}$$

is the *Sobolev class.*

The *Nikol'skij space* $H_p^{\alpha,k}(\mathbf{T}^1)$, $\alpha > 0$, $\alpha = r + \beta$, $0 < \beta \leqslant 1$ is the set of functions $x(\cdot)$ for which there exists an r-th derivative and $\sup\{\|(\Delta_h^k D^r x)(\cdot)\|_{L_p(T^1)}|h|^{-\beta} < \infty : h \in \mathbf{T}^1\}$ (where Δ_h^k is the k-th difference operator). The *Nikol'skij class* $BH_p^\alpha(\mathbf{T}^1)$ $(k = 1)$ is $\{x(\cdot) \in H_p^\alpha(\mathbf{T}^1)$: $\|\Delta_h x^{(r)}(\cdot)\|_{L_p(T^1)} \leqslant |h|^\beta\}$. The class $H_p^\omega(\mathbf{T}^1)$, given by functions of $\omega(\cdot)$ type of modulus of continuity, is defined similarly.

As normed spaces we will usually take the spaces $L_p(T)$, $T = \mathbf{R}^n \vee \mathbf{T}^n$, $1 \leqslant p \leqslant \infty$ or $C(\mathbf{T}^n)$, $BC(\mathbf{R}^n)$.

Let G be a domain in $\mathbf{C}$. A^G denotes the set of functions analytic in G, $BA^G = \{x(\cdot) \in A^G : |x(z)| \leqslant 1, z \in G\}$.

Chapter 1
Classical Approximation Theory

After a brief first section, where we recall the "prehistoric" period of approximation theory, in this chapter we consider the epoch of approximation theory, which began in the last century under the influence of the work and ideas of P.L. Chebyshev, Weierstrass, S.N. Bernstein, Jackson and A.N. Kolmogorov.

In §§ 2 and 3 we discuss the problems of approximation of individual elements (the investigation of the quantity $d(x, A, X)$, finding elements of best approximation, properties of the operator πA etc.). The topic of § 4 is Weierstrass' theorem and its generalizations (that is, the cases when $d(x(\cdot), \mathscr{P}_N, C(T)) \to 0$). In § 5 we discuss the constructive theory of functions, that is, the possibility of describing the properties of functions by the asymptotic behaviour of quantities like $d(x(\cdot), \mathscr{T}_N, C(\Delta))$, and quantities of the type $\varphi(C, \mathscr{T}_N, C(\mathbf{T}^1))$, $d(C, \mathscr{T}_N, C(\mathbf{T}^1)$, $b(C, \mathscr{T}_N, C(\mathbf{T}^1))$ etc.

§ 1. Prehistory

1.1. The Calculation of π. We begin with a small excursion into history (for more details see Archimedes, et al. [1911]). In the introduction it was said that methods of approximation of numbers, functions and mappings were introduced in the seventeenth and eighteenth centuries. A very ancient reference on our theme belongs to time immemorial. In a textbook on the mathematics of the ancient Egyptians—the Rhind papyrus (2000–1700 B.C.)—it is said that the area of a circle is equal to the area of a square with "sides equal to the diameter of the circle, lessened by 1/9 of its length". This gives an approximate value for π of $256/81 = 3.1604\ldots$. Here we immediately come across two problems—the approximation of a number (in this case π) and "quadratures", that is, the replacement of the area under a curve by the area of an equivalent square. First we comment on the calculation of π. Archimedes gave the first two-sided estimate for π: $223/71 = 3.14084\cdots < \pi = 3.14159\cdots < 3.14285\cdots = 22/7$. Vieta was the first to give an analytic formula for π. He calculated π to nine places. Then Huygens gave a series of approximate formulae and calculated π with still greater accuracy. Brouncker decomposed π into a continuous fraction, and Gregory (1670) and Leibniz (1673) expanded π in a series ($\pi = 4 - 4/3 + 4/5 - \cdots$). A set of expressions for π in the form of series were given by Euler in the eighteenth century. Here an attempt was made to calculate π with the greatest possible accuracy. With the help of the formula $\pi = 16 \arctan 1/5 - \arctan 1/239$ Machin calculated the number to roughly 100 places. By themselves such computations have no special value, but they are evidence of the potential of the computational mathematics of the period. "By hand" π was computed to roughly 700 places. In the forties the era of the electronic computer approached. The history of the calculation of π by computer began in 1949 when it was calculated to 2037 places. At the end of the fifties the IBM series machines began to operate. On the IBM 704 π was calculated to 10^4 places, and on the IBM 7090 to 10^5 places. In 1983 π had been calculated to 10,013,395 places.

The ancient mathematicians posed the problem of quadrature of the circle, that is, the construction by ruler and compasses of a square equivalent to a circle. This problem disturbed scholars for more than two millenia and was solved only in the nineteenth century.

The author decided to remind the reader of the history of π in order to show the oddity of the development of mathematics. Starting off, as a rule, from practical needs, mathematicians then often turn to the solution of problems far removed from the original aim. Their attention is drawn to accompanying questions, which lead off to obscure and tempting distant places (recall the quadrature of the circle). Sometimes a return to practicality results, sometimes not. This has happened (and will continue to happen) with many problems of approximation theory. It is not superfluous to remark, that many theoretical problems relating to the "complexity" of π remain open.

We now briefly recall the proper prehistory of approximation theory; the period when the foundations were laid for methods of calculation of numbers, functions and operators.

1.2. Approximation of Functions. At first functions were calculated by special methods. In his letter to Oldenburg (the second letter of 24.10.1676) Newton wrote in detail of how at time of the plague (1665–1666) he had invented "certain means of obtaining logarithms from the area of hyperboli". His enthusiasm for these occupations was very strong. "For me, frankly, it is a shame to say—it is said in the letter—I carry out these calculations to some number of places in my spare time". But shortly after the book by Mercator, "Logarithmotekhnica" (1668), appeared, where, apparently for the first time, series were applied, in particular, the series $\log(1 + x) = x - x^2/2 + x^3/3 - \cdots$. Series were also used in the calculation of numbers. For example, the series $\log 2 = (1/1 \cdot 2) + (1/3 \cdot 4) + \cdots$ (the Brouncker series for π, mentioned above, etc.). Then the series that we call Taylor series appeared in the works of Newton, Leibniz, Taylor, McClaurin, Bernoulli (Newton (1664–71), I. Bernoulli (1694), Taylor (1715)). Trigonometric series arose in the work of Euler (1744), I. Bernoulli and d'Alembert, and what we now call splines arose in the work of Leibniz (1697) and Euler.

1.3. Quadratures. It has been said that the term "quadrature" itself has two and a half thousand years of history and dates from the Ancient Greek civilisation. In modern times the simplest quadrature formulae were used by Kepler, Torricelli (1664) and Simpson (1743). The important Newton-Cotes formula was explained by Newton in a letter to Leibniz (1676), and published by Cotes in 1722. The basic method in all of these quadratures is to change the function into an approximation either by polynomials or by splines.

1.4. Interpolation. The best known interpolation formula was discovered by Lagrange (1795), but already in "The method of differences" (published in 1736) Newton had given a description of two interpolation formula. Interpolation formulae of general form were found in the nineteenth century (Cauchy, Hermite, etc.).

1.5. Approximation of Functionals and Operators. We have already mentioned one of the simplest functionals for the calculation of integrals, that is, quadratures. And already described a basic method of approximation; the replacement of a function by an approximation of it and the approximate representation of a functional or operator by taking the functional or operator on approximate values. This method has been applied for the numerical differentiation and solution of differential equations, beginning in fact with Kepler (1615), and then by Newton, Leibniz, Euler and others.

In all this natural questions arise: how after all can we calculate (anything) by the best possible method? What is the best way to integrate, to approximate a function or functional, to interpolate, to numerically differentiate? And here we must ask: what do we mean by best way?

Such questions, characteristic of approximation theory, will be considered later.

§2. Orthogonal Polynomials and Polynomials Deviating Least from Zero

2.1. Introduction. Orthogonal polynomials and expansions with respect to them (as a means of approximation of functions) arose in the eighteenth century. The earliest were the Legendre polynomials—in the works of Legendre (1785) and Laplace. Subsequently sets of orthogonal polynomials of other types arose.

The role of orthogonal systems of functions in mathematics and its applications (in particular, mathematical physics) turned out to be very significant.

Orthogonal systems of functions can usually be linked to some equation of Sturm-Liouville type. A large number of systems of orthogonal functions are connected with harmonic analysis (see § 2 of Chap. 4), that is, the presence of an invariant structure on a manifold. Hence there extends a thread from the eighteenth century up to the present day: Legendre polynomials appeared, in particular, in the description of spherical functions, and their "relatives" the polynomials of Gegenbauer, Jacobi, Laguerre, Hermite and others, appeared in the construction of harmonic analysis, arising in the representation of groups of rotations of n-dimensional space, groups of unitary matrices, groups of complex triangular matrices of third order and groups of motions of n-dimensional Euclidean space. And, perhaps, the most important orthogonal system, of sines and cosines, emerged in the harmonic analysis of functions on the circle $\mathbf{T}^1$.

In addition, orthogonal polynomials arose as solutions in problems of polynomials least deviating from zero in the L_2 metric. This closely relates their theory to the circle of ideas, which were to take shape after the work of P.L. Chebyshev. Here we consider problems connected with orthogonal polynomials from just this point of view.

The theme "Polynomials least deviating from zero" in fact began with the memoir of P.L. Chebyshev [1947, p 109–143], presented to the Academy of Science in 1853. The mathematical content of the memoir begins with the words: "Soit fx une fonction donnée, U un polynome du degré n avec des coefficients de manière à ce que la difference $fx - U$, depuis $x = a - h$, jusqu'a $x = a + h$, reste dans les limites les plus raprochées de 0, la différence $fx - U$ jouira, comme on le sait, de cette propriété: « Parmi les valeurs les plus grandes et les plus petites de la différence $fx - U$ entre les limites $x = a - h, x = a + h$, on trouve au moins $n + 2$ fois la même valeur numerique »".

(The historians of science remain unclear as to whom, except the author, this result was "known", it being the forerunner of a remarkable result in approximation theory, the theorem on alternation.) In the words of P.L. Chebyshev the problem of the polynomial, deviating least from a given function in the metric on C, has been explicitly posed.

The above result of P.L. Chebyshev applies immediately to the function x^{n+1}, and leads to the polynomial, deviating least from zero in the metric of C. But the possibility of a geometric approach allows us to pose a more general problem.

Let $\hat{I} = [-1, 1]$, $\mu(\cdot)$ a positive measure on $\hat{I}$, $\rho(\cdot)$ a positive continuous function on $\hat{I}$.

Consider the extremal problems

$$\int_{\hat{I}} \left| t^N + \sum_{k=1}^{N} x_k t^{k-1} \right|^p d\mu \to \inf, \qquad 1 \leq p < \infty, \tag{$z_{Np\mu}$}$$

$$\max_{t \in \hat{I}} \left| t^N + \sum_{k=1}^{N} x_k t^{k-1} \right| \rho(t) \to \inf. \tag{$z_{N\infty\rho}$}$$

(If X is a set, C a subset of it, and $f: X \to \bar{\mathbf{R}}$ a functional, where $\bar{\mathbf{R}} = \mathbf{R} \cup \{+\infty\} \cup \{-\infty\}$, then we will write the problem: "find the minimum (maximum) of f on C" as: $f(x) \to \inf(\sup)$, $x \in C$, and attach to this problem the index z.) In other words, in $(z_{Np\mu})$ and $(z_{N\infty\rho})$ the question is of finding the distance of the function $x_N(t) = t^N$ from the space $\mathscr{P}_{N-1}$ of polynomials of degree $N-1$ in the metrics of $L_p(\hat{I}, \mu)$ and $C(\hat{I}, \rho)$.

The solutions of these problems for $\hat{\mu}(t) = t$ and $\hat{\rho}(t) \equiv 1$ are denoted by $T_{Np}(\cdot)$, $1 \leqslant p \leqslant \infty$, and we omit the indices $\hat{\mu}$ and $\hat{\rho}$. P.L. Chebyshev (in the work cited) investigated $(z_{N\infty})$, he and also A.N. Korkin and E.I. Zolotarev investigated (z_{N1}). The solution of (z_{N2}) had essentially been studied in the eighteenth century by Legendre and Laplace. The solutions of these problems are the classical *Chebyshev polynomials of the first and second kind* and the *Legendre polynomials*. These can be briefly expressed by the formulae (where $x_N(t) = t^N$):

$$T_{N\infty}(t) = 2^{-(N-1)}\cos(N\arccos t), \qquad d(x_N(\cdot), \mathscr{P}_{N-1}, C(\hat{I})) = 2^{-(N-1)},$$

$$T_{N1}(t) = 2^{-N}\sin((N+1)\arccos t)/(1-t^2)^{1/2},$$

$$d(x_N(\cdot), \mathscr{P}_{N-1}, L_1(\hat{I})) = 2^{-(N-1)},$$

$$T_{N2}(t) = (N!)((2N)!)^{-1}\left(\frac{d}{dt}\right)^N (t^2-1)^N,$$

$$d(x_N(\cdot), \mathscr{P}_{N-1}, L_2(\hat{I})) = \frac{2^N(N!)^2}{(2N)!}\left(\frac{2}{2N+1}\right)^{1/2}.$$

In the above formulae we have the exact solutions of three extremal problems. The theme of "Approximation theory and the theory of extremal problems" is mentioned in the last chapter. There we endeavour to delineate reasons, why some problem or other has obtained a solution (as a rule, this is connected with some specificity and uniqueness; there are very few exact solutions). All three problems $(z_{N\infty})$, (z_{N1}) and (z_{N2}) have a uniform standard solution. Here we quote the second and third.

2.2. Legendre Polynomials and Orthogonal Polynomials

Theorem 1 (Rodrigues formula (1814)). *The following formula holds*:

$$T_{N2}(t) = (N!)((2N)!)^{-1}\left(\frac{d}{dt}\right)^N (t^2-1)^N.$$

Theorem 1 gives a description of the polynomials, deviating least from zero in the metric of $L_2(\hat{I})$. The *Legendre polynomials* are defined as follows:

$$P_N(t) = (N!)^{-1}2^{-N}\left(\frac{d}{dt}\right)^N (t^2-1)^N.$$

From comparison of these formulae it is clear that $P_N(t) = ((2N)!)2^{-N}(N!)^{-2}T_{N2}(t)$. We quote the form of the Legendre polynomials for small degrees: $P_0(t) \equiv 1$,

$P_1(t) = t$, $P_2(t) = (3t^2 - 1)/2$, $P_3(t) = (5t^3 - 3t)/2, \ldots$. In theorem 1 an exact solution was found for the extremal problem (z_{N2}). We treat this problem according to the standard plan (see the problem book: Alekseev, Galeev & Tikhomirov [1984]), by dividing the solution into four stages: 1. formalization, 2. writing down necessary conditions, 3. solution of the equations of stage 2, and 4. investigation of the solutions.[1]

1. The problem is formalized as follows:

$$f(x) = \int_{\hat{I}} \left(t^N + \sum_{k=1}^{N} x_k t^{k-1} \right)^2 dt \to \inf \qquad (\text{for } x \in \mathbf{R}^N)$$

This is the problem without constraints where f is continuous and convex and, moreover, quadratic (this is the "specificity", which ultimately allows us to solve the problem). From the fact that $f(x) \to \infty$ as $|x| \to \infty$, and the Weierstrass theorem it follows that a solution $\hat{x} = (\hat{x}_1, \ldots, \hat{x}_N)$ exists, that is

$$T_{N2}(t) = t^N + \sum_{k=1}^{N} \hat{x}_k t^{k-1}.$$

2. A necessary (and in this case, from the convexity of f, a sufficient) condition for a minimum is that $f'(\hat{x}) = 0$ (Fermat's theorem).

3. We will solve the equation $f'(\hat{x}) = 0$ by differentiating under the integral sign:

$$\int_{\hat{I}} T_{N2}(t) t^{k-1}\, dt = 0, \qquad k = 1, \ldots, N. \tag{i}$$

4. Put

$$g(t) = ((N-1)!)^{-1} \int_{\hat{I}} (t - \tau)_+^{N-1} T_{N2}(\tau)\, d\tau. \tag{ii}$$

This is the well known *Lagrange formula for the inversion of the operator* of n-times *differentiation* (see later, in Chap. 2, § 1). From (ii) it follows immediately that

a) $g^{(N)}(t) = T_{N2}(t)$,
b) $g^{(k)}(-1) = 0$, $k = 0, 1, \ldots, N-1$.

And from (i) it follows immediately that

c) $g^{(k)}(1) = 0$, $k = 0, 1, \ldots, N-1$.

From a) it follows that $g(\cdot) \in \mathscr{P}_{2N}$, and from b) and c) b) and c) that $g^{(k)}(\pm 1) = 0$, $k = 0, 1, \ldots, N-1$. This means that

$$g(t) = C(t^2 - 1)^N$$

---

[1] One of the aims of the theory of extremal problems is to give uniform methods for the investigation of problems on maxima and minima of any kind. Approximation theory gives very meaningful material for the verification of the effectiveness of general methods. If we wish to pose this problem, we can also give some solutions.

with a coefficient C defined from a)

$$C\left(\frac{d}{dt}\right)^N (t^2-1)^N = t^N + \cdots \Rightarrow C = (N!)((2N)!)^{-1}.$$

The theorem is proved.

We mention several properties of the Legendre polynomials, which follow at once from the definition and Theorem 1.

α) All zeroes of Legendre polynomials are real and lie in the interval $(-1, 1)$.

β) The Legendre polynomials are orthogonal on $[-1, 1]$ with unit weight

$$\int_{\hat{I}} P_n(t)P_m(t)\,dt = 0, \qquad n \neq m.$$

This fact, which follows immediately from Rodrigues formula, can be used to prove that $T_{N2}(\cdot)$ is the polynomial deviating least from zero in $L_2(\hat{I})$ (we recall that the condition $f'(\hat{x}) = 0$ is a sufficient condition for minimality). This is how it is usually done.

γ) The Legendre polynomials satisfy a differential equation (the *Legendre equation*):

$$(1-t^2)\ddot{x} - 2t\dot{x} + N(N+1)x = 0.$$

This equation is an equation of *Sturm-Liouville type*.

δ) For Legendre polynomials there are uniform recurrence formulae, for example:

$$(N+1)P_{N+1}(t) - (2N+1)tP_N(t) + NP_{N-1}(t) = 0.$$

ε) Legendre polynomials can be defined using *generating functions*

$$F(r, t) = \frac{1}{(1-2rt+r^2)^{1/2}} = \sum_{k \geqslant 0} r^k P_k(t).$$

These formulae link the theory of Legendre polynomials with harmonic analysis on the sphere $\mathbf{S}^2$, that is, with the theory of Laplace spherical functions; here we have written the decomposition of the Green's function for the Laplace operator relative to the zonal spherical functions on $\mathbf{S}^2$. We will show, how to deduce an explicit form for the generating functions (by the *Darboux method*). Let Γ be a contour surrounding $\hat{I}$.

The Rodrigues formula can be given in the form:

$$\begin{aligned} P_n(z) &= 2^{-n}(n!)^{-1}\left(\frac{d}{dz}\right)^n (z^2-1)^n \\ &= 2^{-n}(n!)^{-1}(2\pi i)^{-1}\int_\Gamma (\zeta^2-1)^n \left(\frac{d}{dz}\right)^n \left(\frac{1}{\zeta - z}\right) d\zeta \\ &= 2^{-n}(2\pi i)^{-1}\int_\Gamma (\zeta^2-1)\frac{d\zeta}{(\zeta - z)^{n+1}}. \end{aligned}$$

Using Taylor's formula and the residue theorem, we obtain:

$$F(w, z) =: \sum_{n\geqslant 0} w^n P_n(z) = (2\pi i)^{-1} \int_\Gamma \sum_{n\geqslant 0} \frac{(\zeta^2-1)^n w^n \, d\zeta}{2^n(\zeta - z)^{n+1}}$$

$$= (2\pi i)^{-1} \int_\Gamma \frac{d\zeta}{-\dfrac{w}{2}\left(\zeta^2 - \dfrac{2\zeta}{w} + \left(\dfrac{2z}{w} - 1\right)\right)} = -\frac{2}{w} \frac{w}{-2\sqrt{1-2wz+w^2}}$$

$$= \frac{1}{\sqrt{1-2wz+w^2}}.$$

A similar approach plays a major role in Padé approximation theory.

Legendre polynomials were the first example of a *system of orthogonal polynomials*. There then followed a *series of classical polynomials*: Chebyshev, Jacobi, Hermite and Laguerre. All of them *orthogonal* systems of *polynomials* on an interval $\hat{I}$, the line $\mathbf{R}$ or the half-line $\mathbf{R}_+$ with a weight $h(\cdot)$, that is, they satisfy the orthogonality relation

$$\int_\Delta p_n(t) p_m(t) h(t) \, dt = 0, \qquad n \neq m,$$

where $\deg p_n(\cdot) = n$, and $h(\cdot)$ is non-negative on Δ. In the case of the Chebyshev polynomials (the topic of subsequent subsections) $\Delta = \hat{I}$, $h(t) = (1-t^2)^{1/2}$ (T_n the *polynomial of the first kind*); $\Delta = \hat{I}$, $h(t) = (1-t^2)^{1/2}$ (U_n the *polynomial of the second kind*), for the *Jacobi polynomials* $P_n(\cdot, \alpha, \beta)$, $\Delta = \hat{I}$, $h(t) = (1-t)^\alpha(1+t)^\beta$, $\alpha > -1$, $\beta > -1$; for the *Hermite polynomials* $H_n(\cdot)$, $\Delta = \mathbf{R}$, $h(t) = \exp(-t^2)$; for the *Laguerre polynomials* $L_v(\cdot, \alpha)$, $\Delta = \mathbf{R}_+$, $h(t) = t^\alpha \exp(-t)$, $\alpha > -1$. The weights h for the classical polynomials satisfy a special differential equation of the first order (the *Pearson equation*):

$$\frac{h'(t)}{h(t)} = \frac{a_0 + a_1 t}{b_0 + b_1 t + b_2 t^2} = \frac{A(t)}{B(t)}$$

and the boundary conditions

$$\lim_{t\to t_0+0} h(t)B(t) = \lim_{t\to t_1-0} h(t)B(t) = 0.$$

The classical polynomials $K_N(\cdot)$ satisfy the *generalised Rodrigues formula* (c_N are normalisation coefficients):

$$K_N(t) = \frac{c_N}{h(t)}\left(\frac{d}{dt}\right)^N (h(t)B^N(t))$$

and also satisfy a differential equation of Sturm-Liouville type

$$B(t)\ddot{x} + (A(t) + \dot{B}(t))\dot{x} - N(a_1 + (N+1)b_2)x = 0.$$

The theory of orthogonal polynomials is a special division of mathematical analysis, closely bordering approximation theory and we have barely touched on its problems. It is widely represented in the monographic literature. We mention the monographs of Jackson [1941], Szegö [1921] ad P.K. Suetin [1979].

2.3. Chebyshev Polynomials of the First Kind

Theorem 2 (P.L. Chebyshev (1947)).

$$T_{N\infty}(t) = 2^{-(N-1)}\cos(N\arccos t).$$

In this theorem there is a description of the polynomials deviating least from zero in the metric of $C(\hat{I})$. The *Chebyshev polynomials* are defined in the following way:

$$T_N(t) =: \cos(N\arccos t) = 2^{(N-1)}T_{N\infty}(t)$$
$$= 2^{-1}((t+\sqrt{t^2-1})^N + (t-\sqrt{t^2-1})^N).$$

We give several Chebyshev polynomials of low degree: $T_0(t) \equiv 1$, $T_1(t) = t$, $T_2(t) = 2t^2 - 1$, $T_3(t) = 4t^3 - 3t, \ldots$.

P.L. Chebyshev deduced Theorem 2, starting from an extremal property of the polynomial least deviating from the function $x_N(t) = t^N$ (which we quoted above). We give P.L. Chebyshev's argument; subsequently, using various modifications of this argument, a number of other similar problems were solved (by P.L. Chebyshev himself and by E.I. Zolotarev). From the properties of the polynomial least deviating from zero, noted by P.L. Chebyshev, it follows at once, that $\dot{T}_{N\infty}(\cdot)$ has $n-1$ zeroes in the interval $(-1, 1)$ and moreover, that $T_{N\infty}(\cdot)$ itself must attain its maximum and minimum value (which we denote by L) on the boundaries, that is, at ± 1. From this it follows immediately that the polynomials of degree $2N$, $(1-t^2)\dot{T}_{N\infty}(t)$ and $L^2 - T^2_{N\infty}(t)$ have simultaneous zeroes, that is, are proportional. Equating leading coefficients gives us the differential equation

$$\frac{dx}{\sqrt{L^2-x^2}} = N\frac{dt}{\sqrt{1-t^2}},$$

integration of which leads to the answer: $x(t) = T_{N\infty}(t) = 2^{-(N-1)}\cos(N\arccos t)$.

We mention several properties of Chebyshev polynomials which follow from their definition.

α) All zeroes of Chebyshev polynomials lie in the interval $(-1, 1)$ and have the form

$$\tau_{kN} = \cos(2k-1)\pi/2N, \qquad k = 1, \ldots, N.$$

β) The Chebyshev polynomials $T_N(\cdot)$ are orthogonal on $\hat{I}$ with the weight $(1-t^2)^{-1/2}$.

This situation has already been mentioned in the previous subsection: the system $\{T_N(\cdot)\}_{N\in\mathbf{N}}$ is related to a number of classical orthogonal polynomials.

γ) The Chebyshev polynomials $T_N(\cdot)$ satisfy the differential equation

$$(1-t^2)\ddot{x} - t\dot{x} + N^2x = 0.$$

δ) The Chebyshev polynomials can be defined using a generating function

$$\frac{(1-r^2)}{1-2rt+r^2} = 1 + 2\sum_{k\geqslant 1} r^k T_k(t).$$

Replacing t by $\cos\varphi$ ($\Leftrightarrow \varphi = \arccos t$), we get the well-known Fourier series decomposition.

$$\frac{(1-r^2)}{1-2r\cos\varphi+r^2} = 1 + 2\sum_{k\geqslant 1} r^k \cos k\varphi.$$

This formula links the theory of Chebyshev polynomials with harmonic analysis on the circle: here we have the decomposition of the *Poisson kernel* (that is, the Green's function for the Laplace operator on the disc) as a Fourier cosine series.

Chebyshev polynomials play an enormous role in approximation theory. They find constant application in computational mathematics—in interpolation problems and in questions of approximation of functions. Decompositions into series with respect to Chebyshev polynomials are very popular. Apart from anything else, Chebyshev polynomials are optimally adapted for the approximation of functions on $[-1, 1]$ which admit an analytic extension to ellipses with foci at the points ± 1. This theme will be discussed later.

2.4. Chebyshev Polynomials of the Second Kind

Theorem 3 (P.L. Chebyshev (1859), A.N. Korkin-E.I. Zolotarev (1873)).

$$T_{N1}(t) = \dot{T}_{N+1}(t)/(N+1) = 2^{-N}\sin((N+1)\arccos t)/(1-t^2)^{1/2}.$$

Theorem 3 describes polynomials deviating least from zero in the $L_1(\hat{I})$ metric. *Chebyshev polynomials of the second kind* are defined by the equalities:

$$U_N(t) =: \sin((N+1)\arccos t)/(1-t^2)^{1/2} = 2^N T_{N1}(t).$$

For low degrees we have: $U_0(t) \equiv 1$, $U_1(t) = t$, $U_2(t) = 4t^2 - 1$, $U_3(t) = 8t^3 - 4t$.

Theorem 3 follows immediately from the easily verified orthogonality of the functions $t \to \operatorname{sign}(\dot{T}_{N+1}(t))$ from all the powers t^k, $0 \leqslant k \leqslant N-1$ on $\hat{I}$. It follows from this orthogonality that $0 \in \partial f(\hat{x})$, where $f(x)$ is the norm in $L_1(I)$ of the function $t \to t^N + \sum_{k=1}^N x_k t^{k-1}$, and $\hat{x}$ is the solution of the problem $f(x) \to \inf$.

The term "polynomial of the second kind" is used in the general theory of orthogonal polynomials which arise in the expansion of the functions $t \to \int_\Delta (h(\tau)/(t-\tau))\,d\tau$ in continued fractions (or, equivalently, Padé approximation of this function at infinity). The polynomials $T_N(\cdot)$ and $U_N(\cdot)$ appeared in these expansions for $\Delta = \hat{I}$ and $h(\tau) = (1-t^2)^{\pm 1/2}$.

We mention several properties of Chebyshev polynomials of the second kind, which follow at once from the definition and Theorem 3.

α) All zeroes of the polynomials $U_N(\cdot)$ are real and lie in the interval $(-1, 1)$.

β) The polynomials $U_N(\cdot)$ are orthogonal with the weight $(1-t^2)^{1/2}$.

(We have noted this earlier.) Thus, the system $\{U_N(\cdot)\}_{N\geq 0}$ is related to a number of classical orthogonal polynomials.

γ) The functions $(1-t^2)^{1/2}U_{N-1}(\cdot)$ are second linearly independent solutions of the equations which are satisfied by the Chebyshev polynomials $T_N(\cdot)$.

δ) The polynomials $U_N(\cdot)$ can be defined with the help of the generating function

$$\frac{1}{1-2rt+r^2} = \sum_{k\geq 0} r^k U_k(t).$$

This formula unites the theory of the polynomials U_N with harmonic analysis on the three-dimensional sphere.

2.5. Gaussian Quadratures. In § 1 we mentioned the Newton approach to the integration of functions.

Gauss proposed considering a *quadrature formula* of the form

$$\int_{\hat{I}} x(t)\,dt = \sum_{k=1}^{N} p_k x(\tau_k), \tag{i}$$

taking the weights p_k and the nodes τ_k so that this formula is exact on polynomials of maximum degree. Since there are $2N$ parameters in all here, one would expect that degree to be equal to $2N-1$ ($\dim P_{2N-1} = 2N$). This turns out to be the case.

Theorem 4 (Gauss, Werke [1986]). *Let $\{\hat{\tau}_1, \ldots, \hat{\tau}_N\}$ be the zeroes of the Legendre polynomial P_N. Then there are numbers $\hat{p}_1, \ldots, \hat{p}_N$, such that formula* (i) *is exact for polynomials of degree $2N-1$.*

In fact, consider the linear system

$$\sum_{j=1}^{N} p_j \hat{\tau}_j^s = \int_{\hat{I}} t^s\,dt, \qquad s = 0, 1, \ldots, N-1. \tag{ii}$$

Its determinant—the Vandermond determinant—is not zero (we recall, that all zeroes of P_N are distinct and lie in the interval $(-1, 1)$). This means that the system has a unique solution $\hat{p}_1, \ldots, \hat{p}_N$. Let (ii) be proved for $p_j = \hat{p}_j$ for $0 \leq j \leq N-1-l$, $0 \leq l \leq N-1$. We will prove it for $s = N+l$. We have, using the orthogonality relation (for the definition and this property see subsection 2.2):

$$\int_{\hat{I}} t^{N+l}\,dt = \int_{\hat{I}} \left(T_{N2}(t) - \sum_{k=1}^{N} \hat{x}_k t^{k-1}\right) t^l\,dt = -\int_{\hat{I}} \sum_{k=1}^{N} \hat{x}_k t^{k-1+l}\,dt$$

$$\overset{\text{(ii)}}{=} -\sum_{j=1}^{N} \hat{p}_j \sum_{k=1}^{N} \hat{x}_k \hat{\tau}_j^{k+l-1} = (\hat{\tau}_j \text{ zeroes of } T_{N2}(\cdot))$$

$$= \sum_{j=1}^{N} \hat{p}_j \left(T_{N2}(\hat{\tau}_j)\hat{\tau}_j^l - \sum_{k=1}^{N} \hat{x}_k \hat{\tau}_j^{k+l-1}\right) = \sum_{j=1}^{N} \hat{p}_j \hat{\tau}_j^{N+l}.$$

Theorem 4 is proved.

In Chap. 3 we will discuss other ways to approach optimality of quadratures.

§ 3. The Circle of Ideas Due to P.L. Chebyshev

3.1. Introduction. Earlier we quoted P.L. Chebyshev, where he gave the motivation for his research into approximation theory. He posed the question: what is the best way to approximately represent a given function?

When the function is analytic, then a desire to use Taylor's formula can arise. But, as P.L. Chebyshev established, there is an approximate representation of a function in the form of a polynomial, which gives far greater accuracy. And if the function is not analytic, say $\sqrt{t}$, how is it to be approximated? (What is the best way to extract a root?) At the time of P.L. Chebyshev, in calculations "by hand", it was only possible to use the four operations of arithmetic. Therefore P.L. Chebyshev limited himself to two tools of approximation: polynomials and rational functions. (At the present time there are new arguments in favour of rational approximations of elementary functions.)

P.L. Chebyshev posed and solved the problem of best approximation (by these methods of approximation) for certain concrete functions. He sought exact solutions; rather, he tried to express these solutions using explicit formulae. (As we shall see, the number of exact solutions is very small.)

In its general form P.L. Chebyshev's project admits a geometric treatment which we have already mentioned in the introduction to this part. In studying the question of best approximation of a function, P.L. Chebyshev, as a matter of fact, sought the distance of the function $x(\cdot)$ from some approximating set A (P_N or P_{MN}) in some metric (for P.L. Chebyshev usually in $C(\Delta)$). In general this kind of problem is posed as follows: in a normed space X it is required to find the value $d(x, A, X) = \inf\{\|x - y\| : y \in A\}$, in other words, it is required to solve the extremal problem

$$\|x - y\| \to \inf, \qquad y \in A. \tag{z}$$

Here an investigator encounters the whole complex of problems which accompany the solution of any extremal problem—problems of existence, uniqueness, criteria which the solution must satisfy, etc. Some of these problems P.L. Chebyshev himself did not touch on at all, some he solved in passing, but because of his research great value came to be attached to them. The whole associated entourage of questions forms the *circle of ideas due to P.L. Chebyshev.*

3.2. Existence, Uniqueness and Alternation in the Case of Polynomial Approximations. Let $X = C(\Delta)$, $\Delta = [t_0, t_1]$, $|t_1 - t_0| < \infty$, $A = \mathscr{P}_N$. Does there exist a polynomial of best approximation for a function $x(\cdot) \in C(\Delta)$, is it unique, and how can one characterise it? The answer to all these questions is given by the following theorem, which became the source for many scientific directions.

Theorem 5 (on alternation). *Let $x(\cdot)$ be a continuous function on a finite interval Δ. Then a polynomial $\hat{p}(\cdot) \in \mathscr{P}_N$ of degree N deviating least from $x(\cdot)$ (that is, such* *that* $d(x(\cdot), \mathscr{P}_N, C(\Delta)) = \|x(\cdot) - \hat{p}(\cdot)\|$) *exists, is unique and is characterized by the fact that there are* $N + 2$ *points* $\tau_1 < \tau_2 < \cdots < \tau_{N+2}$, $\tau_i \in \Delta$, *at which the difference* $x(\cdot) - \hat{p}(\cdot)$ *alternates and takes its maximum and minimum values, which are equal in modulus.*

This property of a function of alternately taking its maximum and minimum values, equal in modulus, at k points is called *k-alternation.* If $x(\cdot)$ has k-alternation then we write $\operatorname{Alt} x(\cdot) \geqslant k$.

In what follows we will not be able to use much space on the formulation of theorems. We will use abbreviations. The formulation of Theorem 5 in brief will be:

$$/x(\cdot) \in C(\Delta), |\Delta| < \infty/ \Rightarrow \exists! \hat{p}(\cdot) \in \mathscr{P}_N\colon d(x(\cdot), \mathscr{P}_N, C(\Delta)) = \|x(\cdot) - \hat{p}(\cdot)\|_{C(\Delta)},$$

$$\hat{p}(\cdot) \in \mathscr{P}_N, \qquad d(x(\cdot), \mathscr{P}_N, C(\Delta)) = \|x(\cdot) - \hat{p}(\cdot)\|_{C(\Delta)} \Leftrightarrow \operatorname{Alt}(x(\cdot) - \hat{p}(\cdot)) \geqslant N + 2.$$

Here and later the conditions of a theorem will be enclosed in diagonal lines.

The proof of Theorem 5 will be discussed in Chap. 4, § 1. It proceeds by the standard scheme, which we have already encountered in the proof of Theorem 1. The crux here is that in a given concrete case problem (z) reduces to convex programming, where it is possible to apply the main theorem of convex analysis—the refinement theorem (see part 1 of this volume).

It will not be possible for us to give exhaustive historical information on each result. But here, as we are discussing properly speaking, one of the first results of the entire theory, we will make an exception. In the form in which we have formulated the theorem, it was proved in the monograph of Borel [1905]. P.L. Chebyshev did not consider the existence problem at all, uniqueness he sometimes considered, but in this case failed to mention it. What about the necessity of k-alternation; at that time (we have quoted where) P.L. Chebyshev wrote that this property was "known", and made an attempt (see Chebyshev [1947]) at the proof, an attempt which contained the seed of the refinement theorem. The necessity of alternation in Chebyshev problems was proved by Kirchberger [1903].

To conclude this subsection, we give an analogue for rational functions (it is called *Chebyshev's theorem on existence and uniqueness for rational approximations*).

Theorem 6.

$$/x(\cdot) \in C(\Delta), \Delta = [t_0, t_1], |t_1 - t_0| < \infty, N \in \mathbf{Z}_+, M \in \mathbf{Z}_+/ \Rightarrow \exists! \hat{r}(t) = \hat{p}(t)/\hat{q}(t),$$

$$\hat{p}(\cdot) \in \mathscr{P}_{M-\mu}, \qquad \hat{q} \in \mathscr{P}_{N-\nu}\colon d(x(\cdot), \mathscr{R}_{MN}, C(\Delta)) = \|x(\cdot) - \hat{r}(\cdot)\|_{C(\Delta)},$$

$$\operatorname{Alt}(x(\cdot) - \hat{r}(\cdot)) \geqslant M + N - \min(\mu, \nu) + 2,$$

$$\hat{p}(\cdot) \not\equiv 0; \qquad \operatorname{Alt}(x(\cdot) - \hat{r}(\cdot)) \geqslant N + 2 \text{ if } \hat{p}(\cdot) \equiv 0.$$

For the proof see Akhiezer [1965].

Sometimes useful is

Theorem 7 (Vallée-Poussin). *If the polynomials* $\hat{p}(\cdot)$ *and* $\hat{q}(\cdot)$ *are as in theorem 6 and the difference* $x(\cdot) - (\hat{p}(\cdot)/\hat{q}(\cdot))$ $(\hat{p}(\cdot) \not\equiv 0)$ *takes at s successive points of the interval* Δ *non-zero values* $\lambda_1, -\lambda_2, \ldots, (-1)^s\lambda_s$, $\lambda_i > 0$, *where* $s \geqslant M + N - \min\{\mu, \nu\} + 2$, *then*

$$d(x(\cdot), \mathscr{R}_{MN}, C(\Delta)) \geqslant \min_i \{\lambda_i\}.$$

The same inequality is true if $\hat{p}(\cdot) \equiv 0$, $s \geqslant N + 2$.

For the proof see Akhiezer [1965].

This theme is continued in the final subsection of this section. And now we move on to what forms the circle of ideas of P.L. Chebyshev in the proper sense of the word: exact solutions.

3.3. Exact Solutions. A polynomial from $\mathscr{P}_N$ (rational function from $\mathscr{R}_{MN}$) which is a best approximation to a function $x(\cdot)$ (in the metric of $C(\Delta)$) will be denoted by $p_N(\cdot, x(\cdot))$ $(r_{MN}(\cdot, x(\cdot))$.

Theorem 8 (P.L. Chebyshev (1881), on the approximation of $(t - a)^{-1}$.)

$$/a > 1, x_a(t) = (t - a)^{-1}/ \Rightarrow$$

a) $d(x_a(\cdot), \mathscr{P}_N, C(\hat{I})) = (a - \sqrt{a^2 - 1})^N(a^2 - 1)^{-1}$;
b) $p_N(t, x_a(\cdot)) = (t - a)^{-1} - \Phi_N(t)$,

$$\Phi_N(t) = \frac{M}{2}\left(v^N\left(\frac{\alpha - v}{1 - \alpha v}\right) + v^{-N}\left(\frac{1 - \alpha v}{\alpha - v}\right)\right),$$

$$t = 2^{-1}(v + v^{-1}), \qquad \alpha = a - \sqrt{a^2 - 1} < 1, \qquad M = 4\alpha^{N+2}/(1 - \alpha^2)^2.$$

A remarkable contribution to the theory of best approximations was made by a pupil of P.L. Chebyshev, E.I. Zolotarev. The paper by E.I. Zolotarev [1877] "Application of elliptic functions to question on functions deviating least from zero" began with the words: "I think it is not out of place to consider certain questions on least values, which are resolved using the basic formulae of the theory of elliptic functions. These questions belong to the circle of questions on least values, the methods for solution of which were first given by P.L. Chebyshev". Later the following problem is formulated:

"For an entire function $F(x) = x^n - \sigma x^{n-1} - \sum_{k=2}^{n} p_k x^{n-k}$, where σ is given, find the remaining coefficients so that the greatest deviation from zero of the function $F(x)$, within the limits $x = -1$ to $x = +1$, should be as small as possible".

This problem was completely solved by E.I. Zolotarev.

Theorem 9 (E.I. Zolotarev). *The polynomial* $\hat{p}_{N-2}(\cdot) \in \mathscr{P}_{N-2}$ *which best approximates the function* $t \to t^N - \sigma t^{N-1}$ *on* $[-1, 1]$ *admits an explicit expression in terms of elliptic functions.*

The polynomials $Z_{N\sigma}(t) = t^N - \sigma t^{N-1} - \hat{p}_{N-2}(t)$ are called *Zolotarev polynomials.* We do not give an explicit expression for them because it is unwieldy, so we refer the reader to Akhiezer [1965]. Both Theorems 8 and 9 were obtained initially by P.L. Chebyshev's method, via the solutions of differential equations, without the use of complex analysis. Using complex analysis these two theorems can be proved fairly simply and naturally. (The same can be said of the two preceding theorems.) We will show how the methods of complex analysis are applied in the example of the proof of the original Theorem 2—on the explicit expression of the Chebyshev polynomials.

Consider the function, given parametrically by:

$$f(z) = 2^{-N}(w^{-N} + w^N),$$

$$z = 2^{-1}(w + w^{-1}) \Leftrightarrow f(z) = ((z + \sqrt{z^2 - 1})^N + (z - \sqrt{z^2 - 1})^N)/2^N.$$

It is easy to verify that this polynomial is of degree N with leading coefficient unity. As $\arg w$ varies from 0 to π, so z traverses from $+1$ to -1, and $\arg w^N$ from 0 to $N\pi$. Hence it follows at once that $f(\cdot)$ has $(N + 1)$-alternation on $[-1, 1]$, that is, by Theorem 5, the polynomial $f(z)$ is deviating least from zero $\Rightarrow$

$$T_{N\infty}(t) = f(t) = 2^{-N}((t + \sqrt{t^2 - 1})^N + (t - \sqrt{t^2 - 1})^N)$$
$$= \cos(N \arccos t) \cdot 2^{-(N-1)}.$$

We recall two more exactly solved problems.

Theorem 10 (E.I. Zolotarev, 1868). *The solution of the problem of best approximation by rational functions of the function* $t \to \operatorname{sign} t$ *on the set* $[-1, -\kappa] \cup [\kappa, 1]$, $0 < \kappa < 1$, *by rational functions from* $\mathcal{R}_{NN}$, *admits an explicit representation by elliptic functions.*

A.A. Gonchar immediately deduced, from the explicit expression of Theorem 10, Newman's theorem on the approximation of the function $t \to |t|$ by rational fractions. We will have more to say on this.

Theorem 11 (P.L. Chebyshev). *The solution of the best approximation problem for the function* $t \to \sqrt{(1 - k^2t^2)}$, $0 < k < 1$, *on the interval* $[0, 1]$ *by rational functions from* $\mathcal{R}_{NN}$ *(In the sense of minimization of relative error) has an explicit expression in terms of elliptic functions.*

This theorem gives a method of best "rational" extraction of roots from the point of view of accuracy. However, the method itself turns out to be very complicated. Explicit expressions for the best rational approximations of Theorem 10 and 11 are given in Akhiezer [1965].

3.4. Extremal Properties of Polynomials. Apart from the basic theme of finding exact solutions to approximation problems, an important secondary theme, given by the title of this subsection, was developed in the first stage. The question is of the properties of certain polynomials, which single them out as the solutions of extremal problems. We give several results on this theme.

Let a polynomial $p(\cdot) \in \mathscr{P}_N$ not exceed unity on the interval $\hat{I}$. What is the maximal value it may take at a point τ outside this interval? In this way we formulate the *extrapolation problem for polynomials*[2]. The solution is the Chebyshev polynomial. This is a consequence of the following theorem.

Theorem 12 (P.L. Chebyshev, on extrapolation for polynomials).

$$/p(\cdot) \in \mathscr{P}_N, |\tau| > 1/ \Rightarrow |p(\tau)| \leqslant |T_N(\tau)| \, \|p(\cdot)\|_{C(\hat{I})}$$

A.A. Markov, see Markov [1948], stated the following problem. "Suppose that $-L \leqslant f(x) \leqslant L$ for $a \leqslant x \leqslant b$, where $f(x) = p_0 x^n + p_1 x^{n-1} + \cdots + p_n$. The question arises, what is the upper limit of the numerical value of the derivative $f'(x) = np_0 x^{n-1} + (n-1)p_1 x^{n-2} + \cdots + 2p_{n-2}x + p_{n-1}$ of $f(x)$ relative to x". Such a question was posed by D.I. Mendeleev for $n = 2$ in his work "An investigation of aqueous solutions relative to specific gravity" (§ 86). The answer depends on how x is to be determined. We distinguish two cases: 1) x is a given number, 2) x is an arbitrary number between a and b. The work of A.A. Markov opened up a cycle of research on the problem "*inequalities for derivatives of polynomials and functions*". The problem posed was completely solved in the article cited. The case of k-th derivative ($k \geqslant 2$) was subsequently resolved by V.A. Markov the brother of A.A. Markov.

Theorem 13 (A.A. Markov, 1889). *Let $p(\cdot)$ be a polynomial of degree N. Then*

$$\|\dot{p}(\cdot)\|_{C(\hat{I})} \leqslant |\dot{T}_N(1)| \, \|p(\cdot)\|_{C(\hat{I})}$$
$$= N^2 \|p(\cdot)\|_{C(\hat{I})} \Leftrightarrow b(BW^1_\infty(\hat{I}), \mathscr{P}_N, C(\hat{I})) = N^{-2}.$$

Theorem 14 (V.A. Markov, 1892). *Let $p(\cdot)$ be a polynomial of degree N and $2 \leqslant k \leqslant N-1$. Then*

$$\|p^k(\cdot)\|_{C(\hat{I})} \leqslant |T_N^{(k)}(1)| \, \|p(\cdot)\|_{C(\hat{I})}$$
$$\Leftrightarrow b(BW^k_\infty(\hat{I}), \mathscr{P}_N, C(\hat{I}) = |T_N^{(k)}(1)|^{-1}.$$

Theorem 15 (A.A. Markov). *Let $p(\cdot)$ be a polynomial of degree N, $\tau \in \hat{I}$ and $\|p(\cdot)\|_{C(\hat{I})} \leqslant 1$. Then the maximal value of the quantity $p^{(k)}(\tau)$ is attained either on the Chebyshev polynomial (for $|\tau| = 1$), or on a Zolotarev polynomial, multiplied by a constant so that its norm is equal to unity.*

It is clear from Theorems 12–15 that the Chebyshev polynomials may supply an extremum for essentially different problems, even though searching for the extrema of linear functionals. Thus, for example the linear functionals

[2] This same problem can be interpreted, as the problem of best approximation of zero on $[-1, 1]$ under restrictions on the polynomial at the point τ ($p(\tau) = 1$).

$\langle x_\tau^*, p(\cdot)\rangle = p(\tau)$, $|\tau| > 1$ or $\langle x_k^*, p(\cdot)\rangle = p^{(k)}(1)$, $1 \leqslant k \leqslant N-1$, attain their maximum (on the intersection of the unit sphere of $C(\hat{I})$ with $\mathscr{P}_N$) at the polynomial $T_N(\cdot)$. For "smooth" spaces of type $L_p(\hat{I})$, $1 < p < \infty$, this is impossible. Convex analysis not only allows us to clarify this phenomenon, but also to describe all the possible linear functionals on $\mathscr{P}_N$, for which T_N will supply an extremum (see Tikhomirov [1976]).

3.5. Appendix. We stress here certain themes from the first subsection devoted to the research of P.L. Chebyshev.

3.5.1. Questions of Existence. When is there an element of best approximation? Here is one of the most impressive results on this theme.

Theorem 16 (James, 1964). *In order that every closed convex set A should have the property of existence of an element of best approximation relative to any element x of a Banach space X it is necessary and sufficient that X be reflexive.*

It follows easily from Weierstrass' theorem that in a normed space X finite-dimensional subspaces have the property of existence of an element of best approximation to any $x \in X$.

3.5.2. Questions of Uniqueness. P.L. Chebyshev considered approximation by the space of algebraic polynomials. Subsequently any finite-dimensional (and even infinite-dimensional) subspaces of $C(\Delta)$ (and then even $C(T)$, where T is a compact set) were studied from the point of view of uniqueness of the closest element. Spaces having the property of existence and uniqueness of the element of best approximation are called *Chebyshev*. A criterion for a finite-dimensional subspace to be Chebyshev was given by Haar.

Theorem 17 (Haar [1918]). *Let T be compact. The subspace $L_N \subset C(T)$ generated by $\{x_k(\cdot), 1 \leqslant k \leqslant N\}$, $x_k(\cdot) \in C(T)$, $L_N = \operatorname{lin}\{x_1(\cdot), \ldots, x_N(\cdot)\}$ is Chebyshev if and only if the number of zeroes of any element of L_N (except the zero) is not greater than $N-1$.*

It follows from Haar's theorem, in particular, that $\mathscr{P}_N$ and $\mathscr{T}_N = \operatorname{lin}\{1, \cos t, \sin t, \ldots, \cos Nt, \sin Nt\}$ are Chebyshev on $C([a, b])$ and $C(\mathbf{T}^1)$. On the other hand, for many-dimensional compact sets constructing a Chebyshev system has failed. The reason for this is made clear by the following theorem.

Theorem 18 (Mairhuber [1956]). *Let T be a compact set and for some $N \geqslant 2$ let there exist a Chebyshev subspace $L_N \subset C(T)$. Then T is homeomorphic to a subset of the circle $\mathbf{T}^1$.*

The properties of Chebyshev subspaces were studied by M.G. Krejn, Karlin, Schoenberg and many others. See Chapters 3, 4 of this part.

§4. Weierstrass' Theorem and its Generalizations

4.1. Introduction. In 1885 Weierstrass, (Weierstrass [1885]), proved that every continuous function on a finite interval admits an approximation to any accuracy by algebraic polynomials. It is usual to stress the fundamental role of this discovery in approximation theory S.N. Bernstein, for example, wrote: "The discovery of this theorem, remarkable in its generality, determined the future progress of the development of analysis". In fact Weierstrass' theorem in conjunction with the results and ideas of Chebyshev suggested the search for the mutual connection between the smoothness of a function and its best approximations. In fact, it is precisely this that formed the programme of the constructive theory of functions (the topic of the next section).

Weierstrass' theorem turned out to be a starting point for the problem of approximation of complex functions. And, finally, the generalization by Stone of Weierstrass' theorem was one of cornerstones on which was founded the edifice of the theory of Banach algebras.

All of these topics will be discussed in this section.

4.2. Proofs of Weierstrass' Theorem

Theorem 19 (Weierstrass, 1885). *Every function which is continuous on a finite interval can be uniformly approximated by polynomials to any degree of accuracy.*

This can be expressed by the brief formula:

$$/x(\cdot) \in C(\Delta), |\Delta| < \infty/ \Leftrightarrow \lim_{n\to\infty} d(x(\cdot), \mathscr{P}_n, C(\Delta)) = 0.$$

And another restatement. Denote by $\mathscr{P}(\Delta)$ the union of all the restrictions to Δ of polynomials from the $\mathscr{P}_n$. Weierstrass' theorem asserts that $C(\Delta) = \bar{\mathscr{P}}(\Delta)$. (the overbar denotes closure in the uniform metric).

There are various proofs of Weierstrass' fundamental theorem. Some of them left so great a trace in the further development of the theory that we have decided to briefly discuss them.

4.2.1. Weierstrass' Proof (1885). This is based on the idea of *smoothing*.

Extend $x(\cdot) \in C(\Delta)$ to a function of compact support on **R**. Let $y(\cdot) \in C_0(\mathbf{R})$, $y|_\Delta(\cdot) = x(\cdot)$. Put

$$(Q_\alpha y)(t) = (2\pi\alpha)^{-1/2} \int_R y(t + \tau)\exp(-\tau^2/2\alpha)\, d\tau. \tag{i}$$

The *Poisson kernel* $\Pi_\alpha(\tau) = (2\pi\alpha)^{-1/2}\exp(-\tau^2/2\alpha)$ is the density of the Gaussian distribution, that is, in particular $\int \Pi_\alpha(t)\, dt = 1$; it is also the Green's function for the heat equation, hence $\Pi_\alpha(\cdot)$ is "δ-form", that is,

$$\lim_{\alpha\to 0}\int_{\mathbf{R}} \Pi_\alpha(\tau)z(\tau)\,d\tau \to z(0) \qquad \forall z(\cdot)\in BC(R).$$

Hence it follows at once that:

$$|y(t)-(Q_\alpha y)(t)| =: \left|\int_{\mathbf{R}} (y(t+\tau)-y(t))\Pi_\alpha(\tau)\,d\tau\right|$$

$$\leqslant \int_{\mathbf{R}} \omega(\tau, y(\cdot))\Pi_\alpha(\tau)\,d\tau \to \omega(0, y(\cdot)) = 0$$

(by the continuity of ω) and it only remains, for small $\alpha = \alpha_0$, to replace $\exp(-\tau^2/2\alpha_0)$ by its Taylor polynomial and then formula (i) gives a polynomial $\tilde{Q}_{\alpha_0}y(\cdot)$, which sufficiently well approximates $x(\cdot)$ on Δ.

4.2.2. Bernstein's Proof (S.N. Bernstein, 1912) was based on probabilistic arguments. Let $\Delta = I = [0, 1]$ (clearly it reduces to this case). Put

$$(B_N x)(t) = \sum_{k=0}^{N} \binom{N}{k} x(k/N)t^k(1-t)^{N-k}.$$

A random quantity ξ taking the value k/N with probability $p = \binom{N}{k}t^k(1-t)^{N-k}$, is a sum: $\xi = (\xi_1 + \cdots + \xi_N)/N$, where ξ_i are independent of each other and take the values 0 and 1 with probabilities $(1-t)$ and t. In particular, $\sum_{k=0}^{N}\binom{N}{k}t^k(1-t)^{N-k} = 1$. The mathematical expectation of ξ (denoted $M\xi$) is equal to $(\sum_{k=0}^{N} M\xi_i)/N = t$, and the dispersion (denoted $D\xi$) is equal to $(\sum_{k=0}^{N} D\xi_i)/N^2 = t(1-t)/N$. By Chebyshev's inequality the probability $P_N(\delta)$, that $|\xi - M\xi| > \delta$, does not exceed $D\xi/\delta^2$. Now given $\varepsilon > 0$ there is a $\delta > 0$ such that $|(k/N) - t| \leqslant \delta \Rightarrow |x(t) - x(k/N)| \leqslant \varepsilon/2$ and then an N such that $P_N(\delta) \leqslant \varepsilon 4^{-1}\|x(\cdot)\|_{C(I)}^{-1}$ and we obtain:

$$|(B_N x)(t) - x(t)| = \left|\sum_{k=0}^{N}\binom{N}{k} t^k(1-t)^{N-k}(x(k/N)-x(t))\right|$$

$$\leqslant \sum_{|(k/N)-t|\leqslant\delta} + \sum_{|(k/N)-t|>\delta} \leqslant \frac{\varepsilon}{2} + 2\|x(\cdot)\|_{C(I)}P_N(\delta) \leqslant \varepsilon.$$

4.2.3. Lebesgue's Proof (Lebesgue, 1898) is a by-product of the following simple results.

1) A continuous function can be uniformly approximated by a broken line; 2) every broken line can be represented in the form $a_0 + \sum_{k=1}^{n} a_k|t-\tau|$ and therefore it is sufficient to prove that; 3) the function $|t|$ can be uniformly approximated by polynomials on $\hat{I} = [-1, 1]$. The latter assertion follows, for example, from the uniform convergence of the following series:

$$(1+\alpha)^{1/2} = 1 + \alpha/2 - \alpha^2/8 + \cdots,$$

hence we obtain:

$$|t| = (t^2)^{1/2} = (1 - (1 - t^2))^{1/2}$$
$$= \lim_{N\to\infty} \left(1 - \frac{(1-t^2)}{2} - \frac{(1-t^2)^2}{8} + \cdots + (-1)^{N-1}(1-t^2)^N(N!)^{-1} \right.$$
$$\left. \times \frac{1}{2}\left(\frac{1}{2} - 1\right)\cdots\left(\frac{1}{2} - N + 1\right)\right).$$

From which follows Weierstrass' theorem. Lebesgue's plan found its completion in the Stone-Weierstrass theorem.

4.3. The Development and Generalization of Weierstrass' Theorem

4.3.1. The Periodic Case

Theorem 20. *Every continuous function, periodic on a finite interval, can be approximated by trigonometric polynomials with any degree of accuracy*

$$(/x(\cdot) \in C(\mathbf{T}^1)/ \Rightarrow d(x(\cdot), \mathcal{T}_N, C(\mathbf{T}^1)) \to 0 \text{ as } n \to \infty).$$

This result can be proved by a variety of methods: it can be reduced to Weierstrass' theorem for algebraic polynomials; the fact that harmonics form a complete orthogonal system can be used; etc. We give Fejér's proof, which plays a major role in approximation theory.

We recall that the sum $(D_0 + \cdots + D_{N-1})(\cdot)/N$, where $D_k(\cdot)$ is the Dirichlet kernel, is called the *Fejér kernel* and denoted $F_n(\cdot)$[3]. Here $F_N(t) = \sin^2(Nt/2)/2N\sin^2(t/2)$. The function $(\mathcal{F}_N x)(t) = \pi^{-1}\int_{T^1} x(t-\tau)F_N(\tau)\,d\tau$ is called the *Fejér sum*, $(\mathcal{F}_N x)(\cdot) \in \mathcal{T}_N$.

Theorem 21 (Fejér [1904]).

$$/x(\cdot) \in C(T^1)/ \Rightarrow \|x(\cdot) - (\mathcal{F}_N x)(\cdot)\|_{C(T^1)} \to 0, \qquad N \to \infty$$
$$(\lambda(x(\cdot), (\mathcal{F}_N, \mathcal{T}_N), C(T^1)) \to 0).$$

In fact the kernel $F_N(\cdot)$ has the same δ-form property as the Poisson kernel. In addition, it is obvious that $\pi^{-1}\int_{T^1} F_N(t)\,dt = 1$. Therefore

$$|x(t) - (\mathcal{F}_N x)(t)| =: \pi^{-1}\left|\int_{\mathbf{T}^1} (x(t+\tau) - x(t))F_N(\tau)\,d\tau\right|$$
$$\leqslant \pi^{-1}\int_{\mathbf{T}^1} \omega(\tau, x(\cdot))F_N(\tau)\,d\tau \to 0$$

(by the continuity of ω). Which also proves Theorem 21.

4.3.2. The Theorems of Runge, Walsh, Lavrent'ev, Keldysh and Mergelyan. In the same year, 1885, in which Weierstrass proved his theorem ($C(\Delta) = \bar{P}(\Delta)$),

---

[3] The Dirichlet, Fejér and other kernels will also be described in Chap. 2, § 1.

another German mathematician Runge obtained a result of a similar kind in the complex case.

Theorem 22 (Runge [1885]). *Let G be a simple-connected domain and $T \subset G$ a compact set. Then any function analytic in G can be approximated to any accuracy by polynomials on T.*

In other words,

$$/x(\cdot) \in A^G, T \in \operatorname{Comp} G/ \Rightarrow \lim_{N \to \infty} d(x(\cdot), \mathscr{P}_N, C(T)) = 0.$$

Walsh proved that if G is a simply-connected domain with a Jordan boundary and $x(\cdot) \in A^G \cap C(\bar{G})$, then $d(x(\cdot), P_N, C(\bar{G})) \to 0$, $N \to \infty$. The Soviet mathematicians M.A. Lavrent'ev, M.V. Keldysh and S.N. Mergelyan completely solved the problem of approximation of functions by polynomials.

Theorem 23 (M.A. Lavrent'ev, 1934). *If T is a bounded continuum without interior points then $C(T) = \bar{\mathscr{P}}(T)$ if and only if $\mathbf{C} \backslash T$ is connected.*

Theorem 24 (M.V. Keldysh, [1945]). *Let G be a domain. The equality $\bar{\mathscr{P}}(\bar{G}) = A^G \cap C(\bar{G})$ holds if and only $\mathbf{C} \backslash \bar{G}$ is a connected set containing infinity.*

Theorem 25 (S.N. Mergelyan, [1951]). *Let T be a compact set in $\mathbf{C}$. The equality $A^{\operatorname{int} T} \cap C(T) = \bar{\mathscr{P}}(T)$ holds if and only if $\mathbf{C} \backslash T$ is a connected set containing infinity.*

Thus we have here the completion of the theme of approximation of analytic functions by means of algebraic polynomials.

4.4. The Stone-Weierstrass Theorem

Theorem 26. *Let T be compact and $\mathscr{A} \subset C(T)$ a closed subalgebra of functions with identity which separates points: (that is, $\forall t_1, t_2, t_1 \neq t_2, t_i \in T, i = 1, 2, \exists x(\cdot) \in \mathscr{A}: x(t_1) \neq x(t_2)$). Then $\mathscr{A} = C(T)$.*

This theorem in some sense exhausts the theme of approximation of real functions by polynomials. It was proved by Stone [1937], but afterwards the name of Weierstrass became firmly attached to it. This result exerted a decisive influence on the creation of the Gel'fand theory of Banach algebras, Gel'fand, Raikov & Shilov [1960]. It is natural to consider many of the concepts of approximation theory within the confines of this theory. We recall the proof of this classical result, a proof, to which was stretched the thread of Lebesgue's reasoning, see § 4.2.

The proof comprises the following steps.

A) The function $|t|$ is approximated on $[-1, 1]$ by polynomials. We mentioned this in subsection 4.2.3. Hence it follows at once that if $x(\cdot) \in \mathscr{A}$, then $|x(\cdot)| \in \mathscr{A}$. But then $x(\cdot), y(\cdot) \in \mathscr{A} \Rightarrow (x \bigvee y)(t) = \max(x(t), y(t)) = z(t)$ and $(x \bigwedge y)(t) = \min(x(t), y(t)) = \xi(t) \in \mathscr{A}$, since $2z(t) = (x(t) + y(t) + |(x(t) - y(t)|)$, $2\xi(t) = (x(t) + y(t) - |x(t) - y(t)|)$.

B) Let $t_1, t_2 \in T, t_1 \neq t_2$. By assumption there is a function $x(\cdot) = x(\cdot, t_1, t_2) \in \mathscr{A}$ such that $x(t_1) \neq x(t_2)$. Then $\eta_1(t) =: (x(t) - x(t_1))/(x(t_2) - x(t_1))$ has the property that $\eta_1(t_1) = 0, \eta_1(t_2) = 1$. Similarly we can construct $\eta_2(\cdot)$: $\eta_2(t_1) = 1$, $\eta_2(t_2) = 0$. Thus the function $\eta(t) = a_1\eta_2(t) + a_2\eta_1(t)$ is such that $\eta(t_i) = a_i$, $i = 1, 2$.

C) Let $x(\cdot) \in C(T)$. Denote by $\varphi_{t_1t_2}(\cdot)$ a function from $\mathscr{A}$ such that $\varphi_{t_1t_2}(t) = x(t_i)$, $i = 1, 2$. Given $\varepsilon > 0$ and $\theta \in T$, define for $\tau \in T$ a neighbourhood $U(\tau)$ of τ such that $\varphi_{\theta\tau}(t) > x(t) - \varepsilon\ \forall t \in U(\tau)$. Take a finite cover $T = \bigcup_{i=1}^{s} U(\tau_i)$. Let $x_\theta(t) = \bigvee_{i=1}^{s} \varphi_{\theta\tau}(t) (\varepsilon \mathscr{A})$. Then $x_\theta(t) \geqslant x(t) - \varepsilon\ \forall t \in T$ and $x_\theta(\theta) = x(\theta)$. We then find $V(\theta)$, a neighbourhood of θ, such that $x_\theta(t) < x(t) + \varepsilon\ \forall t \in V(\theta)$ and again choose a finite subcover $T = \bigcup_{i=1}^{l} V(\theta_i)$. Put $y(t) = \bigwedge_{i=1}^{l} x_{\theta_i}(t)$. We obtain $\|x(\cdot) - y(\cdot)\|_{C(T)} < \varepsilon$. The theorem is proved.

4.5. Appendix

4.5.1. Muntz' Theorem. Weierstrass proved that the closure of the subspace $\text{lin}(t^k)_{k \in Z_+}$ coincides with $C(\Delta)$. It is natural to put the question: for which powers $\{t^{\alpha_k}\}_{k \in Z_+}$ does the same result hold.

Theorem 27 (Muntz, [1914]). *Let $\Delta = [t_0, t_1]$, $0 < t_0 < t_1 < \infty$, $(0 \leqslant t_0 < t_1 < \infty)$, $\{\alpha_k\}_{k \in Z_+} \alpha_k \geqslant 0$, be a sequence increasing to infinity. In order that any function $x(\cdot)$ be approximated by polynomials $\sum_{k \geqslant 0} a_k t^{\alpha_k}$, it is necessary and sufficient that $\sum_{k \geqslant 0} \alpha_k^{-1} = \infty$ (that $\alpha_0 = 0$ and $\sum_{k \geqslant 1} \alpha_k^{-1} = \infty$).*

4.5.2. Wiener's Approximation Theorem. Let A be some subset of functions summable on the line. We pose the question: when can we approximate any function from $L_1(R)$ by linear combinations of shifts of the functions of this subset?

Theorem 28 (Wiener, 1933). *In order that we can arbitrarily closely approximate any function from $L_1(\mathbf{R})$ by shifts of functions of a subset $A \subset L_1(\mathbf{R})$, it is necessary and sufficient that for any point $\tau \in \mathbf{R}$ there is a function $x(\cdot) \in A$ whose Fourier transform is not zero at τ.*

The connection between Wiener's theorem and Weierstrass' theorem is realized by the theory of Banach algebras (see Gel'fand, Raikov & Shilov, [1960]).

4.5.3. Theorems on Rational Approximation. We pose the same problem as in §4.3.2., but relative to rational approximation. In the solution of the question on rational approximation the notion of analytic capacity plays a fundamental role.

Let T be a compact set in $\mathbf{C}$, $BA^{\mathbf{C}\setminus T}$ the set of functions analytic outside T, zero at infinity and bounded outside T by the constant one. Put $\gamma(T, x(\cdot)) = \lim zx(\cdot)$ as $z \to \infty$. The number $\gamma(T) = \sup\{|\gamma(T, x(\cdot))|: x(\cdot) \in BA^{\mathbf{C}\setminus T}\}$ is called the *analytic capacity* of T. This notion was introduced by Ahlfors in 1947. The fundamental role of analytic capacity is revealed in

Theorem 29 (Ahlfors [1947]). *Let T be a compact set in* $\mathbf{C}$. *In order that every function which is analytic and bounded outside T can be analytically continued to T, it is necessary and sufficient that* $\gamma(T) = 0$.

The question of rational approximation can be solved in terms of analytic capacity.

Theorem 30 (A.G. Vitushkin [1967]). *In order that every continuous function on a compact set T can be uniformly approximated by rational functions it is necessary and sufficient that for any disc* $D_R + z$ *of radius R with centre at z there is the equality*

$$\gamma((D_R + z)\backslash T) = \gamma(D_R + z) = R.$$

4.5.4. Approximation by Bernstein Polynomials and Saturation. From the practical point of view approximation of smooth functions using Bernstein polynomials is inadvisable. The operators B_N are positive polynomial operators and therefore have "*Saturation*", that is, they do not respond to sufficiently small smoothness of the approximated function. The following formula holds: if $x(\cdot)$ is continuous on I and twice differentiable at a point τ then

$$\lim_{N\to\infty} (B_N(x(\cdot))(\tau) - x(\tau))N = \tau(1-\tau)\ddot{x}(\tau)/2 \quad \text{(E.V. Voronovskaya, [1932])},$$

that is, the speed of approximation is no better that N^{-1}. We have already had occasion to meet these phenomena. Let us note, however, that Bernstein polynomials have numerous remarkable qualities.

§ 5. The Constructive Theory of Functions

5.1. Introduction. In the previous two sections we discussed the circle of ideas of P.L. Chebyshev and Weierstrass' theorem. At the beginning of this century several prominent analysts in their papers began to promote problems on the approximation of functions in the spirit of the union of the ideas of Weierstrass and Chebyshev. Here, in the first place, we must mention the monograph of Borel [1905], the series of articles by Vallée-Poussin, in particular [1911], and his monograph [1919] and the works of Lebesgue [1909, 1910]. Also related to this circle of ideas, in essence, is the work of Hadamard [1892] on Padé approximation. This can be most simply illustrated in the space L_2. Consider a periodic function $x(\cdot) \in L_2(\mathbf{T}^1)$. It has an expansion as a Fourier series

$$x(t) = \sum_{k\in Z} x_k e^{ikt}.$$

This gives a mapping $x(\cdot) \to \{x_k\}_{k\in Z}$. We note that the numbers x_k have an approximative meaning:

$$d(x(\cdot), \mathcal{T}_N, L_2(\mathbf{T}^1)) = \left(\sum_{|k|\geqslant N+1} |x_k|^2\right)^{1/2}.$$

We can try to recover information about $x(\cdot)$ from the sequence $\{x_k\}$ or the sequence $\{d(x(\cdot), \mathcal{T}_N, L_2(\mathbf{T}^1))\}_{N\in Z_+}$, in particular, information about smoothness, singularities, etc.

Analogous questions can also be posed, for example, for functions, expanded as a Taylor series:

$$x(z) = \sum_{k\in Z_+} x_k z^k.$$

Obviously, the sequence $\{x_k\}_{k\in Z_+}$ includes all the information about $x(\cdot)$, in particular, about its singularities. Cauchy gave a formula for the calculation of the disc of convergence of a Taylor series, that is, determined the distance from zero of the first singular point of the function $x(\cdot)$. Questions on the recovery of the singularities of $x(\cdot)$ relative to sequence $\{x_k\}_{k\in Z_+}$ were solved by the theory of rational approximations.

It is natural to consider analogous problems for other sequences, for example, $x(\cdot) \to d(x(\cdot), \mathcal{T}_N, X)$, $x(\cdot) \to d(x(\cdot), \mathcal{P}_N, X)$, $x(\cdot) \to d(x(\cdot), \mathcal{R}_{MN}, X)$ etc.

The question of how to recover information on the smoothness of a function from a sequence of best approximations by polynomials was stated in the form of competitions in 1908–9—on the one hand by the Belgian Academy of Sciences (Vallée-Poussin) and on the other hand, by Göttingen University (Landau). In 1911–12 there appeared significant investigations, which laid the foundations for a new direction, given eventually the name: *the constructive theory of functions*. These were the dissertation of the American mathematician Jackson [1911], presented to Göttingen University (1911) and the memoir of S.N. Bernstein [1952 N3], awarded the prize of the Belgian academy. In the preface to his memoir Bernstein defines the aim of his investigation "The present work is an attempt at the approximate calculation of the least deviation $E_n[f(x)]$ $[\Leftrightarrow d(f(\cdot), \mathcal{P}_n, C([a, b]))]$ and an investigation of the connection between the law of decrease of $E_n[f(x)]$ and the differential properties of the function being considered".

Later this ideology was supplemented by research into "approximation in classes", on which S.M. Nikol'skij spoke in the already mentioned Proceedings of the Amsterdam congress. Here, from the formal point of view, the quantities $d(W, \mathcal{P}_N, X)$, $\lambda(W, (\Lambda, A), X)$ and others of similar type, described in the introduction, were being studies.

Moreover, whereas in the first stage of approximation theory, within the confines of the ideas of P.L. Chebyshev, individual approximation of individual elements was studied, here, in the constructive theory of functions, basic attention was given to the asymptotic behaviour of approximative quantities. Here is how one of the problems of Vallée-Poussin was posed. "It would be interesting to know, whether or not it is possible to approximately represent the ordinate of a polygonal line by a polynomial of degree n with error of higher order than n^{-1}" (S.N. Bernstein was occupied with this problem for several decades.)

After these introductory comments we move on to a survey of results.

5.2. Direct Theorems. Approximation of the Function $|t|$. *Direct theorems* include assertions of the following sort: let a function have some smoothness property, what can then be said about the sequence of its best approximations. The first results in this direction are due to Jackson. The following result refines Theorem 20.

Theorem 31 (Jackson's first theorem, 1911). *Let* $x(\cdot)$ *be a continuous periodic function. Then*

$$d(x(\cdot), \mathcal{T}_{2N}, C(\mathbf{T}^1)) \leqslant A\omega(N^{-1}, x(\cdot)).$$

For the proof of this result Jackson applied the operator subsequently named after him. Put

$$J_{2N}(t) = \frac{3}{2(2N^2+1)N}\left(\frac{\sin\frac{Nt}{2}}{\sin\frac{t}{2}}\right)^4$$

This function is called the *Jackson kernel.* By comparison of the expression for the Jackson kernel with the expression for the Fejer kernel given earlier it is clear that $J_{2N}(\cdot) = a_N F_N^2(\cdot)$. Hence it is clear that $J_{2N}(\cdot) \in \mathcal{T}_{2N}$. Here a is chosen so that $\pi^{-1}\int_{\mathbf{T}^1} J_{2N}(t)\,dt = 1$. It is easy to prove that $J_{2N}(t) = 1/2 + (1 - 3/(2N^2+1))\cos t + \cdots$. We denote by $(\mathcal{J}_{2N}x)(\cdot)$ the operator of convolution type with the Jackson kernel

$$(\mathcal{J}_{2N}x)(t) = \pi^{-1}\int_{\mathbf{T}^1} J_{2N}(\tau)x(t+\tau)\,d\tau.$$

Then, by using the property of the modulus of continuity

$$\omega(\lambda\delta, x(\cdot)) \leqslant (1+\lambda)\omega(\delta, x(\cdot)),$$

we obtain:

$$\begin{aligned}|x(t) - (\mathcal{J}_{2N}x)(t)| &=: \pi^{-1}\left|\int_{\mathbf{T}^1}(x(t+\tau) - x(t))J_{2N}(\tau)\,d\tau\right| \\ &\leqslant \pi^{-1}\int_{\mathbf{T}^1}\omega(\tau, x(\cdot))J_{2N}(\tau)\,d\tau \\ &\leqslant \omega(N^{-1}, x(\cdot))\left(1 + \pi^{-1}N\int_{\mathbf{T}^1}|\tau|J_{2N}(\tau)\,d\tau\right). \qquad \text{(i)}\end{aligned}$$

By virtue of the inequalities $\pi^{-1}|\tau| \leqslant \sin(|\tau|/2)$, $0 \leqslant |\tau| \leqslant \pi/2$, of Cauchy-Bunyakovskij, and the equality $\pi^{-1}\int_{\mathbf{T}^1} J_{2N}(\tau)\,d\tau = 1$, we obtain:

$$\pi^{-1}\int_{\mathbf{T}^1}|\tau|J_{2N}(\tau)\,d\tau$$
$$\leqslant \int_{\mathbf{T}^1}\sin(|\tau|/2)J_{2N}^{1/2}(\tau)J_{2N}^{1/2}(\tau)\,d\tau$$
$$\leqslant \pi^{1/2}\left(\int_{\mathbf{T}^1}\sin^2(\tau/2)J_{2N}(\tau)\,d\tau\right)^{1/2} = (\pi 2^{-1})^{1/2}\left(\int_{\mathbf{T}^1}(1-\cos t)J_{2N}(t)\,dt\right)^{1/2}$$
$$= (\pi 2^{-1})^{1/2}\left(\int_{\mathbf{T}^1}(1/2 + (1 - 3/(2N^2+1))\cos t + \cdots)(1-\cos t)\,dt\right)^{1/2}$$
$$= A_1(2N^2+1)^{-1/2} \leqslant AN^{-1} \quad (A < \pi\sqrt{3}/2). \tag{ii}$$

It follows from (i) and (ii) that

$$\|x(\cdot) - (\mathscr{J}_{2N}x)(\cdot)\|_{C(\mathbf{T}^1)} \leqslant A\omega(N^{-1}, x(\cdot)).$$

Theorem 32 (Jackson's second theorem, 1911). *Let $x(\cdot)$ be an r times continuous differentiable periodic function. Then*

$$d(x(\cdot), \mathscr{T}_{2N}, C(T^1)) \leqslant A^{r+1}N^{-r}\omega(N^{-1}, x^{(r)}(\cdot)),$$

where A is the constant from the preceeding theorem.

For the proof of this result the Jackson operator, unfortunately, does not apply: it, like any positive operator of convolution type, has "saturation", and poorly approximates smooth functions. Therefore Jackson applied another operator $\mathscr{J}_{2N,r} = I - (I - \mathscr{J}_{2N})^{r+1}$. The operator $\mathscr{J}_{2N}$ is shift invariant and this means that it commutes with the differentiation operator. Since

$$\omega(\delta, x(\cdot)) \leqslant \delta\|\dot{x}(\cdot)\|_{C(\mathbf{T}^1)}$$

(theorem on Lagrange means), we obtain from Theorem 31:

$$\|x(\cdot) - (\mathscr{J}_{2N}x)(\cdot)\|_{C(\mathbf{T}^1)} \leqslant A\|\dot{x}(\cdot)\|_{C(\mathbf{T}^1)}N^{-1},$$

hence finally

$$\|(I - \mathscr{J}_{2N}, r)x(\cdot)\|_{C(\mathbf{T}^1)}$$
$$=: \|(I - \mathscr{J}_{2N})^{r+1}x(\cdot)\|_{C(\mathbf{T}^1)}$$
$$= \|(I - \mathscr{J}_{2N})(I - \mathscr{J}_{2N})^r x(\cdot)\|_{C(\mathbf{T}^1)} \leqslant AN^{-1}\|(I - \mathscr{J}_{2N})^r\dot{x}(\cdot)\|_{C(\mathbf{T}^1)} \leqslant \cdots$$
$$\leqslant A^rN^{-r}\|(I - \mathscr{J}_{2N})x^{(r)}(\cdot)\|_{C(\mathbf{T}^1)} \leqslant A^{r+1}N^{-r}\omega(N^{-1}, x^{(r)}(\cdot)),$$

as required.

We remark that other approximating operators are possible. In particular, it would be possible to apply the Vallée-Poussin operator. This circle of questions will be clarified in Chap. 2, § 5. In fact if we repeat all of our arguments relative to the operator

$$(\mathscr{J}_{2N}x)(t) = (2\pi)^{-1} \int_{\mathbf{T}^1} (x(t+\tau) + x(t-\tau)) J_{2N}(\tau)\, d\tau,$$

it is possible to obtain such a result.

Theorem 33 (Zygmund, 1945). *Let $x(\cdot)$ be an r times continuously differentiable function. Then*

$$d(x(\cdot), \mathscr{T}_{2N}, C(\mathbf{T}^1)) \leqslant A^{r+1} N^{-r} \omega_2(N^{-1}, x^{(r)}(\cdot)).$$

We have formulated Theorems 31–33 in "individual" form. However, it is possible to give them as theorems on approximation by classes. Let $\omega(\cdot)$ be a function of modulus of continuity type. Denote by $W^r_\infty H^{\omega k}(\mathbf{T}^1)$ the class of r-times continuously differentiable functions $x(\cdot)$, for which $\omega_k(\delta, x^{(r)}(\cdot)) \leqslant \omega(\delta)$. If $k = 1$ then we simply write $W^r_\infty H^\omega(\mathbf{T}^1)$. Then Theorems 31–33 can be expressed in the following way:

$$d(W^r_\infty H^\omega(\mathbf{T}^1), \mathscr{T}_{2N}, C(\mathbf{T}^1)) \leqslant CN^{-r}\omega(N^{-1}),$$

$$d(W^r_\infty H^{\omega,2}(\mathbf{T}^1), \mathscr{T}_{2N}, C(\mathbf{T}^1)) \leqslant CN^{-r}\omega(N^{-1}).$$

Alongside Jackson, and almost simultaneously, S.N. Bernstein was occupied with similar problems. He solved Vallée-Poussin's problem of approximating broken lines (see subsection 5.1).

Theorem 34. a) $d(|t|, \mathscr{P}_N, C(\hat{I})) \asymp N^{-1}$;
b) $d(|t|, \mathscr{P}_{2N}, C(\hat{I}))\cdot 2N \to \mu (N \to \infty)$, $\mu = 0.282 \pm 0.004$;
c) $d(|t|^p, \mathscr{P}_N, C(\hat{I}))N^p \to \mu(p)(N \to \infty) p > 0$.
The first of these results was obtained by S.N. Bernstein in 1912 (Bernstein [1952] Vol. 1, p. 56, the second in 1915 (Bernstein [1952] Vol. 1, p. 157), the third in 1938 (Bernstein [1952] Vol. 2, p. 262). Thus, S.N. Bernstein returned to these problems over a period of more than a quarter of a century. In the last named memoirs he discovered the connection between the approximation of functions on the circle by polynomials and the approximation of functions by entire functions of first degree on the whole axis.

5.3. Inverse Theorems. *Inverse* theorems are those concerning smoothness properties of a function appearing relative to the singularities of its best approximations.

Theorem 35 (S.N. Bernstein (1912)—Vallée-Poussin (1919)). *Let a continuous function $x(\cdot)$ satisfy the inequality*:

$$d(x(\cdot), \mathscr{T}_N, C(\mathbf{T}^1)) \leqslant CN^{-(r+\alpha)}, r \in \mathbf{Z}_+, 0 < \alpha < 1.$$

Then $x(\cdot)$ is r times continuously differentiable and its r-th derivative satisfies a Hölder condition of order α:

$$|x^{(r)}(t) - x^{(r)}(t')| \leqslant A|t - t'|^\alpha, \qquad t, t' \in \mathbf{T}^1 (\Leftrightarrow x(\cdot) \in H^{r+\alpha}_\infty(\mathbf{T}^1)).$$

For the proof of results of this type we need inequalities for trigonometric polynomials, similar to the Markov inequalities for algebraic polynomials.

Theorem 36 (S.N. Bernstein (1912) with refinements by Landau, M and F. Riesz). *If $x(\cdot)$ is a trigonometric polynomial of degree N, then*

$$\|\dot{x}(\cdot)\|_{C(\mathbf{T}^1)} \leqslant N\|x(\cdot)\|_{C(\mathbf{T}^1)}.$$

This inequality is called *Bernstein's inequality*. We have already mentioned this result and said that it is equivalent to the calculation of the quantity $b(BW_\infty^1, \mathcal{T}_N, C(\mathbf{T}^1))$, which turns out to be equal to N^{-1}.

The problem of Bernstein's inequality is entirely within convex programming and therefore we will not dicuss its proof. There is a huge literature devoted to its generalization. We will say more on this question in the following chapters.

But the proof of Theorem 35 is essential for our further discussions, and we give it here.

By assumption there is a trigonometric polynomial $p_k(\cdot) \in \mathcal{T}_{2^k}$ such that $\|x(\cdot) - p_k(\cdot)\|_{C(\mathbf{T}^1)} \leqslant C2^{-k(r+\alpha)}$. Hence it follows that $x(\cdot) = \sum_{k\geqslant 1} x_k(\cdot)$, where $x_1(t) = p_1(t)$, $x_k(t) = p_k(t) - p_{k-1}(t)$, $k \geqslant 2$, and the series converges uniformly. Here

$$\begin{aligned}\|x_k(\cdot)\| &=: \|p_k(\cdot) - p_{k-1}(\cdot)\|_{C(\mathbf{T}^1)} \\ &= \|(x(\cdot) - p_{k-1}(\cdot)) - (x(\cdot) - p_k(\cdot))\|_{C(\mathbf{T}^1)} \\ &\leqslant C(2^{-k(r+\alpha)} + 2^{-k(r+\alpha)}\cdot 2^{-(r+\alpha)}) = C_1 2^{-k(r+\alpha)}.\end{aligned}$$

By Bernstein's inequality (Theorem 36), we obtain

$$\|x_k^{(r)}(\cdot)\|_{C(\mathbf{T}^1)} \leqslant 2^{kr}\|x_k(\cdot)\|_{C(\mathbf{T}^1)} \leqslant C_1 2^{-kr},$$

and this means that $x_k^{(r)}(\cdot) = \sum_{k\geqslant 1} x_k^{(r)}(\cdot)$, moreover the series converges uniformly.

Now, choosing $\delta > 0$, find m from the condition $2^{m-1} \leqslant \delta < 2^m$. Then for $|h| < \delta$ we obtain

$$\begin{aligned}|x^{(r)}(t+h) - x^{(r)}(t)| &=: \left|\sum_{k\geqslant 1} (x_k^{(r)}(t+h) - x_k^{(r)}(t))\right| \\ &\leqslant \sum_{k=1}^{m-1} + \sum_{k\geqslant m} =: X_1(t) + X_2(t).\end{aligned}$$

Using these estimates, Bernstein's inequality and Lagrange's theorem on finite increments, we obtain:

$$|X_1(t)| \leqslant \delta \sum_{k=1}^{m-1} \|x_k^{(r+1)}(\cdot)\|_{C(\mathbf{T}^1)} \leqslant 2^{-(m-1)}C_1 \sum_{k=1}^{m-1} 2^{k(1-\alpha)} \leqslant C_2\delta^\alpha,$$

$$|X_2(t)| \leqslant \sum_{k\geqslant m} 2C_1 2^{-k\alpha} \leqslant C_3\delta^\alpha,$$

that is,

$$|x^{(r)}(t+h) - x^{(r)}(t)| \leqslant C\delta^{\alpha},$$

as required[4].

Thus, the combination of Jackson's direct theorem (Theorem 32) and the inverse theorem of Bernstein-Vallée-Poussin (Theorem 35) allows us to find constructive necessary and sufficiently conditions for a function to belong to the space

$$W_{\infty}^{r}H^{\alpha}(\mathbf{T}^{1}) =: \{x(\cdot) \in C^{r}(\mathbf{T}^{1}) | \exists C > 0: |x^{(r)}(t) - x^{(r)}(t')| \leqslant C|t - t'|^{\alpha}\}$$

for $0 < \alpha < 1$.

The case $\alpha = 1$ was not obtained for a long time. Do the functions, for which $d(x(\cdot), \mathbf{T}_N, C(\mathscr{T}^1)) \leqslant N^{-(r+1)}$ form such a class? The answer to this question is given by

Theorem 37 (Zygmund, [1945]), *In order that the inequality*

$$d(x(\cdot), \mathscr{T}_N, C(\mathbf{T}^1)) \leqslant CN^{-(r+1)}$$

holds for some constant $C > 0$ it is necessary and sufficient that $x(\cdot) \in C^r(\mathbf{T}^1)$ and $\omega_2(\delta, x^{(r)}(\cdot)) \leqslant A\delta$.

In this way the class $W_{\infty}^{r}Z(\mathbf{T}^1)$ and its generalizations became customary in approximation theory.

We give two further results of S.N. Bernstein, which he particularly singled out (see Bernstein [1937]).

Theorem 38. *In order that a function $x(\cdot)$ be analytic on a segment Δ, it is necessary and sufficient that the least deviation $d(x(\cdot), \mathscr{P}_N, C(\Delta))$ decreases,* as N *increases, more quickly than the terms of a geometric progression.*

Theorem 39. *In order that a function $x(\cdot)$ have derivatives of all orders it is necessary and sufficient that the deviation $d(x(\cdot), \mathscr{P}_N, C(\Delta))$ decreases, as N increases, more quickly than any power.*

(We stress that for periodic functions precisely the same results hold for the Fourier coefficients, as follows immediately from the explicit formulae for them.)

Thus, in this and the preceeding subsections, we have given a "constructive" description of functions of finite and infinite smoothness, and even analytic functions.

5.4. The Statement of the Problem on Comparison of Different Methods of Approximation. Approximation in Classes. We have proved that Fejér sums converge uniformly to the function. But why is it impossible to take Fourier sums? Why in the Jackson theorem is it impossible to take Fejér sums? Such questions were posed at the beginning of this century. In 1876 Dubois-Reymond con-

[4] It is amazing that S.N. Bernstein, having both the general method and his inequality, did not fully prove this result (he proved it "up to ε"). In the form given the theorem was due to Vallée-Poussin.

structed an example of a continuous function whose Fourier series did not converge uniformly. This example turned out to be fairly complex. Simpler examples then appeared, but the very essence of the matter was explained by Lebesgue.

The *Fourier operator* is a convolution operator given by the *Dirichlet kernel*

$$(S_N x)(t) = \pi^{-1} \int_{\mathbf{T}^1} x(t+\tau) D_N(\tau)\, d\tau.$$

The norm of this operator as an operator from $C(\mathbf{T}^1)$ to $C(\mathbf{T}^1)$ is obviously equal to the quantity (now called the Lebesgue constant)

$$L(\mathcal{T}_N) = \pi^{-1} \int_{\mathbf{T}^1} |D_N(t)|\, dt.$$

A simple calculation shows that $L(\mathcal{T}_N) \asymp \log N$ (that is, it grows). In this case the operator S_N cannot converge to unity (this fact was subsequently proved in functional analysis in a general form and is called the Banach-Steinhaus theorem).

The Lebesgue constant plays a significant role in various questions. Its asymptotic behaviour has occupied many mathematicians.

Theorem 40. a) $L(\mathcal{T}_N) = 4\pi^{-2} \ln(N+1) + O(1)$,

b) $\sup_N \left(\pi^{-1} \int_{\mathbf{T}^1} |D_N(t)|\, dt - 4\pi^{-2} \ln(N+1) \right) = 1{,}2706\ldots.$

Result a) is given in many books (for example Dzyadyk [1977]), result b) is contained in the work of P.V. Galkin (1971).

We mention one other important *Lebesgue inequality*.

Theorem 41. *Let* $x(\cdot) \in C(\mathbf{T}^1)$. *Then*

$$\|x(\cdot) - (S_N x)(\cdot)\|_{C(\mathbf{T}^1)} =: \varphi(x(\cdot), \mathcal{T}_N, C(\mathbf{T}^1)) \leqslant (1 + L(\mathcal{T}_N)) d(x(\cdot), \mathcal{T}_N, C(\mathbf{T}^1)).$$

The *proof* of this fact is very simple, but uses a very important method of "intermediate approximation". Let

$$\|x(\cdot) - y(\cdot)\|_{C(\mathbf{T}^1)} = d(x(\cdot), \mathcal{T}_N, C(\mathbf{T}^1)),\ y(\cdot) \in \mathcal{T}_N.$$

Then, certainly, $(S_N y)(\cdot) = y(\cdot)$ and this means that

$$\begin{aligned} |x(t) - (S_N x)(t) &= |x(t) - y(t) - S_N(x-y)(t)| \\ &\leqslant d(x(\cdot), \mathcal{T}_N, C(\mathbf{T}^1)) + L_N d(x(\cdot), \mathcal{T}_N, C(\mathbf{T}^1)) \\ &\leqslant (1 + L(\mathcal{T}_N))\, d(x(\cdot), \mathcal{T}_N, (\mathbf{T}^1)). \end{aligned}$$

From Lebesgue's inequality it is clear that for high degrees of smoothness (when $d(x(\cdot), \mathcal{T}_N, C(\mathbf{T}^1)) \ll N^{-r}$ for large r) the Fourier method is nevertheless a good tool of approximation.

Thus, the Fourier operator is not suitable for approximation of continuous functions. But what about in $L_p(\mathbf{T}^1)$ for $p < \infty$? In fact in $L_2(\mathbf{T}^1)$ it is extremal!

The answer to this question was given by M. Riesz and we will consider it in the second chapter. In the same place we will answer the question, why the Fejér and Jackson operators, etc., are not suitable for the approximation of smooth functions.

The statement of the question on optimality of a method of approximation requires some restrictions. One must not dream of an absolutely universal method. In some cases for integration one must apply the rectangle formula, in others the Gauss formula is incomparably better. But how should one precisely pose the problem?

It is natural, that we require not individual calculations, but rather "bulk" calculations, where we must calculate the same thing many times (say, to approximate or integrate). And where we have some information on the object of our approximation or calculation. This information selects a class of objects, say, a class of functions. Then there also arises the requirement to calculate quantities of the type $d(W, \mathcal{T}_N, X)$, $\varphi(W, \mathcal{T}_N, X)$ etc., to investigate their asymptotic behaviour etc.

The first statement on the precise asymptotic behaviour of such quantities is due to A.N. Kolmogorov.

Theorem 42 (A.N. Kolmogorov [1985], *N*27).

$$\varphi(BW_\infty^r(\mathbf{T}^1), \mathcal{T}_N, C(\mathbf{T}^1)) = 4\pi^{-2} \ln N N^{-r} + O(N^{-r}).$$

The *proof* of this theorem (we carry out only the upper estimate in the even case $n = 2s$) is based on the following formula

$$x(t) - (S_N x)(t) = \int_{\mathbf{T}^1} \varphi_{2s,N}(t + \tau) x^{(r)}(\tau)\, d\tau,$$
$$(-1)^s \varphi_{2s,N}(t) = \pi^{-1} \sum_{k \geqslant N+1} k^{-2s} \cos kt, \tag{i}$$

proved directly by integration. Applying the Abel transform twice, we obtain:

$$\begin{aligned}\varphi_{2s,N}(t) &= -D_N(t)(N+1)^{-2s} + \pi^{-1} \sum_{k \geqslant N+1} (k^{-2s} - (k+1)^{-2s}) D_k(t) \\ &= -D_n(t)(N+1)^{-2s} - ((N+1)^{-2s} - (N+2)^{-2s})(N+1)F_N(t) \\ &\quad + \sum_{k \geqslant N+1} (k^{-2s} - 2(k+1)^{-2s} + (k+2)^{-2s})(k+1)F_k(t), \end{aligned} \tag{ii}$$

where D_k is the Dirichlet kernel and F_k the Fejér kernel. Using the relations

$$\Delta k^{-r} \asymp k^{-(r+1)}, \qquad \Delta^2 k^{-r} \asymp k^{-(r+2)}, \qquad F_k \geqslant 0,$$
$$\pi^{-1} \int_{\mathbf{T}^1} F_k(t)\, dt = 1, \qquad \sum_{k \geqslant N+1} k^{-(r+1)} = O(N^{-r}),$$

we obtain

$$\|x(\cdot) - (S_N x)(\cdot)\|_{C(\mathbf{T}^1)} \leqslant \int_{\mathbf{T}^1} |\varphi_{2s,N}(t)|\,dt$$

$$= \pi^{-1}(N+1)^{-2s} \int_{\mathbf{T}^1} |D_N(t)|\,dt + O(N^{-r})$$

$$= 4\pi^2 N^{-r} \ln N + O(N^{-r}),$$

as required.

After this result a number of sharp and asymptotically sharp results were obtained for many classes of functions and methods of approximation. On the basic question here we refer to the following chapter, and here we give only some preliminary theorems.

Theorem 43 (Bohr-Favard inequality, Bohr (1935), Favard (1936)). *Let the function $x(\cdot)$ belong to the class $BW_\infty^r(\mathbf{T}^1)$ and be orthogonal to the space $\mathcal{T}_N$. Then there is the sharp inequality*:

$$\|x(\cdot)\|_{C(\mathbf{T}^1)} \leqslant K_r(N+1)^{-r}.$$

(The constants K_r, defined in the introduction, are called the *Favard constants*). In other words:

$$f(BW_\infty^r(\mathbf{T}^1), \mathcal{T}_N, C(\mathbf{T}^1)) = K_r(N+1)^{-r}.$$

Theorem 44 (on best approximation of the class $BW_\infty^r(\mathbf{T}^1)$ by trigonometric polynomials, Favard [1936], N.I. Akhiezer-M.G. Krejn [1937]).

$$d(BW_\infty^r(\mathbf{T}^1), \mathcal{T}_N, C(\mathbf{T}^1)) = \sup\{\|x(\cdot) - \Lambda x(\cdot)|_{C(\mathbf{T}^1)} | x(\cdot) \in BW_\infty^r(\mathbf{T}^1)\} K_r(N+1)^{-r},$$

$\Lambda x(\cdot) = \dfrac{a_0}{2} + \sum_{k=1}^N \lambda_k(a_k \cos kt + B_k \sin kt)$, *where* (a_k, b_k) *are the Fourier coefficients of* $x(\cdot)$.

Theorem 45 (On the approximation of the conjugate class, N.I. Akhiezer-M.G. Krejn [1937]),

$$d(B\tilde{W}_\infty^r(\mathbf{T}^1), \mathcal{T}_N, C(\mathbf{T}^1)) = \tilde{K}_r(N+1)^{-r},$$

$$\tilde{K}_r = (4/\pi) \sum_{k \in \mathbf{Z}_+} (-1)^{kr}(2k+1)^{-(r+1)},$$

(here $B\tilde{W}_\infty^r(\mathbf{T}^1)$ denotes the class of functions $x(\cdot)$, for which the conjugate functions belong to $BW_\infty^r(\mathbf{T}^1)$).

Theorem 46 (on approximation by polynomial functions analytic in ellipses, N.I. Akhiezer [1938]).

$$d(B_{\mathbf{R}_e} A^{E_R}, \mathcal{P}_{N-1}, C(\hat{I})) = 4\pi^{-1} \sum_{k \geqslant 0} (-1)^k (2k+1)^{-1} \frac{R^{(2k+1)N}}{1 + R^{2(2k+1)N}},$$

where $B_{\mathrm{Re}}A^{E_R}$ is the set of real functions of $\hat{I}$ which admit an analytic continuation to an ellipse, whose sum of semiaxes is equal to R, and with real part bounded therein by unity.

Theorem 47 (on approximation of the Hölder class by Fejér sums, S.M. Nikol'skij, 1940).

$$\lambda(BH^{\alpha}(\mathbf{T}^1), (\mathscr{F}_N, \mathscr{T}_N), C(\mathbf{T}^1)) \asymp \begin{cases} 2\Gamma(\alpha)\pi^{-1}(1-\alpha)^{-1}\sin(\alpha\pi/2)N^{-\alpha}, & 0 < \alpha < 1, \\ 2\pi^{-1}\ln N \cdot N^{-1}, & \alpha = 1. \end{cases}$$

Theorem 48 (on approximation by Fourier sums, (S.M. Nikol'skij, 1945). *Let ω be a concave modulus of continuity. Then*

$$\varphi(W^r_\infty H^\omega(\mathbf{T}^1), \mathscr{T}_N, C(\mathbf{T}^1)) = 2\pi^{-2}\ln N \cdot N^{-r}\int_0^{\pi/2}\omega(4t/(2N+1))\sin t\,dt + O(N^{-r}\omega(1/N)).$$

Theorem 49 (on approximation of functions satisfying a Lipshitz condition, by algebraic polynomials, S.M., Nikol'skij, 1946).

$$d(BW^1_\infty(\hat{I}), \mathscr{P}_N, C(\hat{I})) \asymp \pi 2^{-1}N^{-1}.$$

Theorem 50 (on approximation of functions from $B\tilde{W}^r_\infty(\hat{I})$ by algebraic polynomials, S.N. Bernstein, 1947).

$$d(BW^r_\infty(\hat{I}), \mathscr{P}_N, C(\hat{I})) \asymp K_r N^{-r},$$

K_r the Favard constant, $r \geqslant 2$.

Above we have given some preliminary results, in which the functions φ, λ, d, f and b (as functions of the number of parameters N) have been exactly, or exactly up to order of some quantity, calculated for a number of classes of functions of one variable and for the classical methods of approximation—the Fourier, Fejér methods, spaces of trigonometric and algebraic polynomials. All of these results are grouped around the theme: "smoothness and approximation", to which, basically, this article is devoted.

5.5. The Latest Developments in the Constructive Theory of Functions. In the fifties and sixties a number of classical problems of the constructive theory of functions were completely solved. We will also discuss these here. More details can be found in the mongraphs Dzyadyk [1977], Kornejchuk [1976], Timan [1960] and others.

Jackson's first theorem (Theorem 31) was generalized to modulus of continuity of the k-th order.

Theorem 51 (S.B. Stechkin [1951]).

$$/x(\cdot) \in C(\mathbf{T}^1)/ \Rightarrow d(x(\cdot), \mathscr{T}_N, C(\mathbf{T}^1)) \leqslant A\omega_k(N^{-1}, x(\cdot)).$$

Weierstrass-Vallée-Poussin inverse theorem (Theorem 35) was also significantly developed by S.B. Stechkin. Let $\omega(\cdot)$ be a continuous monotone function of modulus of continuity type, vanishing at zero and satisfying the inequality $\omega(2t) \leqslant c\omega(\varepsilon)$ and (for r $\geqslant$ 1) the condition

$$\int_0^1 \omega(t)t^{-1}\,dt < \infty.$$

Theorem 52. $/r \in \mathbf{Z}_+, d(x(\cdot), \mathscr{T}_N, C(\mathbf{T}^1)) \leqslant AN^{-r}\omega(N^{-1})/ \Rightarrow x(\cdot) \in C^r(\mathbf{T}^1)$ *and* $\forall k \in \mathrm{N}$

$$\omega_k(t, x^{(r)}(\cdot)) \leqslant \begin{cases} A_1 A t^k \displaystyle\int_0^1 \omega(u)/u^{k+1}\,du, r = 0, \\ A_1 A\left(t^k \displaystyle\int_0^1 \omega(u)/u^{k+1}\,du + \int_0^t \omega(u)u^{-1}\,du\right), \quad r \geqslant 1. \end{cases}$$

This result implies also Theorem 35 (Weierstrass-Vallée-Poussin) and Theorem 37 (Zygmund). We mention here the refinements of Jacksons theorem given by N.P. Kornejchuk, N.I. Chernykh, V.V. Zhuk and A.A. Ligun. We denote

$$\kappa_{Nr}(X) = N^r \sup\{d(x(\cdot), \mathscr{T}_N, X)/\omega(\pi/N, x^{(r)}(\cdot), X)\}.$$

Theorem 53. a) $1 - (2N)^{-1} \langle \kappa_{N_0}(C(\mathbf{T}^1)) < 1$ (N.P. Kornejchuk, 1962)

b) $\kappa_{N_0}(L_2(\mathbf{T}^1)) = 1/\sqrt{2}$ (N.I. Chernykh, 1967).

c) $\kappa_{N,2r-1}(C(\mathbf{T}^1)) = K_r/2$ $(r = 1$—V.V. Zhuk (1962), $r = 2j - 1$, $j > 1$—A.A. Ligun (1973)).

We pass on to direct and inverse theorems for approximation by algebraic polynomials This theme was hardly mentioned earlier. It must be said that between the periodic and aperiodic (with the apparatus of approximation $\{\mathscr{P}_N\}$) cases there are essential differences. Direct theorems in the classical "periodic" form cannot be inverted.

The first to delve into the causes of this phenomenon was S.M. Nikol'skij. He discovered that if $x(\cdot) \in W^1_\infty(\hat{I})$, then

$$|x(t) - p_N(t)| \leqslant \frac{\pi}{2}\left(\frac{\sqrt{1-t^2}}{N} + O\left(\frac{\ln N}{N^2}\right)\right)$$

for some sequence $\{p_N(\cdot)\}_{N \geqslant 1}$ of polynomials $p_N(\cdot) \in \mathscr{P}_N$. (He also conjectured that $O(\log NN^{-2})$ could be replaced by $O(N^{-2})$.) Thus, the polynomials $p_N(\cdot)$ approximated better at the ends of the interval $\hat{I}$ than at the interior points. Therefore for the inversion of the approximation theorem we must first strengthen the direct theorems. Here the function $\rho_N(t) = ((1-t^2)^{1/2}N^{-1} + N^{-2})$ came to play an important role in all these questions. It has a simple geometric meaning, it is the quantity, equivalent to the distance from the point t to the ellipse with foci ± 1 and sum of semiaxes $1 + N^{-1}$. The next result is an analogue of Theorem 32.

Theorem 54 (A.F. Timan (1951)). $/x(\cdot) \in C^r(I),\ N \geqslant r/ \Rightarrow \exists A > 0,\ p_N(\cdot) \in \mathscr{P}_N$:

$$|x(t) - p_N(t)| \leqslant A\rho_N^r(t)\omega(\rho_N(t), x^{(r)}(\cdot)).$$

Subsequently V.K. Dzyadyk and Freud extended this result to the second, and N.A. Brudnyj to any, modulus of continuity.

From Theorem 53 it follows at once that if $x(\cdot) \in BW_\infty^r H^\alpha(\hat{I})$ and $0 < \alpha < 1$, then there exists $p_N(\cdot) \in \mathscr{P}_N$ such that

$$|x(t) - p_N(t)| \leqslant A\rho_N^{r+\alpha}(t),$$

and if $x(\cdot) \in W_\infty^r Z(\hat{I})$, then

$$|x(t) - p_N(t)| \leqslant A\rho_N^{r+1}(t).$$

Inverse theorems turned out to be essentially more difficult to obtain. V.K. Dzyadyk was the first to obtain inverse theorems of the type Weierstrass-Vallée-Poussin. A form analogous to Theorem 52 was given to this result of A.F. Timan.

Theorem 55. *Let $\omega(\cdot)$ satisfy the same conditions as in Theorem 52, and let there exist a sequence of algebraic polynomials $\{p_N(\cdot)\}_{N \geqslant 1}$* such that

$$|x(t) - p_N(t)| \leqslant A\rho_N^r(t)\omega(\rho_N(t)).$$

Then $x(\cdot) \in C^r(\hat{I})$ and for any $k \geqslant 1$

$$\omega_k(t, x^{(r)}(\cdot)) \leqslant \begin{cases} A_1 A t^k \displaystyle\int_t^1 \omega(u)u^{-(k+1)}\,du, & r = 0,\\[2ex] A_1 A\left(t^k \displaystyle\int_t^1 \omega(u)u^{-(k+1)}\,du + \int_0^t \omega(u)u^{-1}\,du\right), & r \geqslant 1. \end{cases}$$

It must be said that the inequalities for derivatives of polynomials also may be refined.

Theorem 56 (G.K. Lebed' [1957], Yu.A. Brudnyj 1959) $/x(\cdot) \in P_N$ *and* $\exists s > 0$, $k \in N,\ \omega_\kappa(\cdot) \in C(\mathbf{R}_+)$: $\omega_\kappa \uparrow,\ \omega_\kappa(0) = 0$,

$$\omega_k(2t) \leqslant 2^k\omega_k(t), \qquad |x(t)| \leqslant A\rho_N^s(t)\omega_k(\rho_N(t))/ \Rightarrow \exists A_1 = A_1(s, k, v):$$
$$|x^{(r)}(t)| \leqslant A_1 A\rho_N^{s-r}(t)\omega_k(\rho_N(t)).$$

We cite results analogous to the theorems of Lebesgue (Theorem 40) and A.N. Kolmogorov (Theorem 42), relating to Taylor series. Denote by D_∞^r the class of functions analytic inside the unit circle $D = D_1 = \{z: |z| \leqslant 1\}$, whose r-th derivative is bounded on the circle by the constant unity. Each function of this class can be expanded in a Taylor series $x(z) = \sum_{k \geqslant 0} x_k z^k$. We denote by $\tau_N(\cdot, x(\cdot))$ the function $\sum_{k \geqslant N+1} x_k z^k$.

Theorem 57.

$$\sup_{x(\cdot)\in B^r_\infty} \|\tau_N(\cdot, x(\cdot))\|_{C(D)} = \pi^{-1}\log(N+1)(N+1)^{-r} + O((N+1)^{-r}).$$

For $r = 0$ this result was obtained by Landau (1916), for arbitrary $r \geqslant 1$ by S.B. Stechkin [1953].

In conclusion we will mention how the research on the approximation of the clases $BW^r_\infty(\mathbf{T}^1)$, $BW^r_\infty H^\omega(\mathbf{T}^1)$ by trigonometric polynomials was completed.

After the research of Favard-Akhiezer-Krejn the question of approximation of the classes $BW^r_\infty(\mathbf{T}^1)$ for fractional r remained unsolved. The solution of this problem for $0 < r < 1$ was given by V.K. Dzyadyk (1953). After six years the whole problem was solved by V.K. Dzyadyk and Sun Yun Shen. We formulate their theorem

Theorem 58.

$$d(BW^r_\infty(\mathbf{T}^1), \mathscr{T}_N, C(\mathbf{T}^1)) = K_r(\beta)(N+1)^{-r},$$

where

$$K_r(\beta) = \frac{4}{\pi}\left|\sum_{k\geqslant 0} \frac{\sin((2k+1)\beta - \pi r/2)}{(2k+1)^{r+1}}\right|,$$

where $\beta = 0$ for $0 < r \leqslant 1$ and β is a root of the equation

$$\sum_{k\geqslant 0} \cos((2k+1)\beta - \pi r/2)(2k+1)^{-r} = 0$$

for $r > 1$.

To solve this problem V.K. Dzyadyk had to devise essentially new methods of investigation of approximation by trigonometric polynomials of classes of functions, given by convolutions with kernels which are integrals of absolutely monotone functions.

The approximation problem in the class $BW^r_\infty H^\omega(\mathbf{T}^1)$ for convex $\omega(\cdot)$ was solved by Kornejchuk in a cycle of works in the sixties. We cite the definitive result.

Theorem 59. *Let $\omega(\cdot)$ be a convex modulus of continuity, $r \in \mathbf{Z}_+$. Then*

$$d(W^r_\infty H^\omega(\mathbf{T}^1), \mathscr{T}_N, C(\mathbf{T}^1)) = \|\tilde{x}_{N+1,r,\omega}(\cdot)\|_{C(\mathbf{T}^1)},$$

where the function $\tilde{x}_{n,r,\omega}(\cdot)$ is defined by induction: $\tilde{x}_{n,0,\omega}(\cdot)$ is the $2\pi n^{-1}$-periodic function odd function equal to $\omega(2t)/2$, for $0 \leqslant t \leqslant \pi(2n)^{-1}$ and $\omega(2(\pi n^{-1} - t))/2$ for $\pi(2n)^{-1} \leqslant t \leqslant \pi n^{-1}$,

$$\tilde{x}_{nr\omega}(t) = \begin{cases} \displaystyle\int_{\pi/2n}^t \tilde{x}_{n,r-1,\omega}(\tau)\,d\tau, & r = 1, 3, 5, \ldots, \\ \displaystyle\int_0^t \tilde{x}_{n,r-1,\omega}(\tau)\,d\tau, & r = 2, 4, 6, \ldots. \end{cases}$$

This result can be deduced from the Favard-Akhiezer-Krejn theorem and the following fact on the smoothing of the class $W_\infty^r H^\omega(\mathbf{T})$ by the class $KBW_\infty^{r+1}(\mathbf{T})$.

Theorem 60 (N.P. Kornejchuk).

$$d(W_\infty^r H^\omega(\mathbf{T}^1), \quad KBW_\infty^{r+1}(\mathbf{T}^1), \quad C(\mathbf{T}^1))$$
$$\leqslant 2^{-1} \max_{0 \leqslant a \leqslant \pi} \int_0^a \Phi_{a,r-1}(a-t)(\omega(t) - Kt)\, dt,$$

where $\Phi_{a,0}(t) = 1/2$ *for* $0 \leqslant t \leqslant a$ *and zero for* $t \geqslant a$,

$$\Phi_{a,r}(t) = \begin{cases} 2^{-1} \displaystyle\int_0^{a-1} \Phi_{a,r-1}(\tau)\, d\tau, & 0 \leqslant t \leqslant a, \\ 0, \quad t \geqslant a, \quad r \in \mathrm{N}. \end{cases}$$

Here, if K is chosen so that the maximum relative to a is attained at the points $a = \pi/n$, then equality holds.

These results are amongst the most difficult in approximation theory. In order to obtain them the authors had to closely examine some delicate properties of permutations of functions and investigate by specific methods a series of extremal properties of differentiable functions.

The results of N.P. Kornejchuk on this circle of questions laid the foundation of an entire direction in approximation theory. Among the most active mathematicians working on extremal problems of the constructive theory of functions we name V.F. Babenko, V.L. Velikin, V.G. Doronin, A.A. Zhensykbaev, A.A. Ligun, V.P. Motornij, V.I. Ruban (this list can be extended). In what follows we will only be able partially expose the achievements of this direction. Most of the principal results are presented in the survey articles and monographs Zhensykbaev [1976], Kornejchuk [1976, 1984, 1985], Kornejchuk, Ligun & Doronin [1982].

Chapter 2
Classical Methods of Approximation

In this chapter we will continue the discussion of the questions concerning the approximation of smooth functions by the classical means of approximation. These means are described in § 1. In § 2 we discuss the content of the term "smoothness". § 3 is devoted to the problem of approximation of smooth functions by subspaces generated by harmonics, § 4 to the approximation by summation methods for Fourier series, §§ 5 and 6 to approximation by rational functions and splines.

§ 1. Methods of Approximation

Here we return to the question mentioned in Chap. 1, § 1 and consider the classical means of approximation: algebraic and trigonometric polynomials, rational functions, and also splines.

1.1. Algebraic Polynomials. *Algebraic polynomials* are functions of the form

$$p(t) = \sum_{k=0}^{N} a_k t^k.$$

If $a_N \neq 0$, then N is called the *degree of the polynomial* and is denoted by $\deg p$. The collection of polynomials of degree $\leqslant N$ is denoted $\mathscr{P}_N$.

Polynomials, as a means of approximation, arose at the time of the origin of analysis. They are very simple for calculation, requiring only addition, subtraction and multiplication. We quote an important formula

Theorem 1. *Let* $\Delta = [t_0, t_1], |\Delta| =: |t_0 - t_1| < \infty, x(\cdot) \in C^{r-1}(\Delta)$ *and let* $x^{(r-1)}$ *be absolutely continuous* ($\Leftrightarrow x^{(r-1)}(\cdot) \in AC(\Delta)$). *Then*

$$x(t) = \left(\sum_{k=0}^{r-1} (x^{(k)}(t_0)/k!)(t - t_0)^k \right)$$
$$+ ((r - 1)!)^{-1} \int_{t_0}^{t_1} (t - \tau)_+^{r-1} x^{(r)}(\tau)\, d\tau. \tag{1}$$

Polynomials of the form

$$\sum_{k=0}^{m} (x^{(k)}(t_0)/k!)(t - t_0)^k =: T_m(t, t_0, x(\cdot))$$

are called *Taylor polynomials*. Taylor polynomials as a means of approximation arose in the works of Newton. They are the polynomials having highest order of tangency with $x(\cdot)$ at $t = t_0$.

Formula (1) is obtained by term-term integration. It implies, in particular, that if $u(\cdot)$ is absolutely continuous and

$$x(t) = ((r - 1)!)^{-1} \int_{t_0}^{t} (t - \tau)_+^{r-1} u(\tau)\, d\tau, \tag{1'}$$

then $x(t_0) = \cdots = x^{(r-1)}(t_0) = 0$, $x^{(r)}(t) = u(t)$ almost everywhere. Formula (1′) is called the *inversion formula of r-times differentiation* (it was mentioned in Chap. 1, § 2).

Taylor polynomials are interpolation polynomials, interpolating $x(\cdot)$ with respect to the values $\{x^{(j)}(t_0), 0 \leqslant j \leqslant r - 1\}$. They are a very convenient method of approximation: if it is known that $|x^{(r)}(t)| \leqslant M$, then $\|x(\cdot) - T_r(\cdot, t, x(\cdot))\| \leqslant M|\Delta|^r/r!$. They approximate *analytic functions*, that is functions which can be expanded in a Taylor series, particularly well. Let $x(\cdot)$ be analytic at zero, that is,

$$x(z) = \sum_{k \geqslant 0} x_k z^k, \qquad x_k = ((k-1)!)^{-1} x^{(k)}(0). \tag{2}$$

Then there is a mapping $x(\cdot) \to \{x_k\}_{k \in Z_+}$. An important idea which has been strongly developed in approximation theory is "to invert" similar associations, that is to extract information on $x(\cdot)$ from $\{x_k\}_{k \in Z_+}$ and other similar sequences.

Theorem 2. *If $x(\cdot)$ is analytic at zero, then the sequence $T_N(\cdot, 0, x(\cdot))$ of Taylor polynomials converges to $x(\cdot)$ inside the disc $D = \{z: |z| \leqslant R\}$, where R is defined by* $\limsup_{k \to \infty} |x_k|^{1/k} = R^{-1}$ *(the Cauchy-Hadamard formula). Outside D the series diverges. On the circle $K = \{z: |z| = R\}$, $0 < R < \infty$, the function has a singularity.*

Here is a typical result in approximation theory: from some approximative functional on a function we infer global properties. The usual sequence of functionals for approximation theory is $x(\cdot) \to d(x(\cdot), \mathscr{P}_N, X)$, where X is some function space.

Using polynomials we produce a *smoothing* of a functions, for example, by means of their integration. But here we must keep in mind that for a function of finite smoothness the most flexibility and adaptability is attained not by polynomials, but by piecewise-polynomial structures–*splines*. Polynomials as a means of approximation are optimal only for certain special classes of analytic functions, say, Taylor polynomials for the approximation of functions analytic in a disc and analytically continuable into a large disc. We will constantly return to this theme of comparison of approximation by algebraic polynomials with other methods.

1.2. Rational Functions. A *rational function* is the ratio of two polynomials: $r(\cdot) = p(\cdot)/q(\cdot)$, $p(\cdot) \in \mathscr{P}_m$, $q(\cdot) \in \mathscr{P}_n$. The totality of such function is denoted $\mathscr{R}_{mn}$.

One very popular method of approximation by analytic functions is *Padé approximations*—an analogue in some sense of Taylor approximation.

Let $x(\cdot)$ be analytic at zero and (2) its Taylor polynomial. The Padé approximation of type (mn) of $x(\cdot)$ is the rational function $\pi_{mn}(\cdot) \in \mathscr{R}_{mn}$, having largest order of tangency with $x(\cdot)$ at zero. In other words, $\pi_{mn}(\cdot) = p(\cdot)/q(\cdot)$ is defined by the condition

$$(qx - p)(z) = \alpha_{mn} z^{m+n+1} + \cdots, \qquad \deg p \leqslant m, \qquad \deg q \leqslant n. \tag{3}$$

It is easy to see that for fixed m and n the function $\pi_{mn}(\cdot, x(\cdot)) = \pi_{mn}(\cdot)$ is uniquely defined and its calculation is algorithmically simple—it is necessary to solve a system of linear equations. Thus there arises the *Padé table* $\{\pi_{mn}(\cdot)\}_{m,n}$. Sequences of the form $\{\pi_{mn}(\cdot)\}_{m \geqslant 0}$ are called the *rows of the Padé table*. The zero row consists of the Taylor polynomials: $\pi_{m0}(\cdot, x(\cdot)) = T_m(\cdot, 0, x(\cdot))$. The functions $\pi_{mn}(\cdot)$, like the Taylor polynomials, have the interpolation property. They are also an excellent means of approximation, particularly of the elementary functions.

The general problem of interpolation (in particular, by rational fractions) was first considered by Cauchy (1821). The functions $\pi_{mn}(\cdot)$ were introduced by Padé

(1892), although they can be found earlier in the works of Frobenius, Chebyshev, Stieltjes, etc. devoted to the expansion of a function in continued fractions. In this way we obtain a mapping $x(\cdot) \to \{\pi_{mn}(\cdot, x(\cdot))\}$. Connected with $\mathcal{R}_{mn}$ there is another "table": $x \to \{d(x(\cdot), \mathcal{R}_{mn}, X)\}$.

We denote by $\rho_m(x(\cdot))$ the radius of the maximum disc, to which $x(\cdot)$ can be extended as a meromorphic function with $\leqslant m$ poles. As the Taylor series (the zero row of the Padé table) characterises $\rho_0(x(\cdot))$ (Theorem 2), so the m-th row characterises $\rho_m(x(\cdot))$.

Theorem 3 (Hadamard [1892]). $\rho_m(x(\cdot)) = l_{m-1}/l_m$,

$$l_m = \limsup_{n\to\infty} |\Delta_{mn}|^{1/n},$$

where

$$\Delta_{mn} = \begin{vmatrix} x_m x_{m-1} \dots x_{m-n} \\ x_{m+n} \dots x_m \end{vmatrix}.$$

Theorem 4 (Fabry [1898]). *Let the coefficients of the series* (2) *satisfy*

$$\lim_{m\to\infty} (x_m/x_{m+1}) = \alpha.$$

Then α is a singular point of $x(\cdot)$.

Theorem 5 (Montessus de Ballore [1902]). *Let $x(\cdot)$ be analytic in a neighbourhood of zero and interior to the disc $D_{\rho_n(x(\cdot))}$ have exactly n poles. Then the poles of the functions $\{\pi_{mn}(\cdot, x(\cdot))\}_{m\geqslant 0}$ tend to the poles of $x(\cdot)$, situated in $D_{\rho_n(x(\cdot))}$*

Here are two more examples, where relative to a sequence of approximations it is possible to catch the singularities of functions: in the Fabry theorem the singularity shows from the Padé second row (its denominator is $x_{m+1}z - x_m$), and in the Montessus de Ballore from the Padé m-th row.

It is clear from Theorems 3–5 how natural it is to consider rational fractions as the tool of approximation for meromorphic functions. But in fact their value is immeasurably broader.

Rational functions require for their calculation the execution of only a finite number of arithmetic operations. This is one of the reasons why rational functions play such a notable role in applied analysis. But there is another reason, more closely connected with approximation theory: rational functions very well, and in a number of cases optimally, approximate analytic functions (see Chap. 3, § 3).

1.3. Trigonometric Polynomials and Some Polynomial Operators. The role and origin of trigonometric polynomials, unlike that of the ordinary polynomials and rational fractions, is "nonarithmetic": they are related to the group structure of the circle, namely to the group of its shifts. Special functions, similar to sines and cosines, arise on any manifold with a sufficiently rich group of motions: spheres, tori, projective spaces, Lobachevskii spaces, etc. We will touch on this theme in Chap. 4, § 2.

The circle $\mathbf{T}^1$ can be realised as the interval $[-\pi, \pi]$, with the points $\pm\pi$ identified. $\mathbf{T}^1$ is a compact manifold on which the group of shifts $t \to t + h$ acts, generating a symmetric invariant differential operator d^2/dt^2. The eigenfunctions of this operator are the exponentials $\{\exp(int)\}_{n \in Z}$, the eigenvalues the numbers $\{-n^2\}_{n \geqslant 0}$, where zero is one-multiple, and all the others are two-multiple points of the spectrum. In the real case the points of the spectrum $-n^2$ correspond to a two-dimensional subspace $\text{lin}\{\sin nt, \cos nt\} = L_n$. The systems of functions $\{\exp(int)\}_{n \in Z}$ and $\{\cos nt, \sin nt\}_{n \geqslant 0}$ form complete orthogonal systems and $L_2^C(\mathbf{T}^1)(L_2(\mathbf{T}^1))$ decomposes into a direct orthogonal sum $L_n^C = \text{lin}\{\exp(int)\}$, $n \in \mathbf{Z}$, and L_n, $n \geqslant 0$. This is expressed by

Theorem 6. *Let $x(\cdot) \in L_2^C(T)(L_2(T))$. Then $x(\cdot)$ decomposes into a Fourier series*

$$x(t) = \sum_{k \in \mathbf{Z}} x_k \exp(ikt),$$
$$\left(x(t) = (a_0/2) + \sum_{k \in \mathbf{N}} (a_k \cos kt + b_k \sin kt) \right), \tag{4}$$

$$x_k = (2\pi)^{-1} \int_{\mathbf{T}^1} x(t) \exp(-ikt)\, dt,$$
$$\left(a_k = \pi^{-1} \int_{\mathbf{T}^1} x(t) \cos kt\, dt, \qquad b_k = \pi^{-1} \int_{\mathbf{T}^1} x(t) \sin kt\, dt \right) \tag{5}$$

and Parseval's equality holds:

$$\int_{\mathbf{T}^1} |x(t)|^2\, dt = 2\pi \sum_{k \in \mathbf{Z}} |x_k|^2$$
$$\left(\int_{\mathbf{T}^1} x^2(t)\, dt = \pi \left((a_0^2/2) + \sum_{k \geqslant 1} (a_k^2 + b_k^2) \right) \right). \tag{6}$$

Thus there is a mapping $x(\cdot) \to \{x_k\}_{k \in Z}(\{a_k, b_k\}_{k \in Z_+})$. Relative to the sequence $\{x_k\}_{k \in Z}$, certainly all the singularities of $x(\cdot)$ can be recovered. We have already spoken on this. But nevertheless we state one result (following at once from Theorem 6).

Theorem 7. *In order that $x(\cdot)$ be $(r-1)$-times differentiable $(x^{(r-1)}(\cdot) \in AC(T^1))$ and its r-th derivative belongs to $L_2^C(\mathbf{T}^1)$, it is necessary and sufficient that $\sum_{k \in Z} k^{2r} |x_k|^2 < \infty$.*

This theorem is one verification that in Hilbert spaces, for classes of functions which are shift invariant, the Fourier method is an optimal means of approximation and description of functions.

The choice of harmonics for the first part of the spectrum is quite natural. Thus there arise the spaces of trigonometric polynomials $\mathcal{T}_N^C = \text{lin}\{\exp(ikt)\colon |k| \leqslant N\}$

and $\mathscr{T}_N = \operatorname{lin}\{\cos kt, \sin kt: 0 \leqslant k \leqslant N\}$. It is natural to expect that the spaces $\mathscr{T}_N^{\mathrm{C}}$ and $\mathscr{T}_N$ have good approximation properties relative to classes of functions invariant relative to the shifts.

The operator

$$S_N^{\mathrm{C}}: L_2^{\mathrm{C}}(\mathbf{T}^1) \to \mathscr{T}_N^{\mathrm{C}}(S_N: L_2(\mathbf{T}^1) \to \mathscr{T}_N),$$

acting by the rule

$$(S_N^{\mathrm{C}}x)(t) = \sum_{|k| \leqslant N} x_k \exp(ikt)\left((S_N x)(t) = \sum_{k=0}^{N} (a_k \cos kt + b_k \sin kt)\right),$$

is called the *Fourier operator*.

Theorem 8. *The Fourier operator (in the real case) admits the following representation*:

$$(S_N x)(t) = \pi^{-1} \int_{\mathbf{T}^1} x(t + \tau) D_N(\tau)\, d\tau,$$

where the kernel $D_N(\cdot)$ (called the Dirichlet kernel) is defined by the equality

$$D_N(t) = (1/2) + \sum_{k=1}^{N} \cos kt = \sin((N + (1/2))t)/(2 \sin t/2).$$

The operator

$$(\mathscr{F}_N x)(\cdot) = \left(\sum_{k=0}^{N-1} (S_k x)(\cdot)\right) N^{-1}$$

(the arithmetic mean of the Fourier operators) is called the *Fejér operator*, and

$$(\mathscr{U}_{MN} x)(\cdot) = \left(\sum_{k=M}^{N-1} (S_k x)(\cdot)\right)(N - M)^{-1}, \qquad N > M,$$

the *Vallée-Poussin operator*.

Theorem 9 (the integral representation of the Fejér operator).

$$(\mathscr{F}_N x)(t) = \pi^{-1} \int_{\mathbf{T}^1} F_N(t + \tau) x(\tau)\, d\tau, \tag{8}$$
$$F_N(t) = (\sin^2(Nt/2))/2N \sin^2(t/2).$$

We give the periodic analogue of Theorem 1, that is, we write explicitly the integral operator, inverse on $\mathbf{T}^1$ to the operator $(d/dt)^r$.

Theorem 10. *Let $x(\cdot) \in C^{r-1}(T^1)$, $x^{(r-1)} \in AC(T^1)$. Then*

$$x(t) = c_0 + \int_{\mathbf{T}^1} \varphi_r(t + \tau) x^{(r)}(\tau)\, d\tau =: c_0 + (\varphi_r * x^{(r)})(\cdot) \tag{9}$$

$$\varphi_r(t) = \pi^{-1} \sum_{k \in \mathbf{N}} \cos(kt - \pi r/2) \cdot k^{-r} (r \in \mathbf{N}). \tag{10}$$

The function $\varphi_r(\cdot)$ satisfies the equation

$$\varphi_r^{(r)}(t) = \delta(t) - 1/2\pi,$$

where $\delta(\cdot)$ is the Dirac delta function (from which, properly, formulae (9) and (10) also follow). The restriction of $\varphi_r(\cdot)$ to $[0, 2\pi]$ is a polynomial, called the *Bernoulli polynomial.*

An important role in what follows will be played by the functions $\tilde{x}_{nr}(t) = (\varphi_r * \operatorname{sign}\sin n)(t)$. They are called the *Favard functions.* We give their Fourier series expansion:

Theorem 11.

$$\tilde{x}_{nr}(t) = 4\pi^{-1} \sum_{k \in \mathbf{N}} \sin((2k-1)nt - \pi r/2)(2k-1)^{-(r+1)}; \tag{11}$$

here $\|\tilde{x}_{nr}(\cdot)_{C(\mathbf{T}^1)} = K_r \cdot n^{-r}$, *and*

$$K_r = 4\pi^{-1} \sum_{k \in \mathbf{N}} (-1)^{(k+1)(r+1)}(2k-1)^{-(r+1)},$$

$$K_0 = 1, \qquad K_1 = \pi/2, \quad K_2 = \pi^2/8, \ldots$$

We have already come across the functions $\varphi_r(\cdot)\tilde{x}_{nr}(\cdot)$ and the numbers K_r in the preceeding chapter.

Here we have briefly spoken on harmonic analysis on the circle. In a similar way we may describe the analogues of the functions $\varphi_r(\cdot)$, $\tilde{x}_{nr}(\cdot)$ etc. on other manifolds: $\mathbf{R}^n$, $\mathbf{S}^n$, $\mathbf{T}^n$, etc.

We will be constantly dealing with harmonic analysis throughout this article.

1.4. Splines. The term "spline" is not as common, as "polynomial" or "rational function", but its content is certainly well known to all. A spline is a piecewise polynomial function. Amongst splines are the piecewise-constant, piecewise-linear (polygonal), piecewise-parabolic etc., functions. Splines are as much a natural apparatus of approximation, smoothing and recovery of functions of finite smoothness, as those which we must now regard as the classical means of approximation. This is even justified historically, since splines arose as long ago as the dawn of analysis—by Leibniz and Euler—in the development of direct methods of variational calculus (we recall the "Euler polygon").

The word "spline" denotes a beam or rod. American draughtsmen and engineers have long used flexible rods for approximate interpolation.

A function $x(\cdot)$, given on a finite interval $\Delta = [t_0, t_1]$ is called a *spline*, if there is a partition R_n of Δ into intervals $\Delta_i = [\tau_i, \tau_{i+1}]$, $i = 1, \ldots, n$, $\tau_0 = t_0$, $\tau_{n+1} = t_1$ such that the restriction of $x(\cdot)$ to Δ_i is a polynomial of degree m_i: $x|_{\Delta_i}(\cdot) \in \mathcal{P}_{m_i}$. The points τ_i, $i = 1, \ldots, n$ are called the *nodes of the spline*, and the number $m = \max m_i$ its *order*. The quantity k_i is called the *defect of a spline* $x(\cdot)$ *of order* m *at a point* τ_i, if $x^{(j)}(\cdot)$ is continuous at τ_i for $0 \leqslant j \leqslant m - k_i$, and $x^{(m-k_i+1)}(\cdot)$ has a discontinuity there. The number $k = \max k_i$ is called the *defect of the spline.* The collection of splines of order m of defect k for the partition R_n is denoted

$S_m^k(\Delta, R_n)$. In particular $S_0^1(\Delta, R)$ and $S_1^1(\Delta, R)$ are the piecewise-constant functions and the polygons connected with the partition R_n.

Sometimes the term spline is used only for piecewise-polynomial functions of defect one. These we denote by $S_m(\Delta, R_n)$.

A spline $x(\cdot) \in S_m(\Delta, R_n)$, for which $|x^{(m)}(t)| \equiv 1$, $t \neq \tau_i$ and at each τ_i the sign changes is called *ideal* (or *complete* or *perfect*). A spline for which $x^{(m)}(t) \equiv \text{const}$, $t \neq \tau_i$ is called a *monospline*. In the approximation theory of periodic functions an important role is played by splines of order m of defect one relative to the uniform partition of $\mathbf{T}^1$ into $2n$ parts. We denote them by $\hat{S}_{2n,m}(\mathbf{T}^1)$.

There is the following representation theorem:

Theorem 12.

$$x(t) = \sum_{s=0}^{m} a_s(t - t_0)^s + \sum_{i=1}^{n-1} \sum_{j=0}^{k_i-1} a_{ij}(t - \tau_i)_+^{m-j}, \tag{12}$$

where τ_i are the nodes of $x(\cdot)$ with defects k_i, $0 \leqslant k_i \leqslant m + 1$, $a_s = x^{(s)}(t_0)/s!$, $s = 0, 1, \ldots, m$, $a_{ij} = (x^{(m-j)}(\tau_i + 0) - x^{(m-j)}(\tau_i - 0))/(m - j)!$, $i = 1, \ldots, n - 1$, $j = 0, \ldots, k_i - 1$. Here any function of the form (12) *belongs to $S_m^k(\Delta, R_n)$.*

It follows from Theorem 12 that $\dim S_m^k(\Delta, R_n) = nk + m - k + 1$.

The set of all splines of order m with N parameters we will denote $S_{Nm}(\Delta)$.

If $x(\cdot) \in S_m^k(\Delta, R_n)$, $\Delta = [-\pi, \pi]$ satisfies the conditions $x^{(s)}(-\pi) = x^{(s)}(\pi)$, $s = 0, 1, \ldots, m - k$, then it is called a *periodic spline* from $S_m^k(\mathbf{T}^1, R_n)$. An example of a periodic spline is the *Bernoulli monospline* mentioned above. This function has order r, defect 2 and a unique node in the period at the point 0. We give a theorem on the representation of periodic splines.

Theorem 13. $/x(\cdot) \in S_m^k(\mathbf{T}^1, R_n)/ \Rightarrow$

$$x(t) = c + \sum_{i=1}^{n} \sum_{j=0}^{k_i-1} c_{ij}\varphi_{m-j+1}(t - \tau_j), \tag{13}$$

where τ_i are the nodes of $x(\cdot)$ with defects k_i, $0 \leqslant k_i \leqslant m + 1$, $c_{ij} = x^{(m-j)}(\tau_i + 0) - x^{(m-j)}(\tau_i - 0)$, where $0 \leqslant j \leqslant k_{i-1}$, $1 \leqslant i \leqslant n$,

$$c = (2\pi)^{-1} \int_{\mathbf{T}^1} x(t)\, dt, \qquad \sum_{i=1}^{n} c_{i0} = 0. \tag{14}$$

Here any function of the form (13) *with $c_{ik_i-1} \neq 0$ and satisfying* (14) *belongs to $S_m^k(\mathbf{T}^1, R_n)$.*

We recall that the function $\varphi_r(\cdot)$ was constructed in the previous subsection. The restriction of $\varphi_r(\cdot)$ to $[-\pi, \pi]$ is a monospline ($\varphi_r^{(r-1)}(t) \equiv 1$, $t \neq 0$) and hence the functions $\varphi_r(\cdot)$ are also called *Bernoulli monosplines*. We note that the Favard functions, also constructed in the previous subsection, are ideal splines. In the literature they are sometimes called *Euler ideal splines*.

1.5. Interpolation. Interpolation in the classical sense of the term is the approximate recovery of a function from its values and the values of its derivatives at some given points. Consider a function of one variable.

Given a table $T = \begin{pmatrix} \tau_1, \ldots, \tau_s \\ \nu_1, \ldots, \nu_s \end{pmatrix}$, where $\nu_i \in Z_+, t_0 \leqslant \tau_1 < \cdots < \tau_s \leqslant t_1$ and some set of functions $\mathscr{A}(\mathscr{P}_N, \mathscr{T}_N, \mathscr{R}_{mn}, S_{Nm}(\Delta)$, etc.). To interpolate a given sufficiently smooth function $x(\cdot)$ with respect to T by functions from the given set means to find a function $y(\cdot) \in \mathscr{A}$, which satisfies the relations: $x^{(j)}(\tau_i) = y^{(j)}(\tau_i), 0 \leqslant j \leqslant \nu_i$, $i = 1, \ldots, s$.

Interpolation by algebraic polynomials was applied in very first works on mathematical analysis. The simplest case is when T has the form:

$$T = \begin{pmatrix} \tau_1, \ldots, \tau_n \\ 0, \ldots, 0 \end{pmatrix}.$$

Then it is necessary to find a polynomial $p_n(\cdot) \in \mathscr{P}_n$, satisfying the relations: $p_n(\tau_i) = x(\tau_i), 0 \leqslant i \leqslant n$. The solution of this problem is provided by the *Lagrange polynomial*:

$$p_n(t) = \sum_{k=0}^{n} x(\tau_k) \frac{(t - \tau_0)\ldots(t - \tau_{k-1})(t - \tau_{k+1})\ldots(t - \tau_n)}{(\tau_k - \tau_0)\ldots(\tau_k - {}_{k-1})(\tau_k - \tau_{k+1})\ldots(\tau_k - \tau_n)}.$$

Newton proposed another form of notation for the interpolating polynomial, which turned out to be very suitable in calculations:

$$p_n(t) = x(\tau_0) + (t - \tau_0)x(\tau_0, \tau_1) + (t - \tau_0)(t - \tau_1)x(\tau_0, \tau_1, \tau_2) + \cdots$$
$$+ (t - \tau_0)\ldots(t - \tau_{n-1})x(\tau_0, \tau_1, \ldots, \tau_n).$$

Here $x(\tau_0, \ldots, \tau_k)$ is the k-th divided difference.

Cauchy gave general formulae of interpolating type for rational fractions from $\mathscr{R}_{MN}$. A particular case being the formulae for Padé approximations. In conclusion a few words about interpolation by splines.

Let $\Delta = [t_0, t_1]$ be a finite interval, $T = \begin{pmatrix} \tau_1, \ldots, \tau_s \\ \nu_1, \ldots, \nu_s \end{pmatrix}$, $t_0 < \tau_1 < \cdots < \tau_s < t_1$, $0 \leqslant \nu_1 \leqslant m - k$ and in addition choose two subsets I_0 and I_1 of the set $\mathscr{Z}_{m+1} = \{0, 1, \ldots, m\}$. We pose the problem of finding a spline $s(\cdot)$ from $S_m^k(\Delta, R_n)$, solving the interpolation problem

$$S^{(j)}(\tau_i) = \xi_{ij}, \qquad 1 \leqslant i \leqslant s, \qquad 0 \leqslant j \leqslant \nu_i,$$

and satisfying the boundary conditions:

$$S^{(j)}(t_i) = \xi'_{ij}, \quad j \in I_i, \quad i = 0, 1.$$

More generally the number of interpolating conditions is equal to $\sum_{i=1}^{s} (\nu_i + 1) + l$, where l is the number of boundary conditions. The dimension of $S_m^k(\Delta, R_n)$ is equal, as we recall, to $nk + m - k + 1$. We denote by β_{0i} and β_{1i} the number of interpolation conditions on $s^{(j)}(\cdot), 0 \leqslant j \leqslant m$, in the semi-intervals

$[t_0, t_1)$ and $(\tau_i, t_1]$, and by γ_i the number of interpolation conditions for $s^{(j)}(\cdot)$ on Δ. Then

Theorem 14. *Let $m \in N$ and for given k, $1 \leqslant k \leqslant m$, table T and sets I_0, I_1 let there exist an $N \geqslant 2$ such that:*

$$\sum_{i=1}^{s} (v_i + 1) + l = Nk + m - k + 1,$$

$$\beta_{0i} \geqslant ik, \quad \beta_{1i} \geqslant (N - i)k, \quad i = 1, \dots, s,$$

$$\sum_{j=1}^{\mu} v_j \geqslant \mu + 1, \quad \mu = 0, \dots, m.$$

Then for any set $\Xi = \{\xi_{ij}\}, i = 1, \dots, s, 0 \leqslant j \leqslant v_i - 1, \{\xi'_{ij}\}, i = 0, 1, j \in I_i$, there is a unique spline $s(\cdot) \in S_m^k(\Delta, R_N)$, satisfying the interpolation and boundary conditions, generated by the set Ξ.

A similar result is true for the periodic case.

Interpolation by polynomials, rational functions and splines are very common examples of approximation and recovery of functions.

§2. Spaces of Smooth Functions

In this section we define the spaces and classes of Sobolev and Nikol'skij. It is mainly on these that we will subsequently demonstrate the methods of approximation theory.

2.1. Smoothness and Classes of Smooth Functions. To give an exhaustive answer to the question of what is smoothness is very difficult. There are various approaches to the definition of this notion. If a function $x(\cdot)$ of one variable has a derivative of order $r \in \mathbf{N}$, then it is natural to say that it has smoothness r. Here smoothness is defined as an integer. But in many cases it is essential to know to which space (or to which class of functions) the r-th derivative of $x(\cdot)$ belongs. Then smoothness must be characterised by a pair: the number r and the class of functions to which $x^{(r)}(\cdot)$ belongs. In addition, for many reasons, it is natural to consider not just integer values of smoothness, but is also useful to consider smoothness of fractional order. There are several ways to introduce fractional smoothness.

1) Let

$$x(t) = \sum_{k \in \mathbf{Z}} x_k \exp(ikt), \qquad x_0 = 0,$$

where equality is in the sense of generalised functions, and let $\alpha \in \mathbf{R}$. *The Weil fractional derivative* (see Weil [1917]) *of order* α is defined as

$$(D^\alpha x)(t) = \sum_{k\in\mathbf{N}} (ik)^\alpha x_k \exp(ikt), \qquad (ik)^\alpha = |k|^\alpha \exp(i\pi 2^{-1}\alpha \operatorname{sign} k).$$

In other words, we can take as a starting point the formulae (9) and (10) of the preceeding subsection and put

$$x(t) = \int_{\mathbf{T}^1} \varphi_\alpha(t+\tau)u(\tau)\,d\tau, \qquad \varphi_\alpha(t) = \pi^{-1} \sum_{k\in\mathbf{N}} \cos(kt - \pi\alpha/2)k^{-\alpha}. \tag{16}$$

Then $(D^\alpha x)(t) = u(t)$, that is, $u(\cdot)$ is the Weil α-th derivative. This definition leads to the following class of *Sobolev spaces*

$$\mathring{W}_p^\alpha(\mathbf{T}^1) = \{x(\cdot) | D^\alpha x(\cdot) \in L_p(\mathbf{T}^1)\}.$$

Here the zero over the W means that $\int_{\mathbf{T}^1} x(t)\,dt = 0$. We denote by $B\mathring{W}(\mathbf{T}^1)$ the set

$$\{x(\cdot) \in \mathring{W}_p^\alpha(\mathbf{T}^1) | \, \|D^\alpha x(\cdot)\|_{L_p(\mathbf{T}^1)} \leqslant 1\}.$$

2) Let us define *smoothness* in the *Hölder-Bernstein-Jackson-Nikol'skij sense*. The naturality of this method of introducing smoothness follows from the considerations of the preceeding chapter. Let $\alpha > 0, \alpha = r + \beta, r \in \mathbf{Z}_+, 0 < \beta \leqslant 1$. We say that a function $x(\cdot)$ belongs to $H_p^{\alpha,k}(\mathbf{T}^1)$, if $x^{(r)}(\cdot) \in L_p(\mathbf{T}^1)$ and

$$\|\Delta_h^k x^{(r)}(\cdot)\|_{L_p(\mathbf{T}^1)} \ll |h|^\beta \ \forall h \in \mathbf{R}.$$

Here Δ_h^k is k-th difference operator with step h. The spaces $H_p^{\alpha,k}(\mathbf{T}^1)$ are called *Nikol'skij spaces*. The space $H_p^{\alpha,1}(\mathbf{T}^1)$ we denote simply by $H_p^\alpha(\mathbf{T}^1)$' Here smoothness is defined by pair $(1/p, \alpha) \in I \times \mathbf{R}_+$. By $BH_p^\alpha(\mathbf{T}^1)$ we denote the set

$$\{x(\cdot) | \, \|\Delta_h x^{(r)}(\cdot)\|_{L_p(\mathbf{T}^1)} \leqslant |h|^\beta\}.$$

In our approaches each function has a whole spectrum of smoothness. For example the function $t \to |t|$ belongs not only to $H_\infty^1(\mathbf{T}^1)$, (that is, has smoothness 1), but also to $H_p^{1+p^{-1}}(\mathbf{T}^1)$, $1 < p < \infty$, that is, has smoothness close to 2 if $p \to 1$.

The same approaches are also possible in the many-dimensional case. It is natural to say that a function $x(\cdot)$ on $\mathbf{T}^n$ (or in some sufficiently regular domain in $\mathbf{R}^n$) has smoothness $r \in \mathbf{N}$ if all of its partial derivatives of order r exist. Smoothness is then naturally characterised by the integer r. But it is often the case that a function $x(\cdot)$ has different smoothness in different variables. If for $x(\cdot)$ the derivatives $\partial^{r_i} x/\partial t_i^{r_i}$ exist, then its smoothness is naturally characterised by the vector $(r_1, \ldots, r_n)$. In addition, as in the one-dimensional case, it is useful to know in which class a derivative lies. Here we must bear in mind that we now have an extension of the space $L_p(\mathbf{T}^n)$ with vector norm $p = (p_1, \ldots, p_n)$, defined in the following way:

$$\|x(\cdot)\|_{L_p(\mathbf{T}^n)} = \left((2\pi)^{-1} \int_{\mathbf{T}^1} \cdots \left((2\pi)^{-1} \int_{\mathbf{T}^1} |x(t_1, \ldots, t_n)|^{p_1} dt_1 \right)^{p_2/p_1} dt_2 \ldots \right)^{p_n/p_{n-1}} dt_n \Bigg)^{1/p_n}$$

changing integration into vrai sup if $p_i = \infty$.

Again, as in the one-dimensional case, it is natural to try to introduce smoothness of arbitrary vector order.

The Weil operator D^α of fractional order $\alpha = (\alpha_1, \ldots, \alpha_\nu)$ is defined in a similar way to the one-dimensional case: if

$$x(t) = \sum_{k \in \mathbf{Z}^n} x_k \exp(i\langle k, t\rangle), \qquad \prod_{i=1}^{n} k_i = 0 \Rightarrow x_{k_1 \ldots k_n} = 0,$$

then

$$(D_x^\alpha)(t) = \sum_{k \in \mathbf{Z}^n} (ik)^\alpha x_k \exp(i\langle k, t\rangle), (ik)^\alpha = (ik_1)^{\alpha_1} \ldots (ik_n)^{\alpha_n}. \tag{17}$$

Using D^α we introduce the Sobolev spaces

$$\mathring{W}_p^\alpha(\mathbf{T}^n) = \{x(\cdot) | D^\alpha x(\cdot) \in L_p(\mathbf{T}^n)\}$$

and the corresponding Sobolev classes $B\mathring{W}_p^\alpha(\mathbf{T}^n)$.

As in the one-dimensional case we can define Nikol'skij spaces and classes: $\alpha > 0, \alpha = r + \beta, r \in \mathbf{Z}_+^n, 0 < \beta \leqslant 1$,

$$H_p^{\alpha,k}(\mathbf{T}^n) = \{x(\cdot) | \, \|\Delta_h^k D^r x(\cdot)\|_{L_p(\mathbf{T}^n)} \ll h|^\beta \; \forall h \in \mathbf{R}^n\}.$$

$$BH_p^{\alpha,k}(\mathbf{T}^n) = \{x(\cdot) | \, \|\Delta_h^k D^r x(\cdot)\|_{L_p(\mathbf{T}^n)} \leqslant |h|^\beta \; \forall h \in \mathbf{R}^n\}.$$

In the initial works on embedding theorems the space $\mathring{W}_p^r(\mathbf{T}^n)$, for which all derivatives of order $r \in N$ belong to $L_p(\mathbf{T}^n)$, occurred. This space is of course the intersection $\bigcap_{i=1}^n W_p^{re_i}(\mathbf{T}^n)$, e_i the canonical basis in $\mathbf{R}^n$. S.M. Nikol'skij considered the so-called spaces with dominating mixed derivatives. These also are intersections of subspaces of the type $H_p^{r1,k}(T^n)$, $r \in N$.

This leads on to a subsequent general statement of the problem. We answer here the question of what the words "smoothness" and "space of smooth functions" will mean in what follows.

By *smoothness* we mean a set $\{(1/\mathbf{p}, \alpha)\}$, usually finite: $A = \{(1/p^i, \alpha^i), i = 1, \ldots, m\}$. In this connection we will say that $x(\cdot)$ has smoothness A, if it is known that its α^i-th derivative, in some sense or other, belongs to L_{p^i}. In this way the spaces $\mathring{W}^A = \bigcap \mathring{W}_{p^i}^{\alpha^i}$ and $H^A = \bigcap H_{p^i}^{\alpha^i}$, and the classes $B\mathring{W}^A = \bigcap BW_{p^i}^{\alpha^i}$ and $BH^A = \bigcap BH_{p^i}^{\alpha^i}$ arise. An individual point $(1/p^i, \alpha^i)$ will be called an *elementary smoothness.*

2.2. Other Approaches to Smoothness. Statement of the Approximation Problem for Smooth Functions. Above we considered only functions on $\mathbf{T}^n$. The Nikol'skij spaces and classes may, of course, also be defined for domains in the spaces $\mathbf{R}^n$, and also for the spaces $\mathbf{R}^n$ themselves. For the introduction of the Sobolev classes and spaces a well defined operator of fractional differentiation is required. In the classical Schwartz space of generalised functions $\mathscr{S}'(\mathbf{R}^n)$ this operator is not defined. Therefore for its construction we introduce a broader space $\Phi' \supset \mathscr{S}'(\mathbf{R}^n)$ (P.P. Lizorkin). Φ' is the space conjugate to the space $\Phi \subset \mathscr{S}(\mathbf{R}^n)$ of functions $y(\cdot) \in \mathscr{S}(\mathbf{R}^n)$ such that

$$\int_{\mathbf{R}} t_j^k y(t_1,\ldots,t_{j-1},t_j,t_{j+1},\ldots,t_n)\,dt_j = 0, \qquad j=1,\ldots,n, \qquad k\in\mathbf{Z}_+.$$

Sobolev and Nikol'skij classes on other homogeneous spaces, in particular, on n-dimensional spheres $\mathbf{S}^n$ are introduced in a similar way.

But in addition to the two ways of introducing smoothness that we have mentioned there are many others. We have defined smoothness as a subset $A \subset I^n \times \mathbf{R}^n$. But it is possible to consider classes of type $W^r H_p^\omega$:

$$W^r H_p^\omega(\mathbf{T}^1) = \{x(\cdot) | x^{(r)}(\cdot) \in L_p(\mathbf{T}^1), \omega(\delta, x^{(r)}(\cdot), L_p(\mathbf{T}^1)) \leqslant \omega(\delta)\}.$$

Here smoothness is defined by a function parameter $\omega(\cdot)$. There are various methods of defining smoothness via harmonic analysis—relative to the behaviour of Fourier coefficients or the Fourier transform. Here is one of the most important classes of functions—the *Besov class*. It is defined by a triple $(1/\mathbf{p}, \alpha, \theta)$. We give the definition in the one-dimensional case—on the circle $\mathbf{T}^1 (\theta \in \mathbf{R}_+)$:

$$B_{p,\theta}^\alpha(\mathbf{T}^1) = \left\{x(\cdot) \left| \left(\sum_{s\in\mathbf{N}} \|\delta_s x^{(\alpha)}(\cdot)\|_{L_p(\mathbf{T}^1)}^\theta\right)^{1/\theta} \leqslant 1\right.\right\},$$

$$(\delta_s x)(t) = \sum_{2^{s-1}\leqslant |k| < 2^s} x_k \exp(ikt), \qquad \text{if } x(t) = \left(\sum_{k\in\mathbf{Z}} x_k \exp(ikt)\right).$$

It is possible to define classes using integral representations (Bessel kernel, Macdonald kernel, etc.) and many others. But in the majority of cases, nevertheless, it is possible to choose basic parameters of smoothness, similar to those which we have described, and then the results on best approximations will look the same as those which we will give later.

We will investigate the question of best approximation of functions of a given smoothness. What smoothness means was clarified above. Now we must discuss the questions: with what do we approximate and how do we approximate. In passing we must also consider the problem: what means of approximation do we regard as classical, in particular, what is the analogue of the space $\mathcal{T}_N$ of trigonometric polynomials.

Fundamentally, we restrict ourselves to the following four problems:

1) what is the best way to approximate a given class of smooth functions by the Fourier method;

2) what is the best N-dimensional linear method of approximation of a given class;

3) which N-dimensional subspace best approximates our class;

4) what is the Aleksandrov width of the given class.

In addition, we will try to find the best subspaces for a given class relative to the Boor-Favard constant and the Weierstrass-Nikol'skij constant.

However, as a preliminary we must investigate the question of embedding of the classes in the spaces in which we plan to approximate.

2.3. Basic Theorems of Harmonic Analysis in $\mathbf{T}^n$ and $\mathbf{R}^n$ and Representation Theorems. One of the most significant applications of the ideas and methods of approximation theory in classical analysis is the enlistment of the methods of harmonic analysis to the problems of embedding theorems. This approach goes back to the work of S.M. Nikol'skij (see Nikol'skij [1958], [1969]). First we state the basic theorems of harmonic analysis—the theorems which can serve as cornerstones for the whole theory of embedding, and approximation.

We denote by $\square_s = \{k \in \mathbf{Z}^n | 2^{s_j-1} \leqslant |k_j| < 2^{s_j}, j = 1, \ldots, n\}$, $s = (s_1, \ldots, s_n)$, $s_i \in \mathbf{N}$, and by p_s: $\mathscr{S}'(\mathbf{T}^n) \to \mathscr{S}'(\mathbf{T}^n)$ the operator defined by the formula

$$(\delta_s x)(t) = \sum_{k \in \square_s} x_k \exp(i\langle k, t\rangle),$$

$$(Px)(t) = \left(\sum_{s \in \mathbf{N}^n} |\delta_s x|^2(t)\right)^{1/2}.$$

Theorem 15 (Littlewood-Paley). $/1 < \mathbf{p} < \infty, x(\cdot) \in \mathscr{P}'(\mathbf{T}^n)/ \Rightarrow$

$$x(\cdot) \in L_p(\mathbf{T}^n) \Leftrightarrow (Px)(\cdot) \in L_p(\mathbf{T}^n) \& \exists C_1 > 0, C_2 > 0:$$

$$C_1 \|(Px)(\cdot)\|_{L_p(\mathbf{T}^n)} \leqslant \|x(\cdot)\|_{L_p(\mathbf{T}^n)} \leqslant C_2 \|P(x)(\cdot)\|_{L_p(\mathbf{T}^n)}$$

$$(\Leftrightarrow : \|x(\cdot)\|_{L_p(\mathbf{T}^n)} \asymp \|(Px)(\cdot)\|_{L_p(\mathbf{T}^n)}).$$

In some sense Theorem 15 satisfies the role of a Parseval equality for $L_p(\mathbf{T}^n)$. Let $\mu = \{\mu_k\}_{k \in \mathbf{Z}_n}$ and $M(\cdot) = \sum_{k \in \mathbf{Z}_n} \mu_k \exp(i\langle k, \cdot\rangle)$ a generalized function. The operator $\mathscr{M}_\mu$: $\mathscr{S}'(\mathbf{T}^n) \to \mathscr{S}'(\mathbf{T}^n)$, defined by the formula

$$(\mathscr{M}_\mu x)(\cdot) =: (M * x)(\cdot) =: \sum_{k \in \mathbf{Z}_n} \mu_k x_k \exp(i\langle k, \cdot\rangle),$$

is called a *multiplier transform*. It is said that $\mathscr{M}_\mu$ is a *(p, p)-multiplier* if $\|\mathscr{M}_\mu\|_{L_p(\mathbf{T}^n) \to L_p(\mathbf{T}^n)} < \infty$. We denote $\Delta = \Delta_n \circ \cdots \circ \Delta_1$, $\Delta_j\{\mu_k\} = \{\mu_{k+e_j} - \mu_k\}$, $j = 1, \ldots, n, e_1, \ldots, e_n$ the canonical basis in $\mathbf{R}^n$, $\mu_{ks} =: \mu_k$, if $k \in \square_s$ and 0 if $k \neq \square_s$.

Theorem 16 (Marcinkewicz). *Let $1 < p < \infty$ and for a sequence $\{\mu_k\}_{k \in \mathbf{Z}^n}$ let there exist a C such that $|\mu_k| \leqslant C$, $\sum_{k \in \square_s} |\Delta \mu_{ks}| \leqslant C\ \forall s \in Z^n_+$. Then $\mathscr{M}$ is a (p, p) multiplier and $\|\mathscr{M}_\mu x\|_{L_p(\mathbf{T}^n)} \leqslant C A_p \|x(\cdot)\|_{L_p(\mathbf{T}^n)}$, where the constant A_p depends only on p.*

Corollary (M. Riesz). *The Fourier operators S_N are uniformly bounded in norm as operators from $L_p(\mathbf{T}^1)$ to $L_p(\mathbf{T}^1)$ for $p \in (1, \infty)$.*

Fom this corollary and the Banach-Steinhaus theorem it follows at once that for each $x(\cdot) \in L_p(\mathbf{T}^1)$ there is convergence in $L_p(\mathbf{T}^1)$:

One of the first theorems really in embedding theory, was

Theorem 17 (Hardy-Littlewood). $/1 < p < q < \infty, \alpha = 1/p - 1/q, T = \mathbf{T}^n \vee \mathbf{R}^n/ \Rightarrow$

$$\exists C > 0: \|D^{-\alpha} x(\cdot)\|_{L_q(T)} \leqslant C \|x(\cdot)\|_{L_p(T)}\ \forall x(\cdot) \in L_p(T).$$

The fundamental principle at the foundation of the application of harmonic analysis to embedding and approximation problems is that a smooth function $x(\cdot)$ may be decomposed into a series relative to trigonometric polynomials or entire functions and, for the latter, there are definite estimates, corresponding to the smoothness of $x(\cdot)$. We encounter similar phenomena in the discussion of direct and inverse theorems in the constructive theory of functions.

We quote a theorem, ideologically going back to S.M. Nikol'skij, where this principle is clearly visible.

Theorem 18. $/\alpha \in \mathbf{R}^n_+,\ 1 < \mathbf{p} < \infty/ \Rightarrow$

a) $x(\cdot) \in W^\alpha_p(\mathbf{T}^n) \Leftrightarrow (\delta^\alpha P x)(\cdot) =: (\sum |2^{\langle \alpha, s\rangle}(\delta_s x)(\cdot)|^2)^{1/2} \in L_p(\mathbf{T}^n)$ *and*

$$\|x(\cdot)\|_{W^\alpha_p(\mathbf{T}^n)} \asymp \|(\delta^\alpha P x)(\cdot)\|_{L_p(\mathbf{T}^n)},$$

b) $x(\cdot) \in H^\alpha_p(\mathbf{T}^n) \Leftrightarrow \sup 2^{\langle \alpha, s\rangle} \|\delta_s x(\cdot)\|_{L_p(\mathbf{T}^n)} < \infty$ *and*

$$\|x(\cdot)\|_{H^\alpha_p(\mathbf{T}^n)} \asymp \sup 2^{\langle \alpha, s\rangle} \|\delta_s x(\cdot)\|_{L_p(\mathbf{T}^n)}.$$

2.4. Embedding Theorems for the Intersection of Spaces. There is a special article devoted to embedding theorems. Therefore we only touch on this theme briefly and then only the part which we need for later comparisons. For Sobolev classes and Nikol'skij classes the question of embedding of functions of a given smoothness has been almost completely resolved for $1 < \mathbf{p} < \infty$.

We say that a linear topological space X, whose topology is given by a system of seminorms $\{p_\alpha(\cdot)\}_{\alpha \in \mathscr{A}'}$ is (continuously) embedded in a linear topological space $Y \supset X$, whose topology is given by a system of seminorms $\{q_\beta(\cdot)\}_{\beta \in \mathscr{B}'}$ and write $X \hookrightarrow Y$, if for any $\beta \in \mathscr{B}$ there is a finite subset $\{\alpha_1, \ldots, \alpha_s)$, $\alpha_i \in \mathscr{A}$, and a $C > 0$ such that $p_\beta(x) \leqslant C \max_{1 \leqslant j \leqslant s} p_{\alpha_j}(x)$.

We begin the discussion of embedding with the one-dimensional case, where the results can be illustrated with a diagram (see Fig. 1a, b).

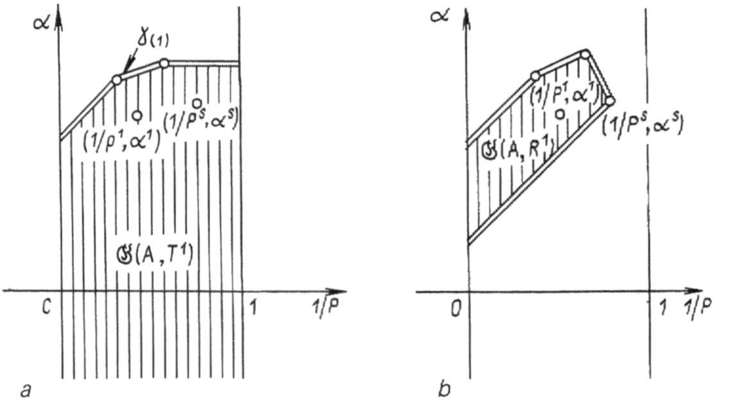

Fig. 1

Let A be a subset of $I \times \mathbf{R}$. Put

$$\mathfrak{G}(A, T) = \begin{cases} \operatorname{co} A + \operatorname{cone}\{-e_1 - e_2\} + \operatorname{cone} e_1, & T = \mathbf{T}^1, \\ \operatorname{co} A + \operatorname{cone}\{-e_1 - e_2\}, & T = \mathbf{R}, \end{cases}$$

$$W^A(T) =: \bigcap_{(1/p,\alpha)} W_p^\alpha(T), \qquad T = \mathbf{T}^1 \vee \mathbf{R}^1.$$

It is natural to introduce the locally convex topology in W^A defined by the family of seminorms $x(\cdot) \to \|D^\alpha x(\cdot)\|_p$, $(1/p, \alpha) \in A$.

Theorem 19. *Let A and $B \subset \operatorname{int} I \times \mathbf{R}$ be nonempty subsets. Then*
a) $B \subset \mathfrak{G}(A, T) \Rightarrow W^A(T \hookrightarrow W^B(T)$.
b) $W^A(T) \hookrightarrow W^B(T) \Rightarrow B \subset \operatorname{cl} \mathfrak{G}(A, T)$.

We see that when A is compact the inclusion $B \subset G(A, T)$ is necessary and sufficient for embedding.

When $T = \mathbf{T}^1$ and A is finite the set $\mathcal{G}(A, T)$ has an upper envelope. It is given by a concave function $1/p \to \gamma(1/p)$, which characterises the possibility for Fourier approximation of spheres in W^A.

We give the basic steps in the proof of the theorem in the periodic case. The proof of sufficiency is based on the following four facts: 1) the embedding $W_q^\alpha(\mathbf{T}^1) \hookrightarrow W_p^\alpha(\mathbf{T}^1)$ for $1/q \leqslant 1/p$ follows from the Hölder inequality, 2) the obvious embedding $W_q^\alpha(\mathbf{T}^1) \hookrightarrow W_q^\beta(\mathbf{T}^1)$ for $\alpha > \beta$, 3) the embedding of the intersection $W_{q1}^{\alpha 1}(\mathbf{T}^1) \cap W_{q2}^{\alpha 2}(\mathbf{T}^1)$ in $W_q^\alpha(\mathbf{T}^1)$, if $\alpha = \theta\alpha^1 + (1 - \theta)\alpha^2$, $1/p = (\theta/p^1) + ((1 - \theta)/p^2)$, which follows from the representation theorem 18; 4) the embedding $W_q^\beta(\mathbf{T}^1) \hookrightarrow W_q^\alpha(\mathbf{T}^1)$, if $\beta = \alpha - (1/p - 1/q)$, $1/p \geqslant 1/q$, which follows from the Hardy-Littlewood theorem. The proof of necessity is based on the consideration of functions of the form $x_{N_1 N_2}(t) = D_{N_1+N_2}(\cdot) - D_{N_1}(\cdot)$ ($D_N(\cdot)$ the Dirichlet kernel) and the theorem on the separation of $(1/p, \beta)$ from $\mathcal{G}(A, T^1)$.

Let us give many-dimensional versions of Theorem 19. If $A \subset I^n \times \mathbf{R}^n$ the set $\mathcal{G}(A, T)$, having the form $\{\operatorname{co} A + \operatorname{cone}\{-e^i - e^{n+i}, e_i, i = 1, \ldots, n\}\}$ for $T = \mathbf{T}^n$ $\{\operatorname{co} A + \operatorname{cone}\{-e^i - e^{n+i}, i = 1, \ldots, n\}\}$ for $T = \mathbf{R}^n$, and the spaces $W^A(T)$ and $H^A(T)$ are introduced in a similar way.

The question of embedding of an intersection of spaces of smooth functions was first posed in general form by N.S. Bakhvalov (Bakhvalov [1963]). He proved the following result.

Theorem 20. $/A \subset I^n \times \mathbf{R}_+^n$, $\operatorname{card} A < \infty$, $H^A(\mathbf{T}^n) =$

$$\bigcap_{(1/p,\alpha)} H_p^\alpha(\mathbf{T}^n), (1/q, \beta) \in \operatorname{int} \mathfrak{G}(A, \mathbf{T}^n)/ \Rightarrow H^A(\mathbf{T}^n) \hookrightarrow H_q^B(\mathbf{T}^n).$$

Here sets of the type $\mathcal{G}(A, \mathbf{T}^n)$ were encountered for the first time.

In a certain sense definitive results have been obtained for embedding of the classes $W^A(T)$. Namely,

Theorem 21. $/A, B \subset (\operatorname{int} I)^n \times \mathbf{R}^n, T = \mathbf{T}^n \bigvee \mathbf{R}^n/ \Rightarrow$

a) $B \subset \mathfrak{G}(A, T) \Rightarrow W^A(T) \hookrightarrow W^B(T)$,

b) $W^A(T) \hookrightarrow W^B(T) \Rightarrow B \subset \operatorname{cl} \mathfrak{G}(A, T)$.

If A is compact (in particular, finite) then $\mathscr{G}(A, T) = \operatorname{cl} \mathscr{G}(A, T)$ and the condition $B \subset \mathscr{G}(A, T)$ is necessary and sufficient for the embedding $W^A(T) \hookrightarrow W^B(T)$. Hence, for example, the known results on embedding of Sobolev classes of functions, classes of functions with dominating mixed derivatives, etc. follow at once.

§ 3. Harmonic Analysis and the Approximation of Classes of Smooth Function

3.1. Introduction. For functions on the n-dimensional torus there is a special, privileged system of functions: these are the harmonics $t \to \exp(i\langle k, t\rangle)$. The experience of classical analysis attests that harmonics are a reasonable tool of approximation for shift invariant classes of functions.

In the one-dimensional case there is a natural and usual ordering of harmonics —relative to the growth of the indices: 1, $\exp(it)$, $\exp(-it)$, $\exp(2it), \ldots$ etc. That is why in approximation theory so much attention is given to approximation by trigonometric polynomials.

But how should one proceed in the many-dimensional case? How should harmonics be ordered? Which subspaces of harmonics should be regarded as analogous to the spaces of trigonometric polynomials?

Attention was drawn to these problems in the sixties. It was clarified that there is no natural means to order harmonics in the many-dimensional case. The ordering must be produced depending on the information available on the functions to be approximated. If, say, a class similar to $\mathring{W}_p^\alpha(\mathbf{T}^n)$ is given with the help of a differential operator D^α, then the harmonics must be ordered according to the growth of $|k|$ and the analogue of the space $\mathscr{T}_N$ of trigonometric polynomials will in this case be the space $\mathscr{T}_N(\alpha)$, consisting of the harmonics $t \to \exp(i\langle k, t\rangle)$, for which $|k|^\alpha \leqslant \gamma$, and γ is chosen so that the number of harmonics is of order N. Such sets are called *hyperbolic crosses*. For smoothnesses, given by subsets of $I^n \times \mathbf{R}^n$, that is, for intersections of classes, we must consider intersections of hyperbolic crosses. However, from technical considerations, connected with the classical theorems of harmonic analysis, it is usual to consider modified, "block" hyperbolic crosses and their intersections. How to approximate similar subspaces of Sobolev and Nikol'skij classes is the topic of this section.

3.2. Best Approximation of Periodic Functions. Functions, given on $\mathbf{T}^n$, are approximated here by trigonometric polynomials. In order not to get tied up in details we restrict ourselves to the case $\mathbf{1} < \mathbf{p}, \mathbf{q} < \infty$. We begin with the one-dimensional version

Theorem 22.

$$/1 < p, q < \infty, \alpha > (1/p - 1/q)_+/ \Rightarrow \varphi(B\mathring{W}_p^\alpha(\mathbf{T}^1), \mathcal{T}_N, L_q(\mathbf{T}^1))$$
$$\asymp \lambda(B\mathring{W}_p^\alpha(\mathbf{T}^1), \mathcal{T}_N, L_q(\mathbf{T}^1))$$
$$\asymp d(B\mathring{W}_p^\alpha(\mathbf{T}^1), \mathcal{T}_N, L_q(\mathbf{T}^1)) \asymp N^{-\alpha+(1/p-1/q)_+}.$$

Let us comment on this result. Here smoothness is given by a "point" $(1/p, \alpha)$ and the speed of convergence of the Fourier method in the metric on $L_q(\mathbf{T}^1)$ turns out to be equal to $N^{-\gamma}$, where $\gamma = \alpha - (1/p - 1/q)$. Here (asymptotically) it is impossible to improve this estimate not only by any other linear method of summation of Fourier series, but even the very deviation of the class $B\mathring{W}_p^\alpha(\mathbf{T}^1)$ from $\mathcal{T}_N$ in $L_q(\mathbf{T}^1)$ has the same order $N^{-\gamma}$. (The inequality $\gamma > 0$ is the condition for compactness of the embedding of $\mathring{W}_p^\alpha(\mathbf{T}^1)$ in $L_q(\mathbf{T}^1)$). Looking ahead, we remark, that the Fourier method is also asymptotically optimal in the sense of the Fourier width, but is far from it in the sense of the Kolmogorov and linear width.

In the many-dimensional case, as already said, it is natural to give smoothness by a set of points and, consequently to define classes as intersections of classes. But in the one-dimensional case it is reasonable to pose the question of the study of approximative properties of the intersection of classes. How such intersections are approximated by the subspace $\mathcal{T}_N$, is given in the following theorem, where the function $\gamma(\cdot)$ was defined in § 2.

Theorem 23.

$$/A = \{(1/p^i, \alpha^i), i = 1, \ldots, m\}, \gamma(1/q) > 0/$$
$$\Rightarrow \varphi(B\mathring{W}^A(\mathbf{T}^1), \mathcal{T}_N, L_q(\mathbf{T}^1)) \asymp \lambda(BW^A(\mathbf{T}^1), \mathcal{T}_N, L_q(\mathbf{T}^1))$$
$$d(B\mathring{W}^A(\mathbf{T}^1), \mathcal{T}_N, L_q(\mathbf{T}^1)) \asymp N^{-\gamma(1/q)}.$$

A similar result holds for the Nikol'skij classes.

In the many-dimensional case the question of selecting the harmonics for the approximation arises immediately. We begin the discussion with the case when the smoothness is given by one point $(1/p, \alpha) \in \operatorname{int} I \times \mathbf{R}^n$, $\alpha = (\alpha_1, \ldots, \alpha_n)$, $\alpha_1 = \alpha_2 = \cdots = \alpha_{l+1} < \alpha_{l+2} \leqslant \cdots \leqslant \alpha_n$. What should we regard here as the analogue of the spaces of trigonometric polynomials $\mathcal{T}_N$?

The problem of approximation of the class $B\mathring{W}_p^\alpha(\mathbf{T}^n)$ in $L_p(\mathbf{T}^n)$ for $p = 1$ or ∞ was first considered by K.I. Babenko, Babenko [1960]. There also hyperbolic crosses appeared

$$\mathcal{T}_N(\alpha) = \operatorname{lin}\{\exp i\langle k, \cdot\rangle | k \in \mathring{\mathbf{Z}}^n, \quad |k|^\alpha \leqslant \gamma\}, \quad \operatorname{card} \mathcal{T}_N(\alpha) \asymp N.$$

S.A. Telyakovskij [1964] suggested considering approximation by "extended" hyperbolic crosses $\mathcal{T}_N(\alpha')$, constructed relative to a point $\alpha' = (\alpha'_1, \ldots, \alpha'_n)$, where $\alpha'_i = \alpha_i$, $i = 1, \ldots, l+1$, $\alpha_1 < \alpha'_i < \alpha_i$. $i > l + 1$.

In addition to the ordinary and extended hyperbolic crosses, the "block" hyperbolic crosses, suggested by the Littlewood-Paley structure theorem, became very common:

$$\mathscr{T}_N(\alpha) = \operatorname{lin}\{\exp i\langle k, \cdot\rangle | k \in \mathring{\mathbf{Z}}^n, k \in \square_s, s \in \mathbf{N}^n, \langle \alpha, s\rangle \leqslant \nu\},$$

$$\operatorname{card} \mathscr{T}_N(\alpha) \asymp N.$$

Theorem 24.

$$/1 < p, q < \infty, \alpha \in \mathbf{R}^n, \quad 0 < \alpha_1 = \cdots = \alpha_{l+1} < \alpha_{l+2} \leqslant \cdots \leqslant \alpha_n/$$

$$\Rightarrow d(B\mathring{W}_p^{\alpha}(\mathbf{T}^n), \mathscr{T}_N(\alpha), L_q(\mathbf{T}^n)) \asymp (N^{-1} \log^l N)^{\gamma},$$

$$\gamma = \alpha_1 - (1/p - 1/q)_+.$$

This is an analogue of Theorem 22. Somewhat later (see Theorem 31) we will give a result on the approximation of a Nikol'skij class by an extended hyperbolic cross, and then discover the different asymptotic forms. We again draw attention to the fact that here the means of approximation ($\mathscr{T}_N(\alpha)$) are defined by the class itself (not true in the one-dimensional case, where they depend only on N).

Therefore it is also reasonable to ask is there a universal subspace? To what extent is our choice reasonable? etc. Such questions can be answered only by returning to the notion of type of widths. A definitive answer to these questions in terms of Kolmogorov and other widths is given in Chap. 3.

For intersections of Sobolev spaces the result turns out to be completely analogous, but for its statement it is necessary to precisely write down the approximating space. Consider the following set of harmonics:

$$\mathscr{T}(A, \mathbf{q}, \mu) = \operatorname{lin}\{\exp(i\langle k, \cdot\rangle) | k \in \square_s, s \in \mathbf{N}^n, \langle \alpha, s\rangle \leqslant \mu, \alpha \in \mathfrak{G}_q(A, \mathbf{T}^n)\},$$

where

$$\mathfrak{G}_q(A, \mathbf{T}^n) = \{\alpha | (1/\mathbf{q}, \alpha) \in \mathfrak{G}(A, \mathbf{T}^n)\}.$$

We see that this is an intersection of "block" hyperbolic crosses, corresponding to smoothnesses from the polyhedron $\mathscr{G}_q(A, \mathbf{T}^n)$. We now choose μ so that the dimension of this subspace is weakly equivalent ($\asymp$) to N. From a theorem of Din' Zung, to be mentioned later, μ must be defined from the condition $2^{\mu M}\mu^l \asymp N$, where M is the value and l the dimension of the set of solutions of the problem

$$\langle s, \mathbf{1}\rangle \to \sup, \quad \langle s, s\rangle \leqslant \mathbf{1}, \quad \alpha \in \mathfrak{G}_q(A, \mathbf{T}^n).$$

For such a μ we will denote $\mathscr{T}(A, \mathbf{q}, \mu)$ by $\mathscr{T}_N(A, \mathbf{q})$.

Theorem 25.

$$/1 < \mathbf{p}^i, \mathbf{q} < \infty, A = \{(1/p^i, \alpha^i), i = 1, \ldots, m\}, \mathfrak{G}_q(A, \mathbf{T}^n) \cap \mathring{\mathbf{R}}_+^n \neq \varnothing/$$

$$\Rightarrow \varphi(BW^A(\mathbf{T}^n), \mathscr{T}_N(A, \mathbf{q}), L_q(\mathbf{T}^n))$$

$$\asymp \lambda(BW^A(\mathbf{T}^n), \mathscr{T}_N(A, \mathbf{q}), L_q(\mathbf{T}^n))$$

$$\asymp d(BW^A(\mathbf{T}^n), \mathscr{T}_N(A, \mathbf{q}), L_q(\mathbf{T}^n)) \asymp (N^{-1} \log^l N)^{1/M}.$$

This is an analogue of Theorem 23.

3.3. Best Approximation of Functions on $\mathbf{R}^n$. Again we begin with the one-dimensional case. How should we approximate functions on $\mathbf{R}^1$? What should we regard as the analogue of $\mathscr{T}_N$ in this case? S.N. Bernstein suggested that such an analogue could be the space of functions, whose "spectrum", that is, the carrier of the Fourier transform, lies in the interval $[-N, N]$. The set of such functions from $L_p(\mathbf{R})$ we denote by $B_{Np}(\mathbf{R})$.

We leave until Chap. 4, § 2, the discussion of why precisely this space is natural to regard as an analogue of $\mathscr{T}_N$, and put the question of how the subspaces $B_{Np}(\mathbf{R})$ approximate the Sobolev classes.

An analogue of Theorem 24 is the following result:

Theorem 26.

$$/A = \{(1/p^i, \alpha^i), i = 1,\dots,m\}, 1 < p^i, q < \infty, i = 1,\dots,m, \gamma(1/q) > 0/$$
$$\Rightarrow \lambda(BW^A(\mathbf{R}), B_{Nq}(\mathbf{R}), L_q(\mathbf{R})) \asymp d(BW^A(\mathbf{R}), B_{Nq}(\mathbf{R}), L_q(\mathbf{R})) \asymp N^{-\gamma(1/q)}.$$

The direct many-dimensional generalization of the space $\mathscr{T}_N(A, \mathbf{q})$ is the space $B_{Nq}(A, \mathbf{R}^n)$, consisting of those functions for which the carrier of the Fourier transform is concentrated in the intersection of the hyperbolic crosses

$$\sum(\xi^j, \mathbf{q}, \mu) = \left\{\sigma \in \mathbf{R}^n \,\middle|\, \prod_{i=1}^n |\sigma_i|^{\xi_i^j} \leqslant \mu, \xi^j = (\xi_1^j,\dots,\xi_n^j), j = 1,\dots,s\right\},$$

where ξ^j are the vertices of the polyhedron $\mathscr{G}_q(A, \mathbf{R}^n)$ and $\operatorname{mes} B_{Nq}(A, \mathbf{R}^n) = N$.

We note that the Lebesgue measure of $B_{Nq}(A, \mathbf{R}^n)$ may be infinite. In particular this will always be the case when the smoothness is given by one point. Because of this situation, it is impossible to give any analogue of Theorem 24. As follows from the theorem of Din' Zung, the inequality $\max_{i,j} \xi_i^j > 0$ (condition (A)) guarantees its finiteness.

Theorem 27.

$$/A = \{(1/p^i, \alpha^i), i = 1,\dots,m\} \subset \operatorname{int} I^n \times \mathbf{R}^n,$$
$$\mathscr{G}_q(A, \mathbf{R}^n) \cap \mathring{\mathbf{R}}_+^n \neq \varnothing \textit{ and condition } (A) \textit{ holds}/$$
$$\Rightarrow \lambda(BW^A(\mathbf{R}^n), B_{Nq}(A, \mathbf{R}^n), L_q(\mathbf{R}^n))$$
$$\asymp d(BW^A(\mathbf{R}^n), B_{Nq}(A, \mathbf{R}^n), L_q(\mathbf{R}^n)) \asymp (N^{-1} \log^l N)^{1/M},$$

where M is the value and l the dimension of the affine hull of the set of solutions of the problem

$$\langle x, 1\rangle \to \sup, \langle \xi^j, x\rangle \leqslant 1, \quad j = 1,\dots,s, \quad x \geqslant 0.$$

We note that we have not mentioned here the quantity $\varphi(BW^A(\mathbf{R}^n), B_{Nq}, L_q(\mathbf{R}^n))$. This is related to the fact that the characteristic functions of sets, in general, are not multipliers. The question of how the ideology of widths is then modified, we postpone until Chap. 4.

The number M and l, occurring in the statement of the theorem, are related to certain asymptotic integrals, which arise naturally in the discussion of approximation of the classes being considered.

Let $\mathscr{P}$ be a set in $\mathbf{R}^n$. Put $\tilde{Z}(\mathscr{P}, \beta) = \sum_{k \in \mathscr{P} \cap \mathbf{Z}_+^n} 2^{\langle \beta, x \rangle}$, $Z(\mathscr{P}, \beta) = \int_{\mathscr{P}} 2^{\langle \beta, x \rangle}\, dx$. The following result is due to Din' Zung.

Let $\mathscr{P}$ be a polyhedral set in $\mathbf{R}^n$, $\operatorname{int} \mathscr{P} \neq \emptyset$, K its recessive cone, $\beta \in \mathbf{R}^n$. Then the quantities $Z(\xi\mathscr{P}, \beta)$ and $\tilde{Z}(\xi\mathscr{P}, \beta)$ are finite for any $\xi > 0$ if and only if $\beta \in \operatorname{int} K^0$ (K^0 the polar of K), and when the latter holds $Z(\xi\mathscr{P}, \beta) \asymp \tilde{Z}(\xi\mathscr{P}, \beta) \asymp 2^{M\xi}\xi^l$, $\xi > 0$, where M is the value and l the dimension of the affine hull of the set of solutions of the problem $\langle \gamma, x \rangle \to \sup$, $x \in P$.

This result gives the estimates of the dimension of $\mathscr{T}(A, \mathbf{q}, \mu)$, and the measure of the intersection of the sets $\sum(\xi^j, \varphi, \mu)$, which we have already used.

3.4. The Bernstein-Nikol'skij Inequality for Sobolev Classes

Theorem 28.

$$/1 \leqslant p, q \leqslant \infty, \alpha > C/ \Rightarrow b(BW_p^\alpha((\mathbf{T}^1), \mathscr{T}_N, L_q(\mathbf{T}^1)) \asymp N^{-(\alpha+(1/q-1/p)_+)}.$$

We recall that $b(W, L, X)$ was defined in the introduction to this article. This result is usually written as follows. Let $1 \leqslant p, q \leqslant \infty$, $x(\cdot) \in \mathscr{T}_N$, $\alpha > 0$. Then:

$$\|D^\alpha x(\cdot)\|_{L_p(\mathbf{T}^1)} \leqslant C N^{\alpha+(1/q-1/p)_+} \|x(\cdot)\|_{L_q(\mathbf{T}^1)}.$$

This inequality (which is sharp up to order) was proved by S.M. Nikol'skij. A particular case ($p = q = \infty$, $\alpha \in \mathbf{N}$) was considered in Chap. 1—the classical Bernstein inequality.

Let us compare two formulae

$b(BW_\infty^1(\mathbf{T}^1), \mathscr{T}_N, C(T^1)) = N^{-1}$ (*Bernstein's inequality*)

$b(BW_\infty^1(\hat{I}), \mathscr{P}_N, C(\hat{I})) = N^{-2}$ (*Markov's inequality*)

In this comparison we must bear in mind that the Bernstein widths here are: $b_{2N+1}(BW_\infty^1(\mathbf{T}^1), C(\mathbf{T}^1)) = \pi/2(N+1)$, $b_{N+1}(BW_\infty^1(\hat{I}), C(\hat{I}^1)) = 1/2N$. Hence it follows that trigonometric polynomials give a true asymptotic form but algebraic polynomials do not. This is one of the essential differences between $\mathscr{T}_N$ and $\mathscr{P}_N$. The exact value of the width is provided by a spline.

We have not mentioned at all the "extreme case", when the quantities p^i and q may be equal to 1 or ∞. In this part of best approximations much has not yet been clarified. The reader will find many details and a detailed bibliography on this circle of questions in a recently published monograph, V.N. Temlyakov [1986], devoted to these problems. Here we note that only for the Bernstein inequalities has a complete result for $p = \infty$ been obtained:

Theorem 29.

$$/\alpha \in \mathbf{R}_+^n, T_N(\alpha) = \operatorname{lin}\{\exp i\langle k, \cdot > | k \in \mathbf{Z}^n, |k|^\alpha \leqslant N\}/$$

$$\Rightarrow b(B\mathring{W}_p^\alpha(\mathbf{T}^n), T_N(\alpha), L_p(\mathbf{T}^n)) \asymp \begin{cases} N^{-1}, & 1 < p < \infty, \\ (N \log^{n-1} N)^{-1}, & p = \infty. \end{cases}$$

3.5. Appendix

3.5.1. Embedding Theorems for the Classes H^{ω}. When smoothness is defined by a functional parameter all the results become, of course, much more difficult. There is a significant literature devoted to the question of embedding classes of the type $H_p^{\omega}(\mathbf{T}^1) = \{x(\cdot): \omega(\delta, x(\cdot), L_p(\mathbf{T}^1)) \leqslant \omega(\delta)\}$ in the space $L_q(\mathbf{T}^1)$. We quote the following result due to P.L. Ul'yanov.

Theorem 30.

$$/1 \leqslant p < q < \infty/ \Rightarrow H_p^{\omega}(\mathbf{T}^1) \hookrightarrow L_q(\mathbf{T}^1) \Leftrightarrow \sum_{k \in \mathbf{N}} k^{((q/p)-2)}\omega^q(k^{-1}) < \infty.$$

The work of V.A. Andrienko, K.K. Golovkin and many others has been devoted to these problems. We wished to mention this circle of problems since it is interesting but is far from completion.

3.5.2. On the Approximation of Nikol'skij Classes. We give here just one result of V.N. Temlyakov and we note the difference in the asymptotic form by comparison with the Sobolev case, and the fact that here an extended hyperbolic cross is considered as the approximating subspace.

Theorem 31.

$$/1 < p \leqslant q < \infty, \alpha \in \mathbf{R}^n, 0 < \alpha_1 = \cdots = \alpha_{l+1} < \alpha_{l+2} \leqslant \cdots \leqslant \alpha_n/$$
$$\Rightarrow d(BH_p^{\alpha}(\mathbf{T}^n), \mathcal{T}_N(\alpha), L_q(\mathbf{T}^n))$$
$$\asymp \begin{cases} ((\log^l N)N^{-1})^{\alpha_1 - 1/p + 1/q}(\log^l N)^{1/q}, & q \leqslant 2, \\ ((\log^l N)N^{-1})^{\alpha_1 - 1/p + 1/q}(\log^l N)^{1/2}, & q \geqslant 2. \end{cases}$$

§ 4. Linear Methods of Summation of Fourier Series

4.1. Introduction. In § 1 we said that the space $\mathcal{T}_N$ is the natural tool for the approximation of periodic functions or shift invariant classes. In this connection the Fourier method has extremal properties in $L_2(\mathbf{T}^1)$. In view of a theorem of M. Riesz, Fourier sums approximate any function from $L_p(\mathbf{T}^1)$, $1 < p < \infty$, that is, the Fourier method is also good in $L_p(\mathbf{T}^1)$, for $1 < p < \infty$. But in the space $C(\mathbf{T}^1)$ this property is lost: we recall that a Fourier sum, in general, does not converge uniformly to a continuous function. On the other hand, Fejér or Jackson sums approximate continuous functions. All of this stimulates us to investigate shift invariant polynomial operators (that is, operators Λ_N: $\mathcal{S}'(\mathbf{T}^1) \to \mathcal{T}_N$ of the form

$$(\Lambda_N x)(t) = (a_0 \lambda_{0N})/2 + \sum_{k=1}^{N} \lambda_{kN}(a_k \cos kt + b_k \sin kt),$$

where (a_k, b_k) are the Fourier coefficients of $x(\cdot)$, with the aim of investigating which methods have the best universality or, conversely, speciality. Many of the operators considered later (Fourier, Fejér, Jackson, Vallée-Poussin) are shift invariant and therefore representable in this form

$$(S_N \Rightarrow \lambda_{kN} = 1, \quad 0 \leqslant k \leqslant N, \quad \lambda_{kN} = 0, \quad k > N;$$

$$\mathscr{F}_N \Rightarrow \lambda_{kN} = 1 - (k/N), \quad 0 \leqslant k \leqslant N-1, \quad \lambda_{kN} = 0, \quad k \geqslant N;$$

$$\mathscr{J}_N \Rightarrow \lambda_{kN} = a_N((2N-k+1)!/(2N-k-2)!)$$
$$- 4(N-k+1)!/(N-k-2)!), \quad 0 \leqslant k \leqslant N-2,$$

$$\lambda_{kN} = a_N((2N-k+1)!/(2N-k-2)!), \quad N-2 \leqslant k \leqslant 2N-2,$$

$$\lambda_{kN} = 0, \quad k \geqslant 2N-1, \quad a_N = (2N(2N^2+1))^{-1};$$

$$\mathscr{U}_{MN} \Rightarrow \lambda_{kN} = 1, \quad 0 \leqslant k \leqslant N-M,$$

$$\lambda_{kN} = 1 - (k-N+M)/M, \quad N-M \leqslant k \leqslant N, \quad \lambda_{kN} = 0, \quad k > N).$$

Some other well-known operators are in the same class *Zygmund*:

$$\lambda_{kN} = 1 - (k/N)^{\beta}, \quad 0 \leqslant k \leqslant N-1, \quad \lambda_{kN} = 0, \quad k \geqslant N, \quad \beta > 0;$$

Riesz means of order δ:

$$\lambda_{kN} = (1-(k/N)^2)^{\delta}, \quad 0 \leqslant k \leqslant N-1, \quad \lambda_{kN} = 0, \quad k \geqslant N;$$

the Korovkin operator:

$$\lambda_{kN} = ((1 - k/(N+2))\cos k\pi/(N+2))$$
$$+ (1/(N+2))(\sin k\pi/(N+2))\operatorname{ctg}\pi/(N+2),$$
$$0 \leqslant k \leqslant N, \quad \lambda_{kN} = 0, \quad k \geqslant N+1;$$

Bernstein-Rogosinskij:

$$\lambda_{kN} = \cos k\pi/2N, \quad 0 \leqslant k \leqslant N-1, \quad \lambda_{kN} = 0, \quad k \geqslant N.$$

How well do these operators approximate, what are their qualities and deficiencies? We will discuss these questions later.

4.2. Conditions for Convergence. For the metrics $p = 1$ and ∞ the problem of convergence of the operators Λ_N arises. A matrix (λ_{kN}) is called *regular*, if $(\Lambda_N x)(\cdot) \xrightarrow{C} x(\cdot)\ \forall x(\cdot) \in C(\mathbf{T}^1)$. The problem of regularity of methods of summation of series has been studied by Hille-Tamarkin, Nagy, S.M. Nikol'skij and many others. By the Banach-Steinhaus theorem the question of convergence (under the condition that $\lambda_{0N} = 1$ and $\lambda_{kN} \to 1$ as $N \to \infty$) reduces to the boundedness of the norms of the Λ_N. The following gives fairly general conditions.

Theorem 32. *For there to exist a constant $C > 0$ such that*

$$\|\Lambda_N\|_{C(\mathbf{T}^1)\to C(\mathbf{T}^1)} \leqslant C(\Lambda_N \Leftrightarrow (\lambda_{kN})),$$

it is sufficient that one of the following three groups of conditions be satisfied:

1. a) $\sup\limits_{k,N} |\lambda_{kN}| < \infty,$ b) $\sup\limits_N \left|\sum\limits_{k=1}^{N} \lambda_{kN}/(N-k+1)\right| < \infty,$

c) $\Delta_2 \lambda_{kN} \geqslant 0$ *or* $\Delta_2 \lambda_{kN} \leqslant 0,\quad k = 0, 1, \dots, N$

(Δ_2 *the second order difference operator with respect to k*) (*Nikol'skij, 1948*);

2. a) $\sup\limits_{k,N} |\lambda_{kN}| < \infty,$ b) $\sup\limits_N \left|\sum\limits_{k=1}^{N-1} \lambda_{kN}/(N-k)\right| < \infty,$

c) $\sup\limits_N \sum\limits_{k=1}^{N-1} k(N-k)N^{-1}|\Delta_2 \lambda_{kN}| < \infty$ (*A.V. Efimov, 1960*).

3. $\sup\limits_N \left(N^{p-1} \sum\limits_{k=1}^{N} |\Delta\lambda_{kN}|^p\right) < \infty,\quad 1 < p \leqslant 2$ (*G.L. Fomin, S.B. Stechkin, 1967*).

4.3. Asymptotic Results. How do we compare the methods discussed in subsection 1.3? One possible approach to this question is to compare how a method approximates some class of functions.

Let us consider, how the Fourier, Fejér and Vallé-Poussin operators approximate Sobolev classes.

Theorem 33. *Put $\gamma = \alpha - (p^{-1} - q^{-1})$. Then if $\gamma > 0$, then*

a) $$\varphi(B\mathring{W}_p^\alpha(\mathbf{T}^1), \mathscr{T}_N, L_q(\mathbf{T}^1)) \asymp \begin{cases} N^{-\gamma}, & 1 < p,\ q < \infty, \\ N^{-\alpha}\ln N, & p = q = \infty, \end{cases}$$

b) $\lambda(B\mathring{W}_p^\alpha(\mathbf{T}^1), (\mathscr{F}_N, \mathscr{T}_N), L_q(\mathbf{T}^1))$

$$\asymp \begin{cases} N^{-\gamma}, & 0 < \gamma < 1,\quad 1 \leqslant p,\quad q \leqslant \infty, \\ N^{-1}, & \gamma > 1,\quad 1 \leqslant p,\quad q \leqslant \infty,\quad \gamma = 1, \\ 1 < p \leqslant q < \infty, & \gamma = 1,\quad 1 \leqslant q < p \leqslant \infty, \\ N^{-1}\ln N, & \gamma = 1,\quad p = q = 1,\quad p = q = \infty,\quad p = 1,\quad q = \infty, \\ N^{-1}(\ln N)^{1/p'}, & \gamma = 1,\quad 1 < p < q = \infty. \end{cases}$$

c) *If* $\lim M/N = \theta, 0 < \theta < 1$, *then*

$$\lambda(B\mathring{W}_p^\alpha(\mathbf{T}^1), (\mathscr{U}_{MN}, \mathscr{T}_N), L_q(\mathbf{T}^1)) \asymp d(B\mathring{W}_p^\alpha(\mathbf{T}^1), \mathscr{T}_N, L_q(\mathbf{T}^1)) \asymp N^{-\gamma}.$$

Theorem 33 a) is a slightly improved result by comparison with Theorem 23, Theorem 33 b) expresses the universality of the Vallée-Poussin sums. Theorem 33 c) (due to A.I. Kamzolov) shows the restricted possibilities of Fejér sums.

This raises the question: why do these differences arise: the "almost universality" of the Fourier method; the restricted possibilities of the Fejér sums; the universality of Vallée-Poussin sums.

All three methods are given in a similar way, with $\lambda_{kN} = \varphi(k/N)$, where φ: $\mathbf{R}_+ \to \mathbf{R}$, where for the Fourier method $\varphi(t) = 1$ for $0 \leqslant t \leqslant 1$, $= 0$ for $t > 1$; for the Fejér case $\varphi(t) = \max(1 - t, 0)$; for the Vallée-Poussin case (with $\theta = 1/2$) $\varphi(t) = 1$ for $t \in [0, 1/2]$, $= 2(1 - t)$ for $t \geqslant 1/2$.

A discontinuity of $\varphi(\cdot)$ at 1 leads to a logarithmic multiplier for approximation of continuous functions. The function $\varphi(\cdot)$ in the Fejér and Vallée-Poussin case continuously approaches unity, but for the Fejér function the derivative at zero is not zero. This prevents the Fejér method approximating the classes $B\mathring{W}_p^\alpha(\mathbf{T}^1)$ with precision any better than N^{-1}. However, if φ is identically equal to unity in a neighbourhood of zero and continuous and monotonely decreasing to zero at unity, then the corresponding matrix (λ_{kN}) will have universal properties similar to those of the Vallée-Poussin operator.

4.4. The Precise Asymptotic Form of the Kolmogorov-Nikol'skij Constant for Fourier Sums. This and the following subsections are devoted to the development of the theme, begun in the work of A.N. Kolmogorov [1985; No. 27] on the precise asymptotic form of $\varphi(BW_\infty^k(\mathbf{T}^1), \mathcal{T}_N, C(\mathbf{T}^1))$ (Theorem 42 of Chap. 1). A significant extension of this theme was obtained by S.M. Nikol'skij in the forties. S.M. Nikol'skij, instead of Fourier sums, investigated upper bounds of the best approximations for Fejér sums, polynomial interpolation operators and many others. Therefore the problem of finding a quantity of the type $\lambda(W, (A_N, \mathcal{T}_N), X) = \sup\{\|x - A_N x\|: x \in W\}$, where A_N is an operator from X to $\mathcal{T}_N$, is sometimes called the *Kolmogorov-Nikol'skij problem*, and the numbers $\lambda(W, (A_N, \mathcal{T}_N), X)$ the *Kolmogorov-Nikol'skij constants*. These problems were intensively developed during the forties to the sixties and continue to develop even up to present. The articles by Efimov [1960, 1966], Stechkin [1980], Telyakovskij [1960, 1963] are devoted to them and a lot of space has been given to them in the monographs by Dzyadyk [1977], Kornejchuk [1976] and, in particular, by Stepanets [1981].

We give some important formulae.

Theorem 34. *Let* $\alpha > 0$. *Then*

$$\varphi(BW_\infty^\alpha(\mathbf{T}^1), \mathcal{T}_N, C(\mathbf{T}^1)) = 4\pi^{-2} \ln N N^{-\alpha} + O(N^{-\alpha}).$$

The case $\alpha \in \mathbf{N}$, as we have said, was analyzed by A.N. Kolmogorov; his method was extended to fractional derivatives by V.T. Pinkevich (1940).

Theorem 35 (S.M. Nikol'skij, 1941). *Let* $r \in \mathbf{N}$, $0 \leqslant \beta \leqslant 1$. *Then*

$$\varphi(BW_\infty^r H^\beta(\mathbf{T}^1), \mathcal{T}_N, C(\mathbf{T}^1))$$
$$= \pi^{-2} 2^{\alpha+1} \ln N N^{-(r+\beta)} \int_0^{\pi/2} t^\beta \sin t \, dt + O(N^{-(r+\beta)}).$$

S.M. Nikol'skij, in his work on the Kolmogorov-Nikol'skij problem in fact introduced the classes $W_\infty^r H^\omega$, which subsequently occupied so important a place in the investigations in approximation theory. The latter result was generalized by A.V. Efimov (1960).

Theorem 36. *Let $\alpha > 0$, and let $\omega(\cdot)$ be an arbitrary modulus of continuity. Then*

$$\varphi(BW_\infty^\alpha H^\omega(\mathbf{T}^1), \mathcal{T}_N, C(\mathbf{T}^1)) = e_N(\omega(\cdot))\pi^{-1}\ln N N^{-\alpha} + O(N^{-\alpha}\omega(N^{-1})),$$

where

$$e_N(\omega(\cdot)) =: \sup_{x(\cdot)\in H^\omega(\mathbf{T}^1)} \pi^{-1}\left|\int_{\mathbf{T}^1} x(t)\cos nt\, dt\right|.$$

The problem of remainder estimates which are uniform in all parameters has been considered. We mention two important results.

Theorem 37 (S.A. Telyakovskij, 1968). *Let $N \geqslant 1$, $r \in \mathbf{N}$. Then, uniformly with respect to N and r,*

$$\varphi(BW_\infty^r(\mathbf{T}^1), \mathcal{T}_N, C(\mathbf{T}^1)) = N^{-r}\left(4\pi^{-2}\ln\frac{N+r}{r+1} + O(1)\right).$$

Theorem 38 (S.B. Stechkin [1980]). *Let $N \geqslant 1$, $r \in \mathbf{N}$. Then, uniformly with respect to N and r,*

$$\varphi(BW_\infty^r(\mathbf{T}^1), \mathcal{T}_N, C(\mathbf{T}^1)) = N^{-r}\{8\pi^{-2}K(\exp(-r/N)) + O(r^{-1})\},$$

where

$$K(x) = \int_0^{\pi/2} (\sqrt{1 - x^2\sin^2 t})^{-1}\, dt, \quad 0 \leqslant x \leqslant 1,$$

is a complete elliptic integral of the first kind.

This appears to be one of the most beautiful formulae of asymptotic analysis and approximation theory.

Let us say a few words about approximation by Fejér sums. S.N. Bernstein first noted that Fejér sums in Hölder classes achieved the necessary (best) order of approximation, and he noted that for functions satisfying a Lipshitz condition the order deteriorated ($\asymp \log N \cdot N^{-1}$). Precise results were obtained by S.M. Nikol'skij.

Theorem 39 (S.M. Nikol'skij, 1940).

$$\lambda(BH_\alpha(\mathbf{T}^1), (\mathcal{F}_N, \mathcal{T}_N), C(\mathbf{T}^1)) = \begin{cases} \dfrac{2\Gamma(\alpha)\sin\alpha\pi/2}{\pi(1-\alpha)N^\alpha} + O(N^{-1}), & 0 < \alpha < 1, \\ \dfrac{2\ln N}{\pi N} + O(N^{-1}), & \alpha = 1. \end{cases}$$

The last formula was essentially improved by Telyakovskij (1969).

Theorem 40. *If $N = 2m$, then the following asymptotic expansion is valid:*

$$\lambda(BW^1_\infty(\mathbf{T}^1), (\mathscr{F}_N, \mathscr{T}_N), C(\mathbf{T}^1)) \approx 2\pi^{-1} \ln N \cdot N^{-1} + 2\pi(1 + C + \ln 2)N^{-1} + \pi^{-1} \sum_{k=1}^{\infty} (2k-1)k^{-1}N^{-(2k+1)} \times (1 + (-1)^N(1 - 2^{2k}))B_{2k},$$

where C is the Euler constant and B_{2k} are the Bernoulli numbers.

Theorem 41 (S.M. Nikol'skij, (1940), Nagy, 1942, 1946)). *Let $r \in \mathbf{N}$. Then* $\lambda(BW^r_\infty(\mathbf{T}^1), (\mathscr{F}_N, \mathscr{T}_N), C(\mathbf{T}^1)) \sim K_{r-1}N^{-1}$.

Theorem 42 (A.I. Stepanets (1981)).

$$/r \in \mathbf{N}/ \Rightarrow \lambda(W^r H^\omega(\mathbf{T}^1), (\mathscr{F}_N, \mathscr{T}_N), C(\mathbf{T}^1)) \sim A(\omega(\cdot))N^{-1},$$

$$A(\omega(\cdot)) = \sup\left\{\|x(\cdot)\|_{C(\mathbf{T}^1)} | x(\cdot) \in \tilde{W}^r H^\omega(\mathbf{T}^1), \int_{\mathbf{T}^1} x(t)\,dt = 0\right\},$$

$$\lambda(H^\omega(\mathbf{T}^1), \quad (\mathscr{F}_N, \mathscr{T}_N), \quad C(\mathbf{T}^1)) = \frac{2}{\pi n}\int_0^{\pi/2} \omega(2t)\frac{\sin^2 nt}{\sin^2 t}\,dt.$$

We give a result on Vallée-Poussin sums. In it the asymptotic form of approximation by sums V_{NM} is calculated under the assumption that $\lim MN^{-1}$ exists and is equal to $\theta \in I$.

Theorem 43 (Telyakovskij (1960)). *Let $r \in \mathbf{N}$. Then*

$$\lambda(BW^r_\infty(\mathbf{T}^1), (V_{NM}, \mathscr{T}_N), C(\mathbf{T}^1)) = \begin{cases} 4\pi^2 N^{-r}\ln(NM^{-1}) + O(N^{-r}), & \theta = 0, \\ c(r, \theta)N^{-r}\ln(NM^{-1}) + O(N^{-r}), & 0 < \theta < 1. \end{cases}$$

The Kolmogorov-Nikol'skij problems have been very significantly developed by the Ukrain school of approximation theory, headed by V.K. Dzyadyk (V.K. Dzyadyk, V.T. Gavrilyuk, A.I. Stepanets and others). A number of important results in this direction have been obtained by N.P. Kornejchuk and his school. Here many-dimensional spaces appeared and a number of classical one-dimensional results obtained many-dimensional generalizations. Since it is not possible for us to pause on this in detail, we refer the reader to the monographs Dzayadyk [1977] and Stepanets [1981], for an exposition of the basic results and a complete bibliography.

4.5. Appendix

4.5.1. On Individual Functions which cannot be Approximated better than Classes as a Whole. In his article (1983), dedicated to A.N. Kolmogorov on his eightieth birthday, S.M. Nikol'skij gave an interesting historical episode on the relation of S.N. Bernstein to the problems of approximation in classes. S.N. Bernstein expressed the doubt that in a class there would exist functions which achieve the asymptotic upper bound for all or infinitely many values of N. However, it turned out that this was not the case.

Theorem 44 (S.M. Nikol'skij, (1946). *There is a function $\bar{x}(\cdot) \in BW_\infty^1(\hat{I})$ such that*

$$\limsup_{N\to\infty} (Nd(\bar{x}(\cdot), \mathscr{P}_N, C(\hat{I})) = \lim_{N\to\infty} \left(N \sup_{x(\cdot)\in W_\infty^1(\hat{I})} d(x(\cdot), \mathscr{P}_N, C(\hat{I}) \right) = \pi/2$$

There are similar phenomena for Fourier sums, but there more complete results have been obtained.

Theorem 45. a) *There exists a function $\bar{x}(\cdot) \in BW_\infty^r(\mathbf{T}^1)$ such that*

$$\limsup_{N\to\infty} (\varphi(\bar{x}(\cdot), \mathscr{T}_N, C(\mathbf{T}^1))/\varphi(BW_\infty^1(\mathbf{T}^1), \mathscr{T}_N, C(\mathbf{T}^1)) = 1 \quad (\textit{Yu.Ya. Doronin});$$

b) $$\max_{x(\cdot)\in BW_\infty^1(\mathbf{T}^1)} \liminf \|x(\cdot) - (S_N x)(\cdot)\|_{C(\mathbf{T}^1)}/\varphi(BW_\infty^1(\mathbf{T}^1), \quad \mathscr{T}_N, C(\mathbf{T}^1)) = \tfrac{1}{2}$$

(*K.I. Oskolkov* [*1975*])

4.5.2. An Extremal Property of the Fourier Operator. A continuous linear operator P: $X \to L$ from a normed space X to a subspace L is called a *projection* onto L if $Px = x$ for every $x \in L$. The Fourier operator is a projection. The following holds

Theorem 46. *For any projection P_N: $C(\mathbf{T}^1) \to \mathscr{T}_N$ the relation $\|P_N\| \geqslant \|S_N\|$ holds, where the norm of the operators is the norm from $C(\mathbf{T}^1)$ to $C(\mathbf{T}^1)$.*

This result is due to S.M. Lozinskij and F.I. Kharshiladze. A proof of it can be based on the idea of averaging of the projection P_N (see [1976]. The quantity $\|S_N\|$, as we recall, increases (see Chap. 1, § 5). Hence, in particular, it follows that the method of trigonometric interpolation is not convergent (there is a continuation of this theme in § 7 of this chapter).

§ 5. Approximation by Rational Functions

While the methods of harmonic analysis have sufficient approximative universality and effectively approximate both smooth and analytic functions, rational functions and splines (the topic of subsequent sections) have their own sphere of preference. Rational functions are most naturally adapted to the approximation of analytic functions, splines to the approximation of functions of finite smoothness. In this connection both have a structural similarity, and this makes a parallel discussion of them quite natural.

Rational approximations include the polynomial, but on comparison noticeable differences emerge. Later we will try to accentuate these differences. For lack of space it is not very appropriate to encumber the exposition with the introduction of a variety of metrics, and throughout here we will investigate the approximation of continuous functions (that is, functions from $C(T)$, where T is a compact set) which, in general, admit an analytic continuation to a domain $G \supset T$. Two fundamental examples are $T = \hat{I} = [-1, 1]$ (and there we will take real functions) and $T = D =: \{z: |z| \leqslant 1\}$. For each function $x(\cdot) \in C(T)$ there is a table $\{R_{mn}\}_{m\geqslant 0, n\geqslant 0}$ made up of the rational functions of best approximation (If T $= I$, then, as we recall, R_{mn} consists of one function $r_{mn}(\cdot)$ and a table $\{\rho_{mn}\}_{m\geqslant 0, n\geqslant 0}$ of best approximations: $\rho_{mn} = \rho_{mn}(x(\cdot))_T = d(x(\cdot), \mathcal{R}_{mn}, C(T))$.

5.1. The Speed of Rational Approximation and Structural Properties of Functions. The basic tenet of the constructive theory of functions is that the structural properties of functions admit a second description (in the periodic case via the sequence $\{d(x(\cdot), \mathcal{T}_N, X)\}$). The systematic study of the relations between structural properties of functions and the behaviour of the table $\{\rho_{mn}(x(\cdot))\}$ was begun in the middle fifties by many people under the influence of A.N. Kolmogorov and S.N. Mergelyan. Amongst the mathematicians most actively developing this theme we must mention first and foremost A.A. Gonchar and E.P. Dolzhenko (see Gonchar [1966], Dolzhenko [1962, 1967]).

The real difference between rational and polynomial approximations is already clear from the following two results of A.A. Gonchar.

Theorem 47. *There is a function $x(\cdot)$, analytic in the interior of the unit disc and continuous on the unit disc, for which the sequence $\{\rho_{nn}\}_{n\geqslant 0}$, $\rho_{nn} = d(x(\cdot), \mathcal{R}_{nn}, C(K))$ tends to zero arbitrarily quickly, but yet it is not analytic at any point of the unit circle.*

Such a function is not difficult to construct, by taking a series $\sum_{k\in N} a_k(z - z_k)^{-1}$, where the $\{z_k\}$ "coil" around the unit circle, and the a_k decrease sufficiently rapidly. Proved in a similar way is

Theorem 48. *For every function $\omega(\cdot)$: $\mathbf{R}_+ \to \mathbf{R}$ and any sequence $\{\varepsilon_n\}_{n\geqslant 0}$, $\varepsilon_n \to 0$, there is a function $x(\cdot) \in C(\hat{I})$ such that $d(x(\cdot), \mathcal{R}_{nn}, C(\hat{I})) \leqslant \varepsilon_n$, $n \geqslant 1$ and at the same time $\limsup_{\delta>0} \omega(\delta, x(\cdot), C(\hat{I}))/\omega(\delta) > 1$.*

Thus not only is it impossible to achieve analyticity, but also no estimate whatever of the modulus of continuity. Compare these results with the polynomial case. The following theorem is due to S.N. Bernstein.

Theorem 49. *In order that a function $x(\cdot) \in C(D)$ $(C(\hat{I}))$ can be analytically continued to the disc of radius $R > 1$ (to an ellipse with foci at the points ± 1 and sum of semiaxes equal to R) it is necessary and sufficient that*

$$\limsup_{n\to\infty} (d(x(\cdot), \mathscr{P}_n, C(K))^{1/n} = R^{-1};$$

$$\left(\limsup_{n\to\infty} (d(x(\cdot), \mathscr{P}_n, C(\hat{I}))^{1/n} = R^{-1}\right).$$

In general, there is nothing surprising in the fact that the usual theorems of the constructive theory of functions do not hold for rational approximations. The possibility of describing the structural properties of functions via best approximations in the trigonometric case is closely connected with shift invariance, the possibility of expressing smoothness by Fourier coefficients etc.

Rational functions are of a different kind. But nevertheless something of the general ideology of the constructive theory remains.

Theorem 50. a) *If* $\rho_{nn} =: d(x(\cdot), \mathscr{R}_{nn}, C(\hat{I})) \ll n^{-(1+\delta)}$, $\delta > 0$, *then* $x(\cdot)$ *is differentiable almost everywhere.*

b) *If* $\rho_{nn} \ll n^{-(\alpha+\delta)}$, *then for every* $\varepsilon > 0$ *there is a closed set* $I_\varepsilon \subset I$, $\operatorname{mes}(I \setminus I_\varepsilon) < \varepsilon$ *such that the restriction* $x(\cdot)|_{I_\varepsilon} \in H^\alpha_\infty(I)$.

Sometimes it is after all possible to decide on membership of some class or other from the speed of rational approximation.

Theorem 51.

$$\Big/ \sum_{n\geqslant 0} d(x(\cdot), \mathscr{R}_{nn}, C(\hat{I})) < \infty \Big/ \Rightarrow x(\cdot) \in AC(\hat{I}).$$

One of the reasons for the impossibility of a constructive (using rational approximations) description of $BW^\alpha_\infty(\hat{I})$ and $BH^\alpha_\infty(\hat{I})$ is that there are no analogues of the Bernstein inequalities here—the derivative of a rational function cannot be estimated simply from the norm of the function in $C(\hat{I})$ and the degree of the rational function. However, Bernstein-Nikol'skij inequalities are possible for different metrics. One of the first inequalities of this kind was as follows.

Theorem 52.

$$/r(\cdot) \in \mathscr{R}_{nn}/ \Rightarrow \|\dot{r}(\cdot)\|_{L_1(T)} \leqslant 2n \|r(\cdot)\|_{C(T)}.$$

Here T is any Lebesgue measurable subset of **R**. One of the reasons for the constructive description of the class $AC(\hat{I})$ is given by inequalities of this kind. Here is another example.

Theorem 53. *For* $x(\cdot)$ *to belong to the Besov class* $B^{1/p}_p(\hat{I})$, $1 \leqslant p \leqslant \infty$, *it is necessary and sufficient that* $\sum_{n\geqslant 0} d(x(\cdot), \mathscr{R}_{nn}, \mathrm{BMO}) < \infty$.

We have restricted ourselves in these illustrations to the differences in the programme of constructive theory for polynomials and rational fractions.

5.2. Comparison of Rational and Polynomial Approximations. We consider the top row of the table $\{\rho_{mn}(x(\cdot))\}$ (that is the sequence $\{\rho_{n0}\}_{n\geqslant 0}$) and the diagonal

$\{\rho_{nn}\}_{n\geqslant 0}$. It is clear that $\rho_{nn} \leqslant \rho_{n0}$. But is it possible for these sequences to coincide? It is not difficult to see that in $L_p(\hat{I})$ for $1 < p < \infty$ such a coincidence is impossible. But what of $p = \infty$?

Theorem 54. a) $/d(x(\cdot), \mathscr{P}_n, C(\hat{I})) = d(x(\cdot), \mathscr{R}_{nn}, C(\hat{I}))\ \forall n \geqslant 0/ \Rightarrow$

$$\exists n_0\colon x(t) = aT_{n_0}(t) + b, \qquad a, b \in \mathbf{R}; \qquad T_n(t) = \cos n \arccos t.$$

b) $/x(\cdot) \in A^{\operatorname{Int} D} \cap C(D), d(x(\cdot), \mathscr{P}_n, C(D)) = d(x(\cdot), \mathscr{R}_{nn}, C(D))\ \forall n \geqslant 0/ \Rightarrow$

$$\exists n_0\colon x(z) = az^{n_0} + b, \qquad a, b \in \mathbf{C};$$

c) $\exists x(\cdot) \in C(\hat{I}), \{n_k\}_{k\in\mathbf{N}}, \quad n_k \in \mathbf{N}$:

$$d(x(\cdot), \mathscr{P}_{n_k}, C(\hat{I})) = d(x(\cdot), \mathscr{R}_{n_k n_k}, C(\hat{I})) \qquad \forall k \in \mathbf{N}.$$

Examples analogous to c) can be constructed even in the space $A^{\operatorname{int} D} \cap C(D)$. Thus if $x(\cdot)$ is not a Chebyshev polynomial plus a constant (on $\hat{I}$) or if $x(\cdot)$ is not a pure power plus a constant (on $A^{\operatorname{int} D} \cap C(D)$) then there is an n such that $\rho_{nn} < \rho_{n0}$. And it is possible to construct functions for which the subsequences $\rho_{n_k n_k} = \rho_{n_k 0}$ coincide.

This situation leads to the fact that in many classes of functions, studied in approximation theory, the "diagonal" ($\mathscr{R}_{nn}$) has no advantage over the first row (P_n). For example

Theorem 55.

$$d(BA^{\operatorname{int} D_R}, \mathscr{R}_{nn}, C(D)) = d(BA^{\operatorname{int} D_R}, \mathscr{P}_n, C(D)) = R^{-n}.$$

Similar results are true also for the classes $W_p^\alpha(\hat{I})$.

Yet at the same time the experience of classical analysis attests to the significant advantages of rational functions $\mathscr{R}_{nn}$ over polynomials $\mathscr{P}_n$. Here is an example (Newman [1964]):

Theorem 56.

$$2^{-1}\exp(-9\sqrt{n}) < d(|t|, \mathscr{R}_{nn}, C(\hat{I})) < 3\exp-\sqrt{n}).$$

This result (we have already noted that it can be deduced from Zolotarev's theorem; see Theorem 10 of Chap. 1) must be compared with the relation $d(|t|, \mathscr{P}_n, C(\hat{I})) \asymp n^{-1}$ (Theorem 34 of Chap. 1).

How does one explain such an impressive difference in the asymptotic forms? Under what circumstances do rational functions have great advantage over polynomials? We will have more to say on this.

5.3. Rational Approximations and Singularities of Functions. Let a function $x(\cdot)$ be analytically continued from a segment $\hat{I}$. As we recall, polynomial approximations allow us to find the "first singularity" of $x(\cdot)$: by Theorem 49 there is a singularity of $x(\cdot)$ on the ellipse E_R. We recall also (§ 1 of this chapter), that the rows of the Padé table "resolve" the poles of a meromorphic function and accurately point to the singularities; recall the theorems of Fabry and Montessus de Ballore. These two classical theorems have recently obtained considerable development.

Theorem 57. *Let the poles of the n-th row of the Padé table* $\{\pi_{mn}(\cdot)\}_{m\in N}$ *converge to* $\zeta_1,\ldots,\zeta_n$. *Then, if* $|\zeta_i| < D_n(x(\cdot))$*—the n-th disc of meromorphicity, then* ζ_i *is a pole of* $x(\cdot)$, *and if* $|\zeta_i| = D_n(x(\cdot))$, *then* ζ_i *is a singular point.*

Theorem 58.

$$\limsup_{m\to\infty} d(x(\cdot), \mathscr{R}_{mn}, C(\hat{I})) = R_n^{-1},$$

where R_n *is the semisum of the axes of an ellipse with foci at the points* ± 1, *to which* $x(\cdot)$ *can be continued as a meromorphic functions with n poles.*

There is nothing surprising about the fact that rational functions are good for the approximation of meromorphic functions. But their merit is much more than this.

5.4. Rational Approximation of Analytic Functions. We return to the general situation: let a function, continuous on a compact set T, admit an analytic continuation to a domain $G \supset T$. Suppose further, that T is compact with a connected complement. Let G be an arbitrary domain, $F = \mathbf{C}\backslash G$. Suppose that G and T form a regular pair in the sense that there is a function $u(\cdot)$: $\mathbf{C}\backslash(T\cup F)$ ($=: D$), solving the Dirichlet problem in D, that is, is harmonic in D, continuous on $\operatorname{cl} D$, takes a given value on ∂T and is unity on ∂F. Let Γ be a system of contours, separating T from F. The value

$$C = C(T, G) = (2\pi)^{-1}\int_\Gamma \frac{\partial u(\zeta)}{\partial n}\, d\zeta$$

is called the *capacity of the pair* (T, G) (or the *capacity of the condenser* (T, F)).

Theorem 59.

$$\lim_{n\to\infty} (d(BA^G, \mathscr{R}_{nn}, C(T)))^{1/n} = \exp(C^{-1}(T, G)).$$

Here the asymptotically best method of approximation is interpolation of $x(\cdot)$ at points uniformly distributed in the measure $(\partial u/\partial n)\, ds$ on ∂T, by the set of rational functions with poles uniformly distributed on ∂F in the same measure.

In Chap. 3 we will see, that this approximation by rational functions is the best in any sense (linear widths, Kolmogorov widths, Aleksandrov widths, etc.). Here we have already uncovered the impressive universality of rational approximations for analytic functions.

5.5. Appendix

5.5.1. Rational Approximation of Elementary Functions. Many specialists in computational mathematics often express their surprise at the considerable approximative qualities of rational approximations of the majority of "formula" functions. These include algebraic, trigonometric, logarithmic functions etc. A fairly broad general class of analytic functions, where rational approximations demonstrate their conspicuous qualities, are described as follows

$$x(z) = \int_{\Delta} d\mu(t)/(t - z), \mu \text{ a positive measure, } \Delta \subset R, |\Delta| \leqslant \infty.$$

This class was studied by the classical mathematicians: P.L. Chebyshev, A.A. Markov, Stieltjes and others. They expanded $x(\cdot)$ into a continued fraction ($\Leftrightarrow$ Padé approximated $x(\cdot)$ at infinity). Explicit expressions for approximations were given by a system of orthogonal polynomials. In this way a chain of interrelations of objects was formed, which will be covered in more detail in another place.

Among the rationally well approximated functions are the piecewise-analytic functions, for example $|t|$. We give here the beautiful formulae:

Theorem 60. a) $\lim_{n\to\infty} (d(|t|, \mathscr{R}_{nn}, C(\hat{I}))^{1/\sqrt{n}} = \exp(-\pi)$;

b) $/1 \leqslant p \leqslant \infty, \quad s \in \mathbf{Z}_+ (s \neq 0) \quad \text{for} \quad p = \infty)/ \Rightarrow$

$$d(t^s \operatorname{sign} t, \mathscr{R}_{nn}, L_p(\hat{I})) \asymp n^{1/2p} \exp(-\pi n^{1/2} \sqrt{s + p^{-1}}).$$

Formula a) was obtained by A.N. Bulanov, b) by N.S. Vyacheslavov.

5.5.2. Rational Approximations of Convex Functions. Here also the difference between rational and polynomial approximations can be seen. In the polynomial case the speed of approximation is dictated by smoothness properties. For rational approximations of convex functions this is not the case. This fact was first observed by A.P. Bulanov. He obtained the lower estimate in the theorem given later. The definitive result, requiring delicate considerations, was obtained by V.A. Popov and P.P. Petrushev.

Theorem 61. *Let* $\mathrm{BCo}(I)$ *be the set of all convex continuous functions* $x(\cdot)$ *on* I, *with norm in* $C(I)$ *not exceeding unity. Then*

$$d(\mathrm{BCo}(I), \mathscr{R}_{nn}, C(I)) \asymp n^{-1}.$$

§6. Splines in Approximation Theory

Splines were briefly mentioned in §1. They are the piecewise-polynomial functions. Their theory runs from the beginning of the forties to the present day, but in analysis they have always been used, even from the time of Leibniz. Splines form a reasonable and convenient tool for the approximation of functions of finite smoothness. We will try to assess this situation, by making comparisons with the more traditional classical means of approximation. In addition, splines possess a number of significant extremal properties.

6.1. Some General Properties of Splines

6.1.1. Another Description of Splines. Splines form a nonlinear set. How can one describe the set of splines, depending on a given number of parameters?

We recall the inversion formula for the differentiation operator on an interval (Theorem 1):

$$x(\cdot) \in W_p^{r+1}(I) \Rightarrow x(t) = \sum_{k=0}^{r} a_k t^k + (r!)^{-1} \int_I (t-\tau)_+^r u(\tau)\, d\tau$$

($u(\cdot) \in L_p(I)$). The function $\varphi_m(t) = t_+^m$ is a spline of order m defect 1. The above formula suggests approximating $x(\cdot)$ by sums

$$\sum_{k=1}^{N} c_k \varphi_m(t - \tau_k)$$

and the closure of these sums.

Theorem 62. *The closure of the sums of the form*

$$\sum_{k=1}^{N} c_k \varphi_m(t - \tau_k), \qquad \tau_k \in \mathbf{R},$$

in the metric of $L_p(I)$ consists of functions of the form

$$\sum_{i=0}^{m} a_i t^i + \sum_{i=1}^{e} \sum_{j=0}^{\nu_e} a_{ij}(t - \tau_j)_+^{m_j}, \qquad \tau_i \in (0, 1);$$

$$\sum_{i=1}^{N} (\nu_i + 1) + m + 1 \leqslant N, \qquad \nu_i \leqslant m.$$

This describes a family of splines with $2N$ parameters, which we denote $S_{2N,m}(I)$. $S_{2N,m}(\Delta)$, for any $\Delta \subset \mathbf{R}$, is defined similarly.

By Cauchy's theorem a function analytic in a domain G is represented by a Cauchy integral:

$$x(z) = (2\pi i)^{-1} \int_{\partial G} x(\zeta)\, d\zeta/(\zeta - z).$$

It is natural for the approximation of $x(\cdot)$ to use sums of the form

$$\sum_{i=1}^{N} C_k/(\zeta_k - z)$$

and their closure. Thus we arrive at $\mathscr{R}_{nn}$. This situation and a number of others indicates some similarity between splines and rational functions.

On the other hand, many spaces of periodic functions are described via a kernel:

$$W_p^k(\mathbf{T}^1) = \left\{x(\cdot) \in \mathscr{S}'(\mathbf{T}^1) | x(t) = \int_{\mathbf{T}^1} K(t-\tau)u(\tau)\,d\tau,\ u(\cdot) \in L_p(\mathbf{T}^1)\right\}.$$

This suggests the consideration of sums of the form

$$\sum_{k=1}^{N} C_k K(\cdot - \tau_k)$$

and their closure. This is one way of obtaining "spline-like functions", replacing them in related problems.

6.1.2. *B*-splines. In § 1, when we discussed splines, a theorem was given, from which it follows that

$$S_m^1(\Delta, R_n) = \operatorname{lin}\{1, t, \ldots, t^m, (t-\tau_1)_+^m, \ldots, (t-\tau_n)_+^m\}$$

$$\left(R_n \Leftrightarrow \bigcup_{i=1}^{n} \Delta_i, \Delta_i = [\tau_i, \tau_{i+1}]\right).$$

We mention another basis in the same space which is more convenient in calculations. Extend R_n to the whole axis, choosing τ_i for $i < 0$ and $i > n+1$ so that the inequality $\tau_i < \tau_{i+1}$ holds everywhere. Choose any such extension and denote it by R, we will call R an extension of R_n. Put

$$\Phi_m(\tau, t) = (\tau - t)_+^m, \qquad b_{im+1}(t, R) = \Phi_m(\tau_i, \ldots, \tau_{i+m+1}, t).$$

Thus $b_{im+1}(t, R)$ is the $(m+1)$-st divided difference of the function $\tau \to \Phi_m(\tau, t)$, for fixed t, relative to the system of points $\{\tau_i, \ldots, \tau_{i+m+1}\}$. We now define a new function

$$B_{i,m+1}(t, R) = (t_{i+m+1} - t_i) b_{im+1}(t, R).$$

The functions $B_{i,m+1}(\cdot, \cdot)$ are called *normalised B-splines*. In particular, if $\hat{R}$ is the partition, given by the integer points then $B_{i1}(t, \hat{R}) = \chi_{[i,1+1]}(t)$, the function $B_{i2}(t, \hat{R})$ zero inside $(i, i+2)$, linear on $[i, i+1]$ and $[i+1, i+2]$, continuous on $\mathbf{R}$ and equal to 1 at $i+1$.

Theorem 63. *The quantity* $\sqrt{n}B_{0n}(t, \hat{R})$ *is equal to the area of the section of the cube* I^n *by the plane* $x_1 + \cdots + x_n = t$.

Theorem 64. a) *Any partition R has the following properties*

(i) $\operatorname{supp} B_{im}(\cdot, R) = [t_i, t_{i+m}]$;

(ii) $B_{im}(\cdot, R) \geqslant 0$;

(iii) $\sum_{i \in \mathbf{Z}} B_{im}(t, R) = 1\ \forall t$.

b) *the functions* $B_{im}(\cdot, R)$, $i = -m, \ldots, n$, *form a basis in* $S_m^1(\Delta, R_n)$ *for any extension R of the partition* R_n.

Assertion (*i*) in a) implies a "local" property of the basis generated by *B*-splines, which is the main advantage of this basis.

Splines were introduced by Shoenberg in 1946. B-splines appeared in the work of Kerri and Shoenberg in 1947.

6.1.3. Approximation of Classes of Smooth Functions by Splines. We denote by $S_{2N,m}(T^1)$ the space of periodic splines of order m defect 1 with respect to the uniform partition. We give an analogue of Theorem 23 for splines.

Theorem 65.

$$/1 \leqslant p, q \leqslant \infty, \alpha > (1/p - 1/q)_+, \alpha = r + \beta, r \in \mathbf{Z}_+, 0 < \beta \leqslant 1/$$
$$\Rightarrow d(BH_p^\alpha(\mathbf{T}^1), S_{2N,r}, L_q(\mathbf{T}^1)) \asymp N^{-(\alpha-(1/p-1/q)_+)}.$$

An analogous result is true for the classes $BW_p^\alpha(\mathbf{T}^1)$. Apparently it is also possible in the many-dimensional case to obtain spline-approximations which are not inferior to harmonic approximations. The theme of interpolation will be continued later,

6.1.4. On Spline Bases. The well-known basis systems: the Haar system; the Faber-Schauder system; and many others, consist of splines. Moreover there are remarkable results due to Cieselski and Figiel according to which on any smooth manifold M^n the Sobolev space $W_p^r(M^n)$ has a basis consisting of piecewise-polynomial functions.

6.2. Extremal Properties of Splines. Again we fix a partition of the segment $\Delta = [t_0, t_1]$ and pose the problem of interpolation at the nodes of the partition.

Theorem 66. *There is a unique spline* $\hat{x}(\cdot) \in S_{2r-1}(\Delta, R_n)$ *such that*

$$\hat{x}(\tau_i) = \xi_i, \quad \hat{x}^{(k)}(t_0) = \hat{x}^{(k)}(t_1) = 0, \quad k = r, \quad r + 1, \ldots, 2r - 2.$$

The proof of this theorem is based on an extremal property: the required spline is a solution of the problem

$$\int_\Delta (x^{(r)}(t))^2\, dt \to \inf, \quad x(\tau_i) = \xi_i, \quad i = 1, \ldots, n, \quad x(\cdot) \in W_2^r(\Delta). \qquad (z_2)$$

This is a quadratic programming problem. A solution $\hat{x}$ exists (by compactness). A necessary (and sufficient) condition for an extremum is given by the Euler equation $\hat{x}^{(2r)}(t) = \sum_{i=1}^n c_i \delta(t - \tau_i)$ and the transversality condition $\hat{x}^{(k)}(t_0) = \hat{x}^{(k)}(t_1) = 0, r \leqslant k \leqslant 2r - 2$, which also gives Theorem 66.

We could consider an even more general problem:

$$\int_\Delta |x^{(r)}(t)|^p\, dt \to \inf, \quad 1 \leqslant p < \infty \quad \left(\operatorname*{vrai\,sup}_{t \in \Delta} |x^{(r)}(t)| \to \inf, \quad p = \infty\right)$$
$$x(\tau_i) = \xi_i, \quad x(\cdot) \in W_p^r(\Delta). \qquad (z_p)$$

This is a convex programming problem. It is called the *Favard problem*. From compactness, for $1 < p \leqslant \infty$, it can be proved that a solution to the problem exists. For $1 < p < \infty$ it is the solution of the equation $(x_{(p)}^{(r)}(t))^{(r)} = \sum_{i=1}^n c_i\delta(t - \tau_i)$ (where $y_{(p)}(t) = |y(t)|^p \operatorname{sign} y(t)$). The functions obtained, although "spline-like", are, for $p \neq 2$, complicated to construct. Up to now their rôle in analysis has not been very great. For $p = \infty$ again splines are obtained which are, moreover, "ideal".

Theorem 67. *Among the solutions of problem* (z_∞) *there is an ideal spline of the form* $\hat{x}(t) = p(t) = c(t^r + 2(\sum_{j+1}^m (-1)^j (t - \theta_j)_+^r)$, $p(\cdot) \in P_{r-1}$, $m \leqslant N - r + 1$, $c \in R$, θ_i *distinct points of* (t_0, t_1).

Analogous results are obtained also for the general interpolation problem $(x^{(j)}(\tau) = \xi_{ij})$ and for the periodic case.

The statement of problem (z_p) (for the general interpolation problem) is due to Favard (1940), its solution to Favard, Shoenberg and (to a greater extent) de Boor (1974, 1976), the case $p = r = 2$ was, as we recall, investigated by Holladay. The splines in Theorem 66 were called *natural splines* by Shoenberg.

We note that if in place of (z_2) we consider the problem

$$\int_\Delta |Px|^2\, dt \to \inf, \quad x(\tau_i) = \xi_i, \quad i = 1, \ldots, n, \quad x(\cdot) \in W_2^r(\Delta), \qquad (z(p))$$

where P is a differential operator with constant coefficients (or a pseudo-differential operator) of order r, then another series of "spline-like functions" is obtained. They satisfy the equation

$$(P^2 x)(t) = \sum_{i=1}^n c_i \delta(t - \tau_i).$$

Now to some extremal problems containing monosplines and ideal splines; many problems on approximation and recovery in functions classes, on quadratures, approximation operators etc. reduce to them.

Theorem 68. *Let* $1 \leqslant p \leqslant \infty$. *Then there exist a (for* $1 < p < \infty$, *a unique) monospline of the form*

$$\hat{M}(t) = t^r + \sum_{k=1}^r \hat{x}_k (t - \hat{\tau}_k)_+^{r-1} + \sum_{k=1}^{r-1} \hat{z}_k t^{k-1},$$

$$t_0 < \hat{\tau}_1 < \cdots < \hat{\tau}_n < t_1, \quad n + r - 1 \leqslant N,$$

yielding a solution of the problem

$$\|t^r - x(t)\|_{L_p(\Delta)} \to \inf, \quad x(\cdot) \in S_{2N,r-1}(\Delta).$$

For $p = \infty$ this theorem was proved by Johnson (1960), for arbitrary p by Barrar and Loeb and has been significantly developed in the works of Boyanov and A.A. Zhensykbaev (see Zhensykbaev [1981]).

The following theorem belongs to the circle of ideas of P.L. Chebyshev: among ideal splines with a given number of changes of sign of the leading derivative there is a spline with least norm in $C(\hat{I})$.

Theorem 69. *For any natural number r and integer $n \geqslant r$ there is a unique, up to multiplication by ± 1, ideal spline $x_{nr}(\cdot)$ having the properties:*

a) $|x_{nr}^{(r)}(t)| \equiv 1 \ \forall t \in \hat{I}$, *except for n points where the sign of $x_{nr}^{(r)}(\cdot)$ changes,*

b) *the function $x_{nr}(\cdot)$ has $(n + r - 1)$-alternation,*

c) $\|x_{nr}(\cdot)\|_{C(\hat{I})} \sim K_r \pi^{-r} n^{-r}$.

The first function in the sequence $\{x_{nr}\}_{n \geqslant r}$ is proportional to the Chebyshev polynomial: $x_{rr}(t) = (r!)^{-1} 2^{-(r-1)} T_r(t)$. The points at which $x_{nr}^{(r)}(\cdot)$ changes sign we denote by v_{nr}^i, $i = 1, \ldots, n$, and the points of alternation by α_{nr}^i, $i = 1, \ldots, n + r + 1$. Theorem 69 was proved by V.M. Tikhomirov (1969) and has subsequently been greatly developed. In the periodic case the rôle of the splines $x_{nr}(\cdot)$ is played by the function $\tilde{x}_{nr}(\cdot)$, introduced in § 1 of this chapter. The points where $\tilde{x}_{nr}^{(r)}$ changes sign, we denote by $\tilde{v}_{nr}^i$, and its points of alternation by $\tilde{\alpha}_{nr}^i$.

We denote by $\hat{S}_{N+m+1,m}(\hat{I})$ the space of splines of order m defect 1 relative to the partition $\hat{R}_N(\hat{I})$, given by the points $\{v_{nr}^i\}_{i=1}^r$. From general interpolation theorems for splines it follows that $\hat{S}_{2N,m}(\mathrm{T}^1)$ ($\hat{S}_{N+m+1,m}(\hat{I})$) interpolates at the points $\alpha_{N+r+1,r}^i$).

We give two more results which are sometimes (however, not completely successfully) called "*fundamental theorems of algebra*" *for monsplines and ideal splines.*

Theorem 70. a) *For any points $t_0 < \tau_1 < \cdots < \tau_k < t_1$ $(k \leqslant N)$ and any natural numbers $v_1, \ldots, v_k (v_k \leqslant r + 1, \sum_{i=1}^k v_i = N)$ there is a unique monospline $M(\cdot)$ such that $M^{(j)}(\tau_i) = 0$, $i = 1, \ldots, k$, $i \leqslant j \leqslant v_k + 1$.*

b) *Let $t_0 < \tau_1 < \cdots < \tau_{N+m+1} < t_1$. Then there is a unique ideal spline $x(\cdot)$ of order m with N nodes such that $x(\tau_i) = 0$, $i = 1, \ldots, N + m - 1$. Details and references in Zhensykbaev [1981].*

6.3. Exact Solutions in Problems of Approximation of Classes of Smooth Functions by Splines. We give analogues of the Bernstein theorem on inequalities for derivatives and the Favard-Akhiezer-Krejn theorem on best approximations.

Theorem 71. a) $b(BW_\infty^r(\mathrm{T}^1), S_{2N,r}(\mathrm{T}^1), \mathrm{C}(\mathrm{T}^1)) = K_r N^{-r}$;

b) $b(BW_1^r(\mathrm{T}^1), S_{2N,r}(\mathrm{T}^1) L_\infty(\mathrm{T}^1)) = 2K_{r+1}\pi^{-1}N^{-(r+1)}$;

c) $b(BW_\infty^k(\mathrm{T}^1), S_{2N,r}(\mathrm{T}^1), L_\infty(\mathrm{T}^1)) = K_r/(N^k K_{r-k})$, $1 \leqslant k \leqslant r$;

d) $b(BW_1^k(\mathrm{T}^1), S_{2N,r}(\mathrm{T}^1), L_1(\mathrm{T}^1)) = K_{r+1}/(N^k K_{r-k+1})$, $1 \leqslant k \leqslant r$.

Relation a) was proved by V.M. Tikhomirov (1969). By comparing it with the trigonometric case it is clear that the splines $S_{2N,r}$ have a large Bernstein constant, hence it follows that (in contrast to approximation) $\mathscr{T}_N$ is not optimal in the sense of Bernstein width. Relations b) and c) were proved by Yu.N. Subbotin, relation d) by A.A. Ligun. Bernstein's inequality can also be obtained by a "passage to the limit" from c) and d) (we will discuss the limit passage shortly).

There is even greater difference (in order) in favour of splines by comparison (in the inequalities) with algebraic polynomials.

Theorem 72. $b(BW_\infty^r(\hat{I}), \hat{S}_{N+r+1,r}(\hat{I}), C(\hat{I})) = \|x_{Nr}\|_{C(\hat{I})}(\sim K_r(\pi N)^{-r})$ *(we recall that* $b(BW_\infty^r(\hat{I}), P_N, C(\hat{I})) \asymp N^{-2r}$*).*

Let us move on to approximation problems. Denote by $I_{2n,m}(\mathbf{T}^1)(I_{N+r,r-1}(\hat{I}))$ the operator, associating a function $x(\cdot) \in C(\mathbf{T}^1)(C(\hat{I}))$ with the interpolated spline from $S_{2N,m}(\mathbf{T}^1)(\hat{S}_{N+r,r-1}(\hat{I}))$, constructed using Theorem 70.

Theorem 73. a)
$$\lambda(BW_\infty^r(\mathbf{T}^1), I_{2N,r-1}(\mathbf{T}^1), L_q(\mathbf{T}^1)) = d(BW_\infty^r(\mathbf{T}^1), S_{2N,r}(\mathbf{T}^1), L_q(\mathbf{T}^1)) = \|\tilde{x}_{Nr}(\cdot)\|_{L_q(\mathbf{T}^1)};$$

b) $\lambda(BW_\infty^r(\hat{I}), \mathscr{I}_{N+r,r-1}(\hat{I}), L_q(\hat{I})) = d(BW_\infty^r(\hat{I}), \hat{S}_{N+r,r-1}(\hat{I}), L_q(\hat{I})) = \|x_{Nr}(\cdot)\|_{L_q(\hat{I})}$.

We remark that the classical results of approximation theory in the periodic case may be obtained by a passage to the limit from the theorem on approximation by splines. We give two results.

Theorem 74. *Let* $x(\cdot) \in C^r(\mathbf{T}^1)$, R_n *a fixed partition of* $\mathbf{T}^1$ *and* $\{y_m(\cdot)\}$ *a sequence of splines from* $S_{2m+1}^k(\mathbf{T}^1, R_n)$, *defined by the conditions* $x^{(j)}(\tau_i) = y_m^{(j)}(\tau_i)$, $j = 0, \dots, k$, $i = 1, \dots, N$, $R_N \Leftrightarrow \{\tau_i\}_{i=1}^N$. *Then*
$$\lim_{m \to \infty} y_m^\nu(t) = p_N^{(\nu)}(t),$$
uniformly in t, *where* $p_N(\cdot)$ *is a trigonometric polynomial from* $\mathscr{T}_{[N/2]}$, *interpolating* $x(\cdot)$ *at the points* τ_i *with the same multiplicity (uniqueness of the interpolation is provided by minimization of the sum* $a_{N/2}^2 + b_{N/2}^2$*).*

b) *Let* N *be odd and* $x(\cdot) \in L_p(\mathbf{T}^1)$, $1 \leqslant p \leqslant \infty$. *Then*
$$\lim_{m \to \infty} d(\cdot), \hat{S}_{2N,m}, L_p(\mathbf{T}^1)) = d(x(\cdot), \mathscr{T}_{[(N-1)/2]}, L_p(\mathbf{T}^1)).$$

Theorem 74 a) was proved by Shoenberg and Golitschek in 1972, although the special case when $x(\cdot) \in C(\mathbf{T}^1)$, is already contained in the work of Quade-Collatz 1938. Theorem 74 b) is due to V.L. Velikin [1981].

In conclusion we give a result of N.I. Kornejchuk, regarding approximation of the class $W_\infty^r H^\omega(T)$, perhaps, the most difficult in this cycle.

Theorem 75. *Let* $\omega(\cdot)$ *be a convex upwards modulus of continuity and* $m \geqslant r$. *Then*
$$d(W_\infty^r H^\omega(\mathbf{T}^1), \hat{S}_{2n,m}, C(\mathbf{T}^1)) = \|\tilde{x}_{nr\omega(\cdot)}(\cdot)\|_{C(\mathbf{T}^1)},$$
where $x_{nr\omega}(\cdot)$ *was defined in Chap. 1.*

We have touched, naturally, only the foothills of a large and interesting theory. The theory of splines is now elucidated in a large number of monographs, see, for example, Ahlberg, Nilsen & Walsh [1967], de Boor [1978], Stechkin & Subbotin [1976], Zav'yalov, Kvasov & Miroshnichenko [1980].

An idea of the size of the literature on splines is given by the bibliographical index of the Ural Scientific Centre [1984].

§7. Appendix

Here we mention several questions connected with classical approximation theory or the theme of "approximation by classes".

7.1. Approximation by Positive Linear Operators. A linear operator $\Lambda: C(\Delta) \to C(\Delta)$ is called *positive* in $C(\Delta)$, if $x(\cdot) \geqslant 0$ implies that $(\Lambda x)(\cdot) \geqslant 0$. It is called *polynomial* if $\operatorname{Im} \Lambda \subseteq \mathcal{T}_N(\mathcal{P}_N)$. In Chap. 1 and this chapter we have repeatedly dealt with positive linear operators. For example, the Fejér and Jackson operators on $C(\mathbf{T}^1)$ and the Bernstein operator (see Chap. 1, §4) on $C(I)$. There are some interesting phenomena connected with positive polynomial operators.

Theorem 76 (Korovkin's three function theorem). *Let $\{\Lambda_n\}_{n\in\mathbf{N}}$ be a sequence of positive operators in $C(\Delta)$, $|\Delta| < \infty$, let $\{x_1(\cdot), x_2(\cdot), x_3(\cdot)\}$ form a Chebyshev system in $C(\Delta)$ and let $(\Lambda_n x_i)(\cdot)$ uniformly tend to $x_i(\cdot)$, $i = 1, 2, 3$. Then $(\Lambda_n x)(\cdot)$ uniformly tends to $x(\cdot)$ for any function from $C(\Delta)$.*

This fact finds its full significance in the theory of ordered spaces and their operators. Not stopping here, we refer the reader to Golovkin [1964] (see the list of references to the previous article).

Theorem 77. *Let $\{\Lambda_n\}_{n\in N}$, $\Lambda: C(\mathbf{T}^1) \to \mathcal{T}_n$ be a sequence of positive linear polynomial operators. Then one of the sequences: $n\|\mathbf{1} - (\Lambda_n \mathbf{1})(\cdot)\|_{C(\mathbf{T}^1)}$ or $n^2 \left\| \left(\Lambda_n \sin^2 \frac{\cdot - \tau}{2}\right)(\cdot)\right\|_{C(\mathbf{T}^1)}$ does not tend to zero ($\mathbf{1}(t) \equiv 1$).*

This theorem, the proof of which is also due to P.P. Korovkin, implies that polynomial operators have a sufficiently low level of saturation: they can only approximate in the asymptotically best way functions of smoothness $\leqslant 2$. What is the problem here?

If we restrict ourselves to shift invariant linear polynomial operators, then, as we recall, we can restrict ourselves to convolution operators

$$(\Lambda_n x)(t) = (p_n * x)(t), \qquad p_n(t) = \pi^{-1}\left(\lambda_{0n} \cdot 2^{-1} + \sum_{k=1}^{n} \lambda_{kn} \cos kt\right).$$

Applying Theorem 76 to the functions 1, sin, cos, we see that for Λ_n to converge to the identity operator it is necessary and sufficient that $\lambda_{in} \to 1$, $i = 0, 1$. The speed of convergence of Λ_n depends (for $\lambda_{0n} = 1$) on the speed of convergence of λ_{1n} to unity. We have

Theorem 78 (M. Riesz).

$$/p_n(t) = \pi^{-1}\left(2^{-1} + \sum_{k=1}^{n} \lambda_{kn} \cos kt\right) \geqslant 0, \qquad t \in \mathbf{T}^1/ \Rightarrow \lambda_{1n} \leqslant \cos \pi/(n+2).$$

This result expresses a certain "indeterminacy principle". The $p_n(\cdot)$ in the theorem is the density of a probability distribution, whose Fourier transform is known to lie in $[-n, n]$. Then the dispersion of $p_n(\cdot)$ cannot be too small. It follows from Riesz's theorem that the entire function $t \to \cos t$ is approximated with speed $O(n^{-2})$. The following result, due to P.P. Korovkin, also follows from this theorem.

Theorem 79.

$$\inf \lambda(Z(\mathbf{T}^1), A_n, C(\mathbf{T}^1)) = 1 - \cos \pi/(n+2).$$

(Here the inf is taken over all positive polynomial n-dimensional operators).

We pose the question: how well do positive operators approximate Sobolev classes? Put

$$\lambda_n^+(\alpha, p, q) =: \inf\{\lambda(W_p^\alpha(\mathbf{T}^1), A_n, L_q(\mathbf{T}^1)) | A_n \in \mathscr{L}(C(\mathbf{T}^1), \mathscr{T}_N), A_n \geqslant 0\}$$

Theorem 80. $/1 < p, q < \infty/ \Rightarrow$

$$\lambda_n^+(\alpha, p, q) \asymp \begin{cases} n^{-\gamma(1/q)}, & 0 < \gamma(1/q) < 2, \\ n^{-2}, & \gamma(1/q) > 2. \end{cases}$$

This result is due to A.I. Kamzolov, who also included all the boundary cases $p, q = 1, \infty, \gamma(1/q) = 2, \gamma(\cdot)$ the function introduced in § 3. From Theorem 80 it is clear how saturation originates under approximation by positive operators.

7.2. Approximation of Functions by Polynomials in the Complex Domain. The general problem here is posed as follows: let $T \subset \mathbf{C}^1$ be a closed set and $\{\mathscr{P}_n\}$ a collection of algebraic polynomials. What do the direct and inverse theorems look like in this situation?

The level of Weierstrass' theorem was dicussed in Chap. 1, § 4 where a theorem due to S.N. Mergelyan was stated which gave necessary and sufficient conditions on T, under which $d(x(\cdot), \mathscr{P}_n, C(T)) \to 0 \; \forall x(\cdot) \in A(T) \cap C(T)$.

Now the "level of Bernstein-Jackson" will be of interest to us.

7.2.1. Faber Polynomials and Representation of Functions Analytic Inside some Level Line. Theorem 49 gave a constructive characteristic of functions analytically continuable from a disc to concentric discs or from a segment to an ellipse E_R. Now we discuss the situations where, in general, polynomials are the best apparatus of approximation.

Let T be a bounded continuum in $\mathbf{C}$ with a simply-connected complement. Then, as is well known from the theory of functions of a complex variable, there is a unique function $\Phi(\cdot)$, Φ: $\mathbf{C}\setminus T \to \mathbf{C}$, which conformally maps $\mathbf{C}\setminus T$ onto the interior of the unit disc with the condition $\lim_{z\to\infty}(\Phi(z)/z) > 0$. The *Faber polynomial* of n-th order for T is the polynomial $F_n(\cdot)$, which consists of the terms with non-negative powers in the expansion of $(\Phi(\cdot))^n$ in a neighbourhood of $z = \infty$. For the disc $D_R = \{z: |z| \leqslant R\}$ the Faber polynomial is the power $(z/R)^n$, for $\hat{I}$ it is the polynomial $2T_n(\cdot)$, where $T_n(\cdot)$ is a Chebyshev polynomial. On Faber polynomials see Suetin [1984].

Later on it is reasonable to suppose that T satisfies certain conditions. When these conditions (described on p 347 of Dzyadyk [1977]) are satisfied we will call

T an *admissible continuum.* The level lines of an admissible continuum are the sets $\{z: |\Phi(z)| = R, R > 1\}$, for $R = 1$ we suppose $\Gamma_1 = \partial T$. The domain bounded by Γ_R, will be called *canonical* and denoted G_R.

Theorem 81 (The Bernstein-Walsh direct and inverse theorem).

$$x(\cdot) \in A^{G_R} \Leftrightarrow \limsup_{n\to\infty} (d(x(\cdot), \mathscr{P}_n, C(T))^{1/n} \leqslant R^{-1}.$$

For an upper estimate expand the function into a series relative of Faber polynomials, for a lower estimate use the following relation (the *Bernstein-Walsh inequality*):

Theorem 82.

$$b(BA^{G_R}, \mathscr{P}_N, C(T)) \asymp R^{-N}.$$

In the next chapter it will be shown that $\mathscr{P}_N$ is extremal or almost extremal for the classes BA^{G_R}. Thus it is made clear that precisely A^{G_R}, where G_R is a canonical domain of some continuum T, is the natural sphere of polynomial approximation. In the remaining cases, polynomials are asymptotically not optimal and we must to apply rational or some other approximation.

7.2.2. Constructive Theory of Functions in the Complex Domain. This theme was remarked in Chap. 1. There we said that from the constructive point of view the periodic case was distinguished from the case of a segment, if the approximation on the segment was accomplished by polynomials. It is natural to raise the question: how will the direct and inverse theorems of constructive theory of functions look if we consider approximation by polynomials in some sufficiently arbitrary domain G of the complex plane? This question is in fact related to complex analysis and, apparently, will be covered in other articles in this series. Here we only briefly touch on this theme.

The creation of the constructive theory of functions in the complex domain is one of the impressive achievements of the Ukrain mathematical school (V.I. Belyi, V.K. Dzyadyk, P.M. Tamrazov, I.A. Shevchuk and others). A special role in this cycle of works was played by the research of V.K. Dzyadyk, who was one of the first to develope this theme, created the school and obtained a number of the principle results. A summary of achievements in the constructive theory of functions in the complex domain is given in a number of monographs and survey works, for example in Dzyadyk [1977], Tamrazov [1975].

The qualitative results are as in the case of the segment. Both there and here a basic role is played by the function $\rho_N(t) = d(N^{-1}, t)$, where $d(\rho, t)$ is the distance of the point $t \in \partial G$ from the ρ-th level line of the Green's function of $\bar{\mathbf{C}} \setminus G$ with a singularity at infinity. For the case of segment a function, equivalent to $\rho_N(t)$, is given explicitly: $((1 - t^2)^{1/2} N^{-1} + N^{-2})$. At corner points of a level lines one is closer to the continuum than at points of smoothness, and this is made clear by the difference in nature of the approximation at corner and non-corner points. Similarly this also explains the difference in the inequalities for derivatives in the case of the disc and in the case of the segment. By Bernstein's inequality, the exponent of N in the inequality for derivatives of polynomials is equal to one, but in the case of a segment it is two. And again this is because the level line approaches the disc with one speed and the segment with another. (This gives two more examples, where the essence of the problem is revealed by the "output in the complex domain".)

In this connection, there are a variety of opinions as to the perspective of complex analysis in the development of real problems in approximation theory. We give two diametrically opposed opinions. In 1912 S.N. Bernstein, inspired by the first successes of the constructive theory of functions wrote: "From that time [when Weierstrass proved his theorem on approximation by polynomials of continuous functions] the theory of functions of a complex variable, attaining its greatest development at that time, gradually receded into the background, while the study of functions of a real variable advanced". V.D. Erokhin (untimely lost and a very talented mathematician, on whom we speak later) asserted somewhat the opposite. He reckoned that it is difficult to find any interesting facts of real analysis, which were not made completely meaningful via complex analysis.

The second point of view is more impressive, but one must not deny the converse influence of real analysis on the complex. Thus, for example, in the approximation of analytic function with singularities, alongside the rational functions (clearly a complex object), the same order of approximation is given by piecewise-polynomial mappings, a new object in complex analysis. Such mappings must (and do) play a special role in approximations questions for functions of finite smoothness. We comment on this in the next subsection.

7.3. Approximation of Functions of Many Variables

7.3.1. Local Best Approximations of Functions. The idea of describing functions of finite smoothness by local approximations is due to S.N. Bernstein, He obtained the following result, (Bernstein [1937]).

Theorem 83. *For a function $x(\cdot)$ to belong to $C^{n+1}(\Delta)$ it is necessary and sufficient that there is a function $\varphi(\cdot) \in C(\Delta)$ such that for any point τ and uniformly as $\tau' \to \tau$, $\tau'' \to \tau$, $\tau' < \tau < \tau''$ there is the relation*:

$$\lim_{\tau' \to \tau, \tau'' \to \tau} d(x(\cdot), \mathscr{P}_n, C([\tau', \tau'']))/(\tau'' - \tau')^{n+1} = \varphi(\tau) \geqslant 0.$$

Here of course

$$\varphi(\tau) = ((n+1)!)^{-1} 2^{-2n-1} |x^{(n+1)}(\tau)|$$

This theme was developed by D.A. Raikov, H. Whitney and others. And in fact, whenever smoothness is given in the standard way, that is, via local approximation by polynomials, it is natural to give it with the help of piecewise-polynomial mappings. This idea was very widely developed by Yu.A. Brudnyj, who constructed an interesting and complete theory of description of the majority of classical spaces of smooth functions via local approximations. On the other hand, there are a huge number of papers on approximations of functions by splines (the evidence for this can be obtained from Ural Scientific Centre [1984]).

It is not excluded that in a short time spline-approximation and piecewise-polynomial approximation of smooth functions will attain the level of harmonic approximation.

7.3.2. Constructive Theory of Smooth Functions of Many Variables in Arbitrary Domains. How well do polynomials approximate continuous (and even arbitrary) functions on sets of sufficiently general form? A series of interesting papers by V.N. Konovalov are devoted to this question. Certainly one cannot expect to obtain results of Jackson type. For example, the relation

$$d(x(\cdot), \mathscr{P}_N, C(T)) \leqslant C\omega(N^{-1}, x(\cdot), C(T))$$

for a T, consisting of two disjoint segments, is impossible. For the statements of direct theorems it is necessary to introduce new characteristics of domains and average the moduli of continuity. The reader may obtain an idea of the form of theorems of Jackson type from Konovalov [1986]. One gains the impression, that in the papers of V.N. Konovalov a new and interesting direction in approximation theory has started. But it has hardly been noticed and at present, perhaps, it is difficult to say how final are the results obtained up to now.

7.4. Polynomial Interpolation and Quadratures. Earlier we mentioned two classical interpolation formulae; Newton and Lagrange. Slightly later we discussed their comparison relative to complexity. Here we discuss the problem of accuracy. We pose the question: how good is polynomial interpolation?

We analyse one ancient example (Runge). Consider the function $\bar{x}(t) = (1 + 25t^2)^{-1}$ and the sequence of interpolating polynomials $\{p_N(\cdot)\}_{N \geqslant 0}$ interpolating $\bar{x}(\cdot)$ on $[-1, 1] =: \hat{I}$ at equidistant points: $\bar{x}(\bar{\tau}_i) = p_N(\bar{\tau}_i)$, $\bar{\tau}_i = -1 + 2(i-1)/N$, $i = 1, \ldots, N+1$. The function $\bar{x}(\cdot)$ is analytic and therefore it would appear there are no dirty tricks here and one can expect that $e_N =: \|\bar{x}(\cdot) - p_N(\cdot)\|_{C(\hat{I})} \to 0$. However, it turns out that $e \to \infty$. What is the problem?

Let $\tau = (\tau_1, \ldots, \tau_{N+1})$ be a system of distinct points on $\hat{I}$. The operator $\mathscr{T}_\tau$, which associates with $x(\cdot)$ the polynomial from $\mathscr{P}_N$ interpolating $x(\cdot)$ relative to the system τ, is a linear operator from $C(\hat{I})$ to $C(\hat{I})$. The norm of this operator is called the *Lebesgue constant* of $\mathscr{T}_\tau$, We denote it by $\lambda_N(\tau)$.

Theorem 83. *For an equidistant system of nodes* $\bar{\tau} = (\bar{\tau}_1, \ldots, \bar{\tau}_{N+1})$, $\bar{\tau}_i = -1 + 2(i-1)/N$ *there is the estimate*:

$$\lambda_N(\bar{\tau}) \geqslant C \exp(N/2).$$

Theorem 84. *For any* τ *the inequality*

$$\lambda_N(\bar{\tau}) \geqslant 2\pi^{-2} \ln N + b_N,$$

holds, where b_N *is a bounded sequence.*

Theorem 85. *For $\hat{t} = (\hat{t}_1, \ldots, \hat{t}_{n+1})$, where $\hat{t}_i$ are the zeroes of the Chebyshev polynomial $T_{N+1}(\cdot)$, there is the inequality*

$$\lambda_N(\hat{t}) \leqslant 2\pi^{-1} \ln N + 4.$$

(All of these facts are contained, for example, in de Boor [1974].) These results prompt us to express the following opinions:

a) The reasons for divergence in Runge's example is that analyticity of the function $\bar{x}(\cdot)$ "is not enough", to cancel the divergence induced by the growth of the Lebesgue constant for a uniform system of points.

b) In polynomial interpolation we must take care (divergence is possible) but if it is used then we must use Chebyshev nodes.

Interpolation of functions with singularities is a separate theme. There, in general, interpolation by rational functions is preferable.

From Theorems 84 and 85 it is possible to deduce the following major result, due to Faber.

Theorem 86. *No matter what are the nodes of interpolation $\{\tau_{iN}\}$, $N \geqslant 0$, $i = 1, \ldots, N+1$, there is a continuous function $x(\cdot)$ such that the sequence of interpolating polynomials relative to the given system diverges.*

Let us say a few words about quadratures. There are several possible approaches to the problem of optimal quadratures. One of these—the Kolmogorov-Nikol'skij approach—will be discussed in the next chapter. The most interesting results in this direction were obtained in the one-dimensional case, and in the many-dimensional case, the approach, going back to Gauss is central. The monograph Mysovskikh [1981] is devoted to this circle of problems. See also Krylov [1959], Sobolev [1974]. We cannot dwell on this. Here we will briefly discuss certain results, on the Gaussian approach in the one-dimensional case.

Firstly an interesting episode, which many years ago attracted much attention. Now it has, perhaps, only historical interest. The episode is characteristic of the creative style of P.L. Chebyshev. Long ago multiplication was not a very convenient operation and P.L. Chebyshev suggested seeking quadrature formulae of the form

$$\int_I x(t)\,dt \approx a\left(\sum_{i=1}^{n} x(\tau_i)\right),$$

with only one multiplication. Following the Gaussian approach, P.L. Chebyshev posed the question: is it possible to choose the $n+1$ parameters so that the quadrature formula is exact for polynomials of degree $\leqslant n$. P.L. Chebyshev succeeded in constructing such formulae for $2 \leqslant n \leqslant 7$. What happened after that was not clear. The question was fully investigated by S.N. Bernstein.

Theorem 87. *The Chebyshev formula is still possible for $n = 9$ and does not exist for $n \geqslant 10$ and $n = 8$.*

Unfortunately, here the investigator expected failure. Gauss' theorem itself, covered in Chap. 1, admits a generalization to the case of spaces with weights.

Theorem 88. *For the quadrature formula*

$$\int_{\hat{I}} p(t)x(t)\,dt \approx \sum_{k=1}^{N} p_k x(\tau_k)$$

to be exact on $\mathscr{P}_{2N-1}$, *it is necessary and sufficient that*

1) $p_k = \int_{\hat{I}} L_k(t)p(t)\,dt$, *where the* $L_k(\cdot)$ *are the basic functions in the Lagrange interpolation formula, constructed with respect to the nodes* $\tau_1, \dots, \tau_N$;

2) *its nodes are the roots of a polynomial of degree* N

$$\omega(t) = \prod_{i=1}^{N} (t - \tau_i),$$

orthogonal, relative to the weight $p(\cdot)$ *and the segment* $\hat{I}$, *to all polynomials from* $\mathscr{P}_{N-1}$.

The proof of this fact is an almost verbatim repeat of that given in Chap. 1. Here is another generalisation of Gauss' theorem due to Turan.

Theorem 89. *Let* $q = 2s$ *be an even number. Then among quadrature formulae of the form*

$$\int_{\hat{I}} x(t)\,dt \approx \sum_{k=1}^{n} \sum_{j=0}^{q} a_{kj} x^{(j)}(\tau_k),$$

exact on all polynomials of degree $n(q+1)$, *there is a unique formula exact on polynomials of degree* $n(q+2)-1$. *The nodes of this formula* $\{\tau_k\}_{k=1}^{n}$ *are the zeroes of a real polynomial of degree* n *with leading coefficient equal to one, which solves the following extremal problem*:

$$\int_{\hat{I}} \left(t^n + \sum_{k=1}^{n} x_k t^{k-1}\right)^{2s} dt \to \inf.$$

In other words, this polynomial is least deviating from zero in the $L_{2s}(\hat{I})$ metric. Thus we have crossed the bridge to our very preliminary discussions (see Chap. 1, § 1).

7.5. Appendix

7.5.1. On a Class of Functions of Compact Support. We have investigated the methods of harmonic approximation and splines. Both have endured the test of time from the point of view of practical feasibility: both trigonometric polynomials and splines are the constant tools of calculators. But is it possible to find anything else that could be more convenient?

In the next chapter we will give some "motivation" for the feasibility of spaces of trigonometric polynomials and splines. In particular, these spaces are extremal in the sense of Kolmogorov width for the Sobolev classes $W_p^r(\mathbf{T}^1)$ in L_p.

Here trigonometric polynomials are universal; they simultaneously serve, for example, all the classes $W^r_\infty(\mathrm{T}^1)$ in L_∞ but the methods of approximation depend on r and are complicated. Splines of order $r-1$ relative to the uniform partition realize the best approximation for simple interpolation but are not universal. There arises the question: is it possible to obtain a combination of the qualities of trigonometric polynomials and splines? One such attempt was made by V.A. Rvachev.

In 1971 V.L. Rvachev and V.A. Rvachev (see Rvachev & Rvachev [1979]) described the properties of a remarkable infinitely differentiable function, which they denoted by the symbol up $(\cdot)$. up $(\cdot)$ is of compact support with support in $\hat{I}$, is a "hill" function enclosing unit area whose derivative is of "hill" or "hole" form and is homothetic to the function itself. It is easy to give an analytic description of such a function. It has many remarkable properties. In particular, it turns out that if we take the space of shifts of up-functions with uniform step, then we obtain a space with extremal, or almost extremal, properties in the sense of Kolmogorov widths for the classes $W^r_p(\mathrm{T}^1)$ in L_p, $p = 2$ or ∞. Here it is difficult to estimate the perspective and practical feasibility of this concrete approach, but it undoubtedly merits attention. The search for similar new methods of approximation deserves to continue.

7.5.2. Approximation Theory in the Hausdorff Metric. The Hausdorff metric came to be used in questions of approximation of functions more than twenty years ago. It first appeared in the works of Penkov and Sendov in 1962. As we recall, for a long time the fundamental metric in the approximation theory was the uniform, Chebyshev, metric. Then came research into approximation in others, mainly integral, metrics. The merit of the Bulgarian mathematician Sendov was the promotion of the feasibility of constructing approximation theory in the Hausdorff metric. Already very preliminary statements showed that here one is uncovering a distinctive unusual new world, with vivid specificity and uniqueness. Over many years a large school of pupils and followers of Sendov formed, who took approximation theory in the Hausdorff metric a long way. Then mathematicians of other countries joined in this research and, in particular, in the USSR, these problems much occupied E.P. Dolzhenko and his pupils. Unfortunately, for lack of space it is impossible to discuss this in detail. We refer the reader to the monograph by Sendova [1979].

7.5.3. Approximations with Restrictions. Thirty years ago (in 1956) there appeared a paper by Ganelius devoted to one-sided approximation. Up to now a huge number of papers have been devoted to this and related themes. Of particular interest are the solutions obtained of the extremal problems which arise. However, the whole of this theme is somewhat to one side of the questions considered in this article. The reader may obtain information on this domain of approximation theory from Babenko [1983] and the monograph Kornejchuk, Ligun & Doronin [1982].

Chapter 3 Best Methods of Approximation and Recovery of Functions

Here we discuss the third stage of approximation theory. We describe methods of best approximation by subspaces, methods of best Fourier, linear and non-linear approximation and methods of optimal recovery of functions in classes.

§ 1. Preliminary Information

1.1. Definitions. In the introduction to this part we gave the definitions of many widths and all the definitions fitted into one scheme. Here we again give the definitions of the basic widths. Namely those which, in the main, will appear later.

Let X be a normed space with unit sphere B, W a convex, centrally-symmetric subspace of X. For the definition of the Fourier width we require that there should be a Hilbert space H everywhere dense in X and $W \subset H$.

The *Fourier N-width* of the class W is the value

$$\varphi_N(W, X) = \inf_{L_N \in \mathrm{Lin}_N(H)} \sup_{x \in W} \|x - \mathrm{Pr}\, L_N x\|, \tag{1}$$

where $\mathrm{Pr}\, L_N x$ is the orthogonal projection of x onto the subspace L_N, and the infimum is taken over all subspaces of H of dimension $\leqslant N$.

The *linear N-width* is the value

$$\lambda_N(W, X) = \inf_{L_N \in \mathrm{Lin}_N(X),\, \Lambda \in \mathscr{L}(X, L_N)} \sup_{x \in W} \|x - \Lambda x\|, \tag{2}$$

where the infimum is taken over all subspaces of X of finite dimension $\leqslant N$ and all linear continuous operators from X to $L_N (\Leftrightarrow \Lambda \in \mathscr{L}(X, L_N))$.

The *Kolmogorov N-width* is the value

$$d_N(W, X) = \inf_{L_N \in \mathrm{Lin}_N(X)} \sup_{x \in W} \inf_{y \in L_N} \|x - y\| \left(\Leftrightarrow \inf_{L_N} \inf_{\varepsilon} (\varepsilon B + L_N \supset W) \right). \tag{3}$$

The *Aleksandrov N-width* is the value

$$a_N(W, X) = \inf \sup_{x \in W} \|x - f(x)\|, \tag{4}$$

where the infimum is taken over all N-dimensional complexes K_N, lying in X and all continuous mappings f from W to K_N.

The *Bernstein N-width* is the value

$$b_N(W, X) = \sup_{L \in \mathrm{Lin}_{N+1}(X)} \sup_{\varepsilon > 0} (\varepsilon B \cap L \subset W). \tag{5}$$

The *Gel'fand N-width* is the value

$$d^N(W, X) = \inf \sup_{x \in W \cap L_{-N}} \|x\|,$$

where the infimum is taken over all subspaces L_{-N} of codimension N. (We also write $b_N(W)$ and $d^N(W)$).

The *ε-entropy* of W in X is the logarithm of the least number of elements of an ε-net for W in X. It is denoted $H_\varepsilon(W, X)$. This is the quantity inverse to the width $\varepsilon_N(W, X)$ mentioned in the introduction.

The widths (1)–(3) characterise the best approximative possibilities of the Fourier method, any N-dimensional linear method, and also simply the possibility of approximations by N-dimensional subspaces. The Aleksandrov width characterises the best accuracy of approximation by continuous operators with N-dimensional image. The Gel'fand width plays a major role in questions of interpolation and recovery of functions. The Bernstein width characterises the best subspace from the point of view of inequalities of the Bernstein-Nikol'skij type, the entropy $\mathscr{H}_\varepsilon(W, X)$ shows the approximative possibilities of finite point sets. But, undoubtedly, of most importance in approximation theory is the Kolmogorov width which is the most closely connected with the basic direction of classical approximation theory.

1.2. Historical Information and Commentary. Widths were introduced for various reasons. They arose first in dimension theory. In 1922 I.S. Uryson introduced the widths named after him. Let W be a metric space. The *Uryson width* $u_N(W)$ is the infimum of those ε for which there exists a covering of W by open sets of diameter $< \varepsilon$ of multiplicity $N + 1$ (that is, such that each point is covered by $\leqslant N + 1$ sets and some point is covered by exactly $N + 1$ sets).

The introduction of $u_N(W)$ was obviously inspired by the Lebesgue-Brouwer definition of dimension and it as it were characterises "deviation from N-dimensionality". The width a_N was introduced by P.S. Aleksandrov in 1933, the width d_N by A.N. Kolmogorov in 1936, the widths λ_N and b_N by V.M. Tikhomirov in 1960. In a discussion of Tikhomirov's paper with its author, I.M. Gel'fand expressed the idea that there must be a dual to the Kolmogorov diameter, which characterizes best interpolation.[1] Such a quantity was introduced in the article of V.M. Tikhomirov (1955) and called the Gel'fand N-width. Independently similar quantities were introduced in 1965 by S.A. Smolyak and I.F. Sharygin in their dissertations. Fourier width was introduced by V.N. Temlyakov (1982).

The notion of ε-entropy arose in the work of A.N. Kolmogorov (1956). It must be said that its apperance was influenced by the ideas of information theory. In order to distinguish one definite element of a finite set C, with $N(C)$ elements, it is sufficient to give $[\log_2 N(C)] + 1$ binary places. "Therefore it is possible to regard the value $H(C) = \log_2 N(C)$, given by A.H. Kolmogorov in this paper, as

[1] The introduction of such a quantity was suggested by the duality theorem of S.M. Nikol'skij.

a measure of the "amount of information" contained in the indication of one definite element of the set C. For infinite sets it is natural to consider methods of approximation of the given elements". The problem of recovery of an element up to ε using the least number of binary places leads to the quantity $\mathscr{H}_\varepsilon(W)$, equal to the binary logarithm of the number $N_\varepsilon(W)$, where $N_\varepsilon(W)$ is the least number of sets of diameter $\leqslant \varepsilon$, covering W. This quantity is called the *absolute ε-entropy*, and its inverse the *entropic co-width*. Regarding all of this see Kornejchuk [1976, 1984], Lorentz [1966], Pinkus [1985], Tikhomirov [1976].

1.3. Relations Between Widths

1.3.1. Some General Properties of Widths. Let X be a normed space, W a convex centrally-symmetric subset. Then the following properties hold:

a) Monotonicity: $W \subset W_1 \Rightarrow d_N(W, X) \leqslant d_N(W_1, X)$; X isometrically embedded in $X_1 \Rightarrow d_N(W, X) \geqslant d_N(W, X_1)$;

b) homogeneity $d_N(\alpha W, X) = |\alpha| d_N(W, X)$;

c) $d_N(\mathrm{cl}\, W, X) = d_N(W, X)$.

These properties are characteristic also for the other diameters φ_N, λ_N, a_N.

1.3.2. Equalities

Theorem 1. a) *If W is compact metric then*

$$a^N(W) = u_N(W).$$

b) *If W is a centrally-symmetric subspace of a normed space X, then*

$$\lambda^N(W) = 2d^N(W, X).$$

c) *Let C and B be convex centrally-symmetric bodies in $\mathbf{R}^n$, $\mathbf{R}^n_C$ and $\mathbf{R}^n_B$ the n-dimensional spaces with unit spheres C and B respectively. Then*

$$b_{n-N-1}^{-1}(C, \mathbf{R}^n_B) = d^N(B, \mathbf{R}^n_C).$$

d) *Let B and C be convex centrally-symmetric bodies in $\mathbf{R}^n$. Then*

$$d^N(B^0, \mathbf{R}^n_{C^\circ}) = d_N(C, \mathbf{R}^n_B),$$

where B° and C° are the polars of the sets B and C.

e) *Under the assumptions of d) $\lambda_N(C, \mathbf{R}^n_B) = \lambda_N(C^\circ. \mathbf{R}^n_{B^\circ})$.*

The quantity a^N, the *Aleksandrov co-width*, was defined in the introduction. It is interesting that two quantities (a^N and u_N), defined from totally different expressions (one of dimensions, the other of coding) should coincide. The same can be said about relation b).

1.3.3. Absolute Widths. In the notations of the widths φ_N, λ_N, d_N, ε_N, a_N both W and X occur. Each of these widths may diminish by isometrically embedding X in an enveloping space. This situation makes it natural to introduce a notion of absolute width. Let p_N be one of the widths φ_N, λ_N, d_N, ε_N, or a_N. We define the *absolute width* $P_N(W)$ as $\inf p_N(W, \tilde{X})$, where the infimum is taken over all Banach spaces $\tilde{X}$, in which it is possible to isometrically embed X. We denote the absolute widths by Φ_N, Λ_N, D_N, E_N and A_N. The notion of absolute width was introduced by R.S. Ismagilov, although the first prototype of it (connected with ε-entropy) was studied by A.G. Vitushkin.

Theorem 2. *a) Let W be compact metric, then*

a) $2E_N(W) = \varepsilon^N(W)$;

b) $2A_N(W) = a^N(W) = u_N(W)$;

c) *Let W be a convex centrally-symmetric subset in a normed space. Then*

$$2\Lambda_N(W) = \lambda^N(W)(= d^N(W, X)).$$

1.3.4. Inequalities Between Widths

Theorem 3. *Let X be a normed space, $W \subset X$ the algebraic sum of a compact and a finite-dimensional space. Then the following inequalities hold:*

a) $a_N(W, X) \leqslant d_N(W, X) \leqslant \lambda_N(W, X) \leqslant \varphi_N(W, X)$, $a^N(W) \leqslant \lambda^N(W)$;

b) $b_N(W) \leqslant d_N(W, X)$, $b_N(W) \leqslant d^N(W, X)$, $b_N(W) \leqslant 2a_N(W, X)$;

c) $\lambda_N(W, X) \geqslant d^N(W)$;

d) $\mathscr{H}_\varepsilon(W) \leqslant \mathscr{H}_\varepsilon(W, X) \leqslant \mathscr{H}_{\varepsilon/2}(W)$.

We note that if X is a Hilbert space, then, of course,

$$\varphi_N(W, X) = \lambda_N(W, X) = d_N(W, X) \quad \text{and} \quad d_N(W, X) \geqslant d^N(W, X).$$

Width may be decreased by isometric immersion of spaces, but the decrease is subject to the following natural estimate (having a multitude of applications in concrete situations).

Projection Lemma. *Let X be a normed space, L a subspace, $W \subset L$ and P: $X \to L$ a continuous linear projection, then*

$$d_N(W, X) \leqslant d_N(W, L) \leqslant \|P\| d_N(W, X).$$

1.4. Some Calculations and Estimates of Widths of Finite-Dimensional Sets. A precise calculation of the values of widths succeeds in a small number of cases—usually this is a fairly difficult extremal problem. A great deal of stress has been placed on the calculation of widths of the so-called *p-ellipsoids*

$$B_p^n(a) = \left\{x \in \mathbf{R}^n \,\middle|\, \sum_{k=1}^{n} (|x_k/a_k|)^p \leqslant 1\right\}, \quad 1 \leqslant p < \infty,$$

$$B_\infty^n(a) = \{x \in \mathbf{R}^n | \max|x_k/a_k| \leqslant 1\}, \quad a = (a_1, \dots, a_n),$$

$$a_1 \geqslant a_2 \geqslant \cdots \geqslant a_n > 0, \quad B_p^n(1) =: B_p^n.$$

We still call $B_1^n(a)$ an octahedron, and $B_\infty^n(a)$ a parallelopiped. We denote by $\Sigma^n = \{x \in \mathbf{R}^{n+1} \colon x \geqslant 0, \sum_{i=1}^{n+1} x_i = 1\}$ the regular n-dimensional simplex. The calculation and estimation of the widths of $B_p^n(a)$ and Σ^n in the metric l_q^n play a significant role in the computations of widths of function classes. We quote some of the most important, in our opinion, results. We begin with the case l_2^n.

Theorem 4. a) $\varphi_N(B_2^n(a), l_2^n) = \lambda_N = d_N = b_N = a_N(B_2^n(a), l_2^n) = a_{N+1}$;
b) $\varphi_N(B_1^n, l_2^n) = \lambda_N = d_N = (1 - (N/n))^{1/2}, 0 \leqslant N \leqslant n,$

$$b_N(B_1^n, l_2^n) = a_N(B_1^n, l_2^n) = (N + 1)^{-1/2};$$

c) $\varphi_N(B_1^n(a), l_2^n) = \lambda_N = d_N = \max\limits_{N \leqslant m \leqslant n} \left((m - N)\left(\sum\limits_{k=1}^{m} a_k^{-2}\right)^{-1}\right)^{1/2}$;

d) $d_N(\Sigma^n, l_2^{n+1}) = (1 - (N + 1)/(n + 1))^{1/2}$, *except for two cases:*

$$d_{2s-1}(\textstyle\sum^{2s}, l_2^{2s+1}) = (2s + 1 - (2s + 1)^{-1})^{1/2}$$

and

$$d_1(\Sigma^{2s}, l_2^{2s+1}) = (1 - (2/(2s + 1)) - (1/8(2s + 1)s).$$

Picture the square with sides of length one, formed by the points $(1/p, 1/q)$ (see Fig. 2). We divide this square into 5 domains:

$$\mathrm{I} = \{(1/p, 1/q) | 1/p \leqslant 1/q\},$$
$$\mathrm{II} = \{(1/p, 1/q) | 1/q \leqslant 1/p, 1/2 \leqslant 1/q\},$$
$$\mathrm{III} = \{(1/p, 1/q) | 1/q \leqslant 1/2, 1 \leqslant 1/p + 1/q\},$$
$$\mathrm{IV} = \{(1/p, 1/q) | 1/2 \leqslant 1/p, 1/p + 1/q \leqslant 1\},$$
$$\mathrm{V} = \{(1/p, 1/q) | 1/q \leqslant 1/p, 1/p \leqslant 1/2\}.$$

It follows from Theorem 5 that exact solutions of the problem of widths of the pairs $(B_p^n(a), l_q^n)$ can be obtained in the "upper triangle" $I = \{(1/p, 1/q): 1/p \leqslant 1/q\}$ for the widths φ_N, λ_N and d_N and in the "lower triangle" (that is the union of the domains II–V) for a_N and b_N.

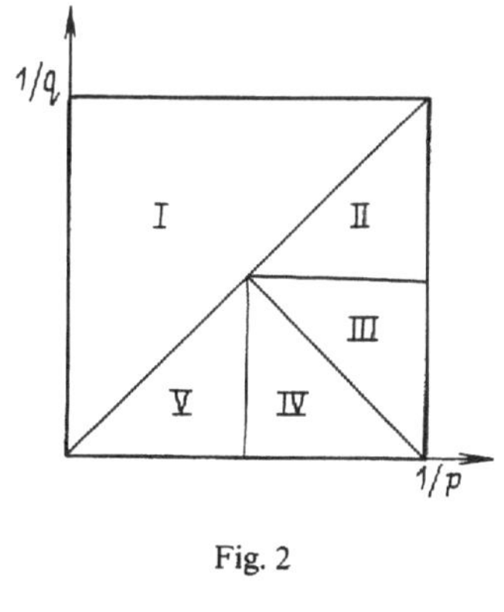

Fig. 2

Theorem 5. a) *For* $1 \leqslant p < q \leqslant \infty$

$$a_N(B_p^n(a), l_q^n) = b_N(B_p^n(a), l_q^n) = \left(\sum_{k=1}^{N+1} a_k^{pq/(p-q)}\right)^{(p-q/pq)}.$$

b) *For* $1 \leqslant q < p \leqslant \infty$

$$\varphi_N(B_p^n(a), l_q^n) = \lambda_N(B_p^n(a), l_q^n) = d_N(B_p^n(a), l_q^n) = \left(\sum_{k=N+1}^{n} a_k^{pq/(p-q)}\right)^{(p-q/pq)}.$$

c) *For* $p = q$

$$\varphi_N(B_p^n, l_q^n) = \lambda_N = d_N = b_N = a_N(B_p^n(a), l_q^n) = a_{N+1}.$$

The values of the widths (that is, the quantities $(\sum a_k^{pq/(p-q)})^{(p-q)/pq}$) have a simple geometric meaning, they are the stationary values of the function $x \to \|x\|_{l_q^n}$ on the manifold $\partial B_p^n(a)$. In the lower triangle (for the widths φ_N, λ_N and d_N) it appears that there should also be some interesting quantities having extremal significance. But none are known, and there are not even any good conjectures, in spite of the presence of concrete material (in particular, $d_2(B_1^4, l_\infty^4) = \sqrt{2} - 1$, $d_2(B_1^6, l_\infty^6) = 2\sqrt{3} - 3$, $\varphi_N(B_1^n, l_\infty^n) = 1 - N/n$ and a number of others have been calculated).

Apparently, the calculation of φ_N, λ_N, d_N in the lower triangle (except the point (1, 1/2), where all have been done; see Theorem 4 c)) is a very difficult problem.

Now we give two theorems on the asymptotic behaviour of the widths λ_N and d_N of the pair (B_p^n, l_q^n). They have played a most important role in the solution of the problem on the asymptotic form of Kolmogorov widths for Sobolev classes. Because of space constraints, unfortunately, we cannot comment on the proofs of the following two very beautiful theorems.

Theorem 6. *Let* $1 \leqslant p, q \leqslant \infty$. *Then*

a) $\varphi_n(B_p^{2n}, l_q^{2n}) \asymp n^{(1/q-1/p)_+}$.

b) $\lambda_n(B_p^{2n}, l_q^{2n}) \asymp \begin{cases} d_n(B_p^{2n}, l_q^{2n}), & 1/p + 1/q \geqslant 1, \\ d_n(B_{q'}^{2n}, l_{p'}^{2n}), & 1/p + 1/q \leqslant 1. \end{cases}$

c) $d_n(B_p^{2n}, l_q^{2n}) \asymp \begin{cases} n^{1/q-1/p} & \text{in I} \cup \text{V}, \\ 1 & \text{in II}, \\ n^{1/q-1/2} & \text{in III} \cup \text{IV}, \end{cases}$

$$1/p + 1/p' = 1/q + 1/q' = 1.$$

Put

$$\Phi(n, N, p, q) = \begin{cases} \max(n^{1/q-1/p}, (1 - N/n)^{(1/p-1/q)/(1/p-1/2)}) & \text{in II}, \\ \max(n^{1/q-1/p}, \min(1, n^{1/q}N^{-1/2}))(1 - N/n)^{1/2} & \text{in III} \cup \text{II}, \\ \min(1, n^{1/q}N^{-1/2})^{(1/p-1/q)/(1/2-1/q)} & \text{in V}; \end{cases}$$

$$\Psi(n, N, p, q) = \begin{cases} \Phi(n, N, p, q), & 1 \leqslant p < q \leqslant p', \\ \Phi(n, N, q', p'), & \max(p, p') < q < \infty. \end{cases}$$

Theorem 7. *Let* $1 \leqslant p \leqslant q < \infty$, *or* $2 \leqslant p < \infty$, $q = \infty$, $N < n$. *Then*

a) $d_N(B_p^n, l_p^n) \asymp \Phi(n, N, p, q)$,

b) $\lambda_N(B_p^n, l_q^n) \asymp \Psi(n, N, p, q)$.

Here we have barely touched the extremal geometric problems, connected with approximation theory. We can hope that there will be very many more brilliant achievements.

§ 2. Widths and Entropies of Classes of Smooth Functions

Here we have the opportunity to compare the best classical methods of optimality (at an asymptotic level). We begin with the one-dimensional case.

2.1. Widths of Sobolev Classes of Functions of One Variable. Here we will give results, relating to the pair $(B\mathring{W}_p^\alpha(\mathbf{T}^1), L_q(\mathbf{T}^1))$. To simplify the statements, we will throughout suppose that $1 < p, q < \infty$.

Theorem 8 (on Fourier-widths). $/\alpha > (1/p - 1/q)_+/ \Rightarrow$

$$\varphi_N(B\mathring{W}_p^\alpha(\mathbf{T}^1), L_q(\mathbf{T}^1)) \asymp \varphi(B\mathring{W}_p^\alpha(\mathbf{T}^1), \mathcal{T}_N, L_q(\mathbf{T}^1)) \asymp N^{-\alpha+(1/p-1/q)_+}.$$

Theorem 9 (on linear widths). $/\alpha > 0$ *in* I, $\alpha > \dfrac{1/p - 1/q}{1 - 2/q}$ *in II*, $\alpha > 1/p$ *in* III $\alpha > 1 - 1/q$ *in* IV, $\alpha > (1/p + 1/q) + \omega$ *in* V$/ \Rightarrow$

$$\lambda_N(B\mathring{W}_p^\alpha(\mathbf{T}^1), L_q(\mathbf{T}^1)) \asymp \begin{cases} \varphi(B\mathring{W}_p^\alpha(\mathbf{T}^1), \mathcal{T}_N, L_q(\mathbf{T}^1)) & in\ \mathrm{I} \cup \mathrm{II} \cup \mathrm{V}, \\ d_N(B\mathring{W}_p^\alpha(\mathbf{T}^1), L_q(\mathbf{T}^1)) & in\ \mathrm{III}, \\ N^{-\alpha+1/2-1/q} & in\ \mathrm{IV}. \end{cases}$$

Theorem 10 (on Kolmogorov widths). $/\alpha > (1/p - 1/q)_+$ *in* I ∪ II, $\alpha > 1/p$ *in* III ∪ IV, $\alpha > \dfrac{1/p - 1/q}{1 - 2/q}$ *in* V$/ \Rightarrow$

$$d_N(B\mathring{W}_p^\alpha(\mathbf{T}^1), L_q(\mathbf{T}^1)) =: d_N(\alpha, p, q) \asymp \begin{cases} N^{-\alpha} & in\ \mathrm{I} \cup \mathrm{V}, \\ N^{-\alpha+1/p-1/q} & in\ \mathrm{II}, \\ N^{-\alpha+1/p-1/2} & in\ \mathrm{III} \cup \mathrm{IV}. \end{cases}$$

Theorem 11 (on Aleksandrov widths and ε-entropy). $/\alpha \in N/ \Rightarrow$

a) $a_N(B\mathring{W}_p^\alpha(\mathbf{T}^1), L_q(\mathbf{T}^1)) \asymp N^{-\alpha}$;

b) $\mathcal{H}_\varepsilon(B\mathring{W}_p^\alpha(\mathbf{T}^1), L_q(\mathbf{T}^1)) \asymp \varepsilon^{-1/\alpha} \quad (1 \leqslant p, q \leqslant \infty)$.

From Theorems 8–10 it is clear that for "large" smoothness (when $\alpha \geqslant \alpha_0$) the speed of approximation is characterised by the four quantities: α, $\alpha - 1/p + 1/q$, $\alpha - 1/p + 1/2$ and $\alpha - 1/2 + 1/q$.

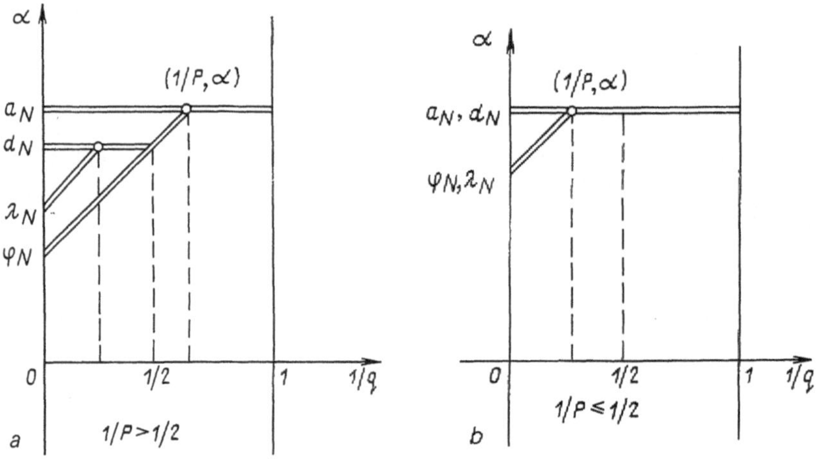

Fig. 3

In Fig. 3 we have tried to show graphically, in which cases and how effective is the corresponding method. We will repeat this, reverting to the previous figure. The Fourier method is optimal in the upper triangle I. In II it is optimal in the sense of the linear and Kolmogorov widths, but concedes to a nonlinear (Aleksandrov) approximation method. In V the Fourier method is optimal in the sense of the linear width, but not in the sense of the Komogorov width (see Theorem 10 and Theorem 23 in Chap. 2): in III ∪ IV ∪ V $d(B\mathring{W}_p^\alpha(\mathbf{T}^1), \mathcal{T}_N, L_q(\mathbf{T}^1))$ and $d_N(B\mathring{W}_p^\alpha(\mathbf{T}^1), L_q(\mathbf{T}^1))$ have different asymptotic forms.

We give another formula for "small smoothness":

$$d_N(B\mathring{W}_p^\alpha(\mathbf{T}^1), L_q(\mathbf{T}^1)) \asymp N^{-q/2(\alpha-1/p+1/q)} \quad 1/p - 1/q < \alpha < 1/p, \quad \text{in III} \cup \text{IV},$$

$$1/p - 1/q < \alpha < \frac{1/p - 1/q}{1 - 2/q} \quad \text{in} \quad V,$$

$$\lambda_N(B\mathring{W}_p^\alpha(\mathbf{T}^1), L_q(\mathbf{T}^1)) \asymp N^{-q/2(\alpha-1/p+1/q)}, \quad \alpha < 1/p, \quad \text{in III}, \asymp N^{-p'/2(\alpha-1/p+1/q)}$$

in IV for $\alpha < 1 - 1/q$.

The discovery of the phenomenon of different asymptotic forms for small smoothnesses is due to B.S. Kashin [1981] (see also the work of E.D. Kulanin [1983]). And here is the situation with the Bernstein-Nikol'skij inequality.

Theorem 12 (on Bernstein widths).

$$b_N(B\mathring{W}_p^\alpha(\mathbf{T}^1), L_q(\mathbf{T}^1)) \asymp \begin{cases} N^{-\alpha}, & 1/q \leqslant 1/p \quad or \quad 1/2 \leqslant 1/p \leqslant 1/q, \quad \alpha > 0 \\ N^{-\alpha+1/p-1/q}, & 1/p \leqslant 1/q \leqslant 1/2, \quad \alpha > 1/p, \\ N^{-\alpha+1/p-1/2}, & 1/p \leqslant 1/2 \leqslant 1/q, \quad \alpha > 1/p \end{cases}$$

It is clear from this result that the space of trigonometric polynomials $\mathcal{T}_N$, for $1/p \leqslant 1/2 \leqslant 1/q$ (cf. Theorem 12 and Theorem 28 of Chap. 2), is not optimal in the sense of the width b_N.

We cannot dwell in detail on the proof of these results. We say only that the proof proceeds by reduction to the finite-dimensional case (Theorems 5 and 6) via discretization. One convenient method is to use a result such as:

Theorem 13 (Marcinkiewicz). $/x(\cdot) \in \mathcal{T}_N$, $1 < p < \infty$, $\xi = (x(0), x(2\pi/(2N+1)), \ldots, x(4\pi N/(2N+1)))/ \Rightarrow \exists C_1, C_2$:

$$C_1 N^{-1/p} \|\xi\|_{l_p^{2N+1}} \leqslant \|x(\cdot)\|_{L_p(\mathbf{T}^1)} \leqslant C_2 N^{-1/p} \|\xi\|_{l_p^{2N+1}}.$$

The idea of discretization was first used in the problem of widths by R.S. Ismagilov, and the method was developed by V.E. Majorov, B.S. Kashin and E.D. Glushkin. We will prove, for example, how the estimate from below in Theorem 10 is obtained. Consider the subspace

$$L_{2N} = \operatorname{lin}\{\exp(ik\cdot), N < k \leqslant 3N\} \quad (\dim L_{2N} = 2N).$$

Later we will use a) the projection lemma, b) the theorem of Marcinkiewicz, c) Theorem 6.

We have

$$d_N(B\mathring{W}_p^\alpha(\mathbf{T}^1), L_q(\mathbf{T}^1)) \overset{\text{a)}}{\gg} d_N(B\mathring{W}_p^\alpha(\mathbf{T}^1) \cap L_{2N}, L_q(\mathbf{T}^1) \cap L_{2N})$$
$$\overset{\text{b)}}{\gg} N^{-\alpha+1/p-1/q} d_N(B_p^{2N}, l_q^{2N}) \overset{\text{c)}}{\gg} d_N(\alpha, p, q).$$

2.2. Trigonometric Widths. We have already said that in the domains II–V (Fig. 2) $\mathcal{T}_N$ is not asymptotically optimal. But we do not like to think, that harmonics are somehow not good enough. It is natural to pose the question: is it possible despite this to choose N harmonics (not necessarily in succession) such that the subspace spanned by them is asymptotically optimal. This leads to the definition of trigonometric (harmonic) widths (R.S. Ismagilov, 1974):

$$d_N^\Gamma(W, X) = \inf_{\{j_1, \ldots, j_N\}} \sup_{x \in W} d(x(\cdot), \operatorname{lin}\{\exp(ij_1), \ldots, \exp(ij_N)\}, X).$$

The widths φ_N^Γ, λ_N^Γ etc. are introduced in a similar way. The theme of "trigonometric widths" is far from exhausted and to many is intriguing. We give just one definitive result:

Theorem 14. $/\alpha > 1/ \Rightarrow$

$$\lambda_N^\Gamma(B\mathring{W}_p^\alpha(\mathbf{T}^1), L_q(\mathbf{T}^1)) \asymp \lambda_N(B\mathring{W}_p^\alpha(\mathbf{T}^1), L_q(\mathbf{T}^1)).$$

For the widths φ_N^Γ Theorem 8 settles everything, but the question of comparison of the asymptotic forms of b_N and b_N^Γ, d_N and d_N^Γ is at present open, although a number of special facts have accumulated.

This cycle of research must include works on the "best Lebesgue constant". In 1948 Littlewood made the following conjecture: there is an absolute constant $C > 0$ such that

$$\left\| \sum_{k=1}^{N} \exp(ij_k\cdot) \right\|_{L_1(\mathbf{T}^1)} \geqslant C \ln N$$

for any set $\{j_1, \ldots, j_N\}$, $j_k \in Z$.

The Littlewood conjecture was proved by S.V. Konyagin [1981] and almost simultaneously (and more simply) by McGehee, Pigno and Smith [1981]. It is natural to somewhat generalize Littlewood's statement of the problem and investigate the asymptotic forms of quantities of the form

$$L_N(\alpha, q) = \inf_{\{j_1, \ldots, j_N\}} \left\| \left(\sum_{k=1}^{N} \exp(ij_k\cdot) \right)^{(\alpha)} \right\|_{L_q(\mathbf{T}^1)}.$$

The solution of Littlewood's problem is equivalent to the following result.

Theorem 15.

$$L_n(0, 1) \asymp \ln N.$$

The asymptotic value of $L_N(\alpha, q)$ for any α and q has not yet been found. We give one (not definitive) result

Theorem 16. a) $/\alpha \geqslant 1/q, 0 \leqslant 1/q \leqslant 1/2$ or $\alpha \geqslant 0, 1 > 1/q > 1/2/ \Rightarrow$

$$L_N(\alpha, q) \asymp N^{\alpha + 1/q'}, \quad L_N(\alpha, 1) \asymp N^{\alpha} \ln N, \quad \alpha \geqslant 0$$

b) $/0 < \alpha < 1/q, q = 2(k+1), k \in \mathbf{N}/ \Rightarrow L_N(\alpha, q) \asymp N^{(\alpha q + 1)/2}$.

c) $/0 < \alpha < 1/q,\ 0 < 1/q < 1/2 \Rightarrow L_N(\alpha, q) \asymp N^{(\alpha q + 1)/2}$ (Belinsky, Konyagin, 1987), $\alpha = 1/q,\ 0 < 1/q < 1/2 \Rightarrow L_N(\alpha, q) \asymp N/\log^{1/q'} N$, $\alpha = 0,\ 0 < 1/q < 1/2$, $L_N(0, q) \asymp N^{1/2}$ (Belinsky, 1987). Thus even here there are still many unanswered questions.

2.3. Widths of Intersections of Sobolev and Nikol'skij Classes in the Many-Dimensional Case. In § 3 of Chap. 2 we discussed how to approximate Sobolev and Nikol'skij classes and their intersections by the Fourier method and by trigonometric polynomials with "natural" choices of harmonics. Here we consider the question of optimality of these methods. The first work on this theme is due to K.I. Babenko. B.S. Mityagin first applied the apparatus of harmonic analysis to similar problems. Essential progress in these problems was made recently by V.N. Temlyakov. A number of definitive results on the approximation of the intersection of classes are due to E.M. Galeev.[2]

We retain the notations of § 3 of Chap. 2.

Theorem 17 (on the Fourier-widths of the intersection of Sobolev classes). $/1 < q < \infty, (1/p, \alpha^i) \in \operatorname{int} I \times \mathbf{R}^n, A = \{(1/p, \alpha^i), i = 1, \ldots, m\}$,

$$\mathfrak{G}_q(A, \mathbf{T}^n) \cap \mathring{\mathbf{R}}^n_+ \neq \varnothing / \Rightarrow \varphi_N(B\mathring{W}^A, L_q(\mathbf{T}^n)) \asymp (N^{-1} \log^l N)^{1/M},$$

[2] In the one-dimensional case E.M. Galeev obtained the weak asymptotic form of the Kolmogorov width for the intersection of classes.

where M is the value, and l the dimension of the set of solutions of the problem

$$\langle s, \mathbf{1}\rangle \to \sup, \langle \alpha, s\rangle \leqslant 1, \alpha \in \mathfrak{S}_q(A, \mathbf{T}^n).$$

It follows from this theorem that the intersection of the stepped hyperbolic crosses constructed in Chap. 2, § 3 is actually optimal relative to order in the sense of Fourier-widths.

Theorem 18 (on linear widths of a Sobolev class). $/1 < p, q < \infty, \alpha = (\alpha_1, \ldots, \alpha_n), \alpha_1 = \cdots = \alpha_{l+1} < \alpha_{l+2} \leqslant \cdots \leqslant \alpha_n, \alpha_1 > (1/p - 1/q)_+ / \Rightarrow$

$$\lambda_N(B\mathring{W}_p^\alpha(\mathbf{T}^n), L_q(\mathbf{T}^n)) \asymp \begin{cases} (N^{-1}\log^l N)^{\alpha_1 - (1/p-1/q)_+} \text{ in I} \cup \text{II} \cup \text{V}, \\ (N^{-1}\log^l N)^{\alpha_1 - 1/p + 1/2} \text{ in III}, \alpha_1 > 1/p, \\ (N^{-1}\log^l N)^{\alpha_1 - 1/2 + 1/q} \text{ in IV}, \alpha_1 > 1 - 1/q. \end{cases}$$

Theorem 19 (on Kolmogorov widths of a Sobolev class). $/1 < p, q < \infty, \alpha = (\alpha_1, \ldots, \alpha_n), \alpha_1 = \cdots = \alpha_{l+1} < \alpha_{l+2} \leqslant \cdots \leqslant \alpha_n, \alpha_1 > (1/p - 1/q)_+ / \Rightarrow$

$$d_N(B\mathring{W}_p^\alpha(\mathbf{T}^n), L_q(\mathbf{T}^n)) \asymp \begin{cases} (N^{-1}\log^l N)^{\alpha_1} \text{ in I} \cup \text{V}, \ \alpha_1 > 1/2, \\ (N^{-1}\log^l N)^{\alpha_1 - 1/p + 1/2} \text{ in III} \cup \text{IV}, \ \alpha_1 > 1/p, \\ (N^{-1}\log^l N)^{\alpha_1 - (1/p-1/q)_+} \text{ in II}. \end{cases}$$

Problems on widths of intersections of Sobolev classes and even the usual Nikol'skij classes are not yet completely solved. We state some particular results in the form of two theorems.

Theorem 20. $/A = \{(1/p^i, \alpha^i), i = 1, \ldots, m\}, \alpha^i \in \mathbf{R}^n, 1 < p^i < \infty, 2 \leqslant q < \infty, \mathfrak{G}_2(A, \mathbf{T}^n) \cap \mathring{\mathbf{R}}^n_+ + \tfrac{1}{2} \neq \varnothing / \Rightarrow$

$$d_N \mathrm{BW}^A(\mathbf{T}^n), L_q(\mathbf{T}^n)) \asymp (N^{-1}\log^l N)^{1/M},$$

where M is the value and l the dimension of the set of solutions of the problem:

$$\langle s, \mathbf{1}\rangle \to \sup, \quad \langle \alpha, s\rangle \leqslant 1, \quad \alpha \in \mathfrak{G}_2(A, \mathbf{T}^n).$$

Theorem 21. $/1 < p, q < \infty, \alpha \in \mathbf{R}^n, 0 < \alpha_1 = \cdots = \alpha_{l+1} < \alpha_{l+2} \leqslant \cdots \leqslant \alpha_n / \Rightarrow$

$$d_N(\mathrm{BH}_p^\alpha(\mathbf{T}^n), L_q(\mathbf{T}^n)) \asymp \begin{cases} (N^{-1}\log^l N)^{\alpha_1}\log^{l/2} N, & q \leqslant p, 2 \leqslant p \text{ or} \\ & 2 \leqslant p < q, \alpha_1 > 1/2, \\ (N^{-1}\log^l N)^{\alpha_1 - 1/p + 1/2}\log^{1/2} N, & p \leqslant 2 \leqslant q, \\ & \alpha_1 > 1/p. \end{cases}$$

In conclusion a result on ε-entropy.

Theorem 22. $/1 \leqslant p, q \leqslant \infty, l/n > (1/p - 1/q)_+, l \in \mathbf{N} / \Rightarrow$

$$\mathscr{H}_\varepsilon(W_p^1(\mathbf{T}^n), L_q(\mathbf{T}^n)) \asymp \varepsilon^{-n/l}.$$

2.4. Some Discussion. Here we find ourselves at an interim stage in the theme "finite smoothness and approximation". This theme began, we recall, with the work of S.N. Bernstein and Jackson. For its natural completion it is desirable to obtain answers to the original questions: What is smoothness? How should one approximate smooth functions? Here we quote some of the progress on asymptotic results.

There is a feeling that fairly soon much of this will be finally cleared up. For the present we can only push off from the experience obtained from research into the approximation of Sobolev classes of one variable. Once again we name certain important phenomena, which, apparently, will be characteristic for general many-dimensional problems. For definiteness let the smoothness be given by a polyhedron (or more generally a subset) in $I^n \times \mathbf{R}^n$. Associated with this polyhedrom and the metric q, in which the approximation is realised, must be at least the four numbers, characterising the best Fourier-approximation, linear N-dimensional approximation, approximation by N-dimensional subspaces and the nonlinear approximation by N-dimensional manifolds; four functionals of the smoothness and q. Apparently they will shortly be named, and by the same token this part of the problem will be mastered. But what about the other branches of the old schemes? In particular, does the ideology of "inverse theorems" make sense in many-dimensional problems? Here we do not have the audacity to look into the future.

§ 3. Widths of Classes of Analytic Functions

Here also we have an opportunity to compare the classical and non-classical results. However, in questions of approximation for analytic functions the classical position looks much more sound: rational approximations (at the asymptotic level) stand, as a rule, concurrent with the remaining methods. Apart from approximation by rational functions, there is only one effective method of best approximation. This is connected with the construction of special bases, generalising the bases of Taylor, Chebyshev, Laurent etc. Both methods are very instructive and interesting. Illustrations will mainly be in the example of classes BA^G.

3.1. ε-Entropy of the Pair (BA^G, C(T)). Let G be a domain (in general, multiply-connected) in $\mathbf{C}$. We recall that A^G denotes the set of functions $x(\cdot)$ analytic in G and satisfying the inequality

$$\sup\{|x(z)| \mid z \in G\} < \infty.$$

(Another possible notation for the class A^G is $H_\infty(G)$). BA^G the set of functions $x(\cdot) \in A^G$, for which

$$\sup\{|x(z)|\,|\,z \in G\} \leqslant 1.$$

Let $T \subset G$ be a set consisting of a finite number of continua.

The problem of ε-entropy for classes of analytic functions on a compact set T, which admit an analytic continuation to a domain $G \supset T$, was posed by A.N. Kolmogorov (1956). He proved

Theorem 23. *Let Δ be a segment of the line* $\mathbf{R}$, *G a domain, $\Delta \subset G$. Then*

$$\mathscr{H}_\varepsilon(BA^G, G(\Delta)) \asymp \log^2 1/\varepsilon.$$

We note the essential difference between the asymptotic forms in the smooth and the analytic cases (in the smooth case $\varepsilon^{-1/\alpha}$, where α is the smoothness, in the analytic case $\log^2 1/\varepsilon$).

Almost immediately after this result improvements appeared. In a number of cases, where there are natural bases in BA^G which well approximate functions on $T \subset G$, A.G. Vitushkin [1959] calculated the entropy up to strong equivalence. We give as an example one of these results. Let $BA_h(\mathbf{T}^1)$ be the set of periodic functions, which admit an analytic continuation to the strip $|\operatorname{Im} z| < h$ of width $2h$ satisfying the inequality $|x(z)| \leqslant 1\ \forall z\colon |\operatorname{Im} z| < h$, $h > 0$.

Theorem 24.

$$\mathscr{H}_\varepsilon(BA_h(\mathbf{T}^1), C(\mathbf{T}^1)) \backsim 2(\log^2 1/\varepsilon)(h \log_2 e)^{-1}.$$

The *proof* of this theorem is based on the properties of the Fourier series (basis of exponentials). On expansion of $x(\cdot)$ as a Fourier series $x(t) = \sum_{k \in Z} x_k \exp(ikt)$, a consequence of the Cauchy inequality is the estimate: $|x_k| \leqslant C_1 \exp(-|k|h)$. And then an ε-net for the class $BA_h(\mathbf{T}^1)$ is formed by the functions of the form

$$y(t) = C_2 \sum_{|k| \leqslant n} (m_{1k} + i m_{2k}) \varepsilon (2n+1)^{-1} \exp(ikt),$$

where $n = (\log 1/\varepsilon)(h \log_2 e)^{-1} + O(1)$, and m_{ik}, $i = 1, 2$ are integers satisfying the inequality $|m_{ik}| \leqslant C_2(2n+1)\exp(-|k|h)\varepsilon^{-1} =: N_k$. In other words, it is necessary "to remember" the Fourier coefficients x_k for $|k| \leqslant n$ with an accuracy of $C_2 \varepsilon (2n+1)^{-1}$. A comparatively simple calculation (see Kolmogorov & Tikhomirov [1959]) shows, that the power of the ε-net turns out to be that required up to order. For the lower estimate we use the function

$$z(t) = 2 \sum_{|k| \leqslant m} (s_{1k} + i s_{2k}) \varepsilon \exp(ikt), \quad |s_{ik}| \leqslant C_4 \exp(-|k|h)(m\varepsilon)^{-1}.$$

$$m = (\log_2 1/\varepsilon)(h \log_2 e)^{-1} + O(1).$$

C_4 can be chosen so that all these functions belong to $BA_h(\mathbf{T}^1)$ (by the Bernstein-Walsh inequalities on the growth of polynomials in $|\operatorname{Im} z| < h$). In addition they all differ by 2ε and the power of the set of these functions has the necessary order. Thus it is possible to prove Theorem 24. A.G. Vitushkin considered other cases where the function could be expanded in a "natural" series (Taylor in the disc,

Laurent in the ring, Chebyshev on a segment, under the condition that the functions could be continued to an ellipse, etc.). Throughout analogues of the inequalities of Cauchy and Bernstein were used.

After the work of A.G. Vitushkin, A.N. Kolmogorov in his lectures (1957) and various speeches began to actively promote the problem of finding the strong asymptotic form of the quantity $\mathscr{H}_\varepsilon(BA^G, C(T))$, for sufficiently arbitrary pairs (T, G). Kolmogorov's problem in the case when T is a simply-connected continuum, not dividing the simply-connected domain G, was solved simultaneously by K.I. Babenko [1958] and V.D. Erokhin [1968]. Under these assumptions they proved.

Theorem 25.

$$\mathscr{H}_\varepsilon(BA^G, C(T)) \backsim (\log_2 \rho^{-1})^{-1} \log^2 1/\varepsilon,$$

where ρ is the inner radius of the ring $\{w\colon \rho < |w| < 1\}$, to which $G \setminus T$ is mapped conformally (the conformal radius of the doubly-connected domain $G \setminus T$).

The solutions were based on different ideas. K.I. Babenko used rational approximations, V.D. Erokhin constructed a special basis for the pair (T, G). As a result, later research became established, as it were, into two branches "basis" and "rational".

To formulate the answer for more complex pairs (T, G), it was necessary to introduce the notion of the capacity $c(T, G)$ of a compact set T relative to a domain G (the capacity of the condensor (T, G)). Suppose that $G \setminus T$ is regular with respect to the Dirichlet problem and let $u(\cdot)$ be a function harmonic in $G \setminus T$, continuous in $\overline{G \setminus T}$ and solving the Dirichlet problem in $G \setminus T$ with boundary conditions, equal to zero on T and unity on ∂G. Let Γ be a system of smooth contours, separating T from ∂G and let v be the normal to Γ directed from T to ∂G. The *capacity of the condensor* is the value

$$c = c(T, G) = (2\pi)^{-1} \int_\Gamma \frac{\partial u(\xi)}{\partial v}\, d\xi.$$

The value

$$c(T) = \lim_{R \to \infty} R \exp(-(c(T, D_R))^{-1}),$$

where D_R is the disc of radius R, is called the *logarithmic capacity.*

3.2. Rational Approximations of the Classes BA^G in the Space $\mathrm{C(T)}$

Theorem 26. *Let ∂G have positive logarithmic capacity, and let $\mathbf{C} \setminus G$ have a countable set of connected components. Then*

$$\lim_{n \to \infty} (d_n(BA^G, C(T)))^{1/n} = e^{-(-(c(T,G))^{-1})},$$

$$\lim_{\varepsilon \to 0} (\mathscr{H}_\varepsilon(BA^G, C(T))) \log_2^{-2}(1/\varepsilon) = c(T, G).$$

Theorem 26 contains the result of Theorem 25 (where $c = (\log_2 \rho^{-1})^{-1}$). This theorem is due to Widom [1972]. The first detailed result for the general situation of a pair (T, G) (under significantly more severe assumptions) was obtained by A.L. Levin and V.M. Tikhomirov (1968). The asymptotically best method of approximation here is the natural one (it appeared earlier in many investigations, beginning with Walsh, and was mentioned in Chap. 2). The pair (T, G) generates a well defined measure on T and ∂G (connected with the distribution of electricity under the charge T and insulator ∂G). It turned out that we must construct rational functions with zeroes, "uniformly distributed" in measure on T, and with poles, "uniformly distributed" in measure on ∂G. In Babenko's case one takes the preimages of points, uniformly distributed relative to neighbourhoods bounded by the ring under a conformal mapping of $G \setminus T$ onto this ring.

Widom's theorem and this method of approximation were discussed in a number of papers (Ganelius (1976), Fisher and Micchelli (1980), A.A. Gonchar (1984) and others).

Fisher and Micchelli [1980] proved that the exact value of the n-width can be expressed (under fairly broad assumptions about the pair (T, G)) though the value of an extremal problem connected with the Blaschke product. We give their theorem in the case when $G = \{z: |z| < 1\}$ is the unit disc (that is, when $BA^G = BH_\infty$), and T is compact in G.

Theorem 27. $d_n(BH_\infty, C(T)) = \inf \|B(\cdot)\|_{C(T)}$, *where the infimum is taken over all Blaschke products*:

$$B_n(z) = e^{i\theta} \prod_{i=1}^{n} (z - z_i)/(z\bar{z}_i - 1).$$

In a number of cases the answer can be given explicitly (for example, in terms of elliptic functions).

3.3. The Construction of Special Bases for the Pair (BA^G, $C(T)$). V.D. Erokhin proved Theorem 25 by constructing unusual bases associated with the structure of the pair (T, G). At the heart of this construction lies the following.

Theorem 28. *Let T be a continuum, not dividing the plane, G a simply-connected domain $T \subset G$, Γ_1 and Γ_2 the boundary continua of the doubly connected domain $G \setminus T$ listed in no particular order. Then the mapping $F\colon G \setminus T \to \mathbf{C}$ on the ring $c < |w| < 1$ can be written as a composition $F = F_1 \circ F_2$, where F_1 is a conformal mapping of the simply-connected domain with boundary Γ_1, and F_2 of the domain with boundary $F_2(\Gamma_2)$.*

If ∂G is a "level line" for T (that is the preimage of the circle $|w| = R$ under the mapping $\Phi\colon \mathbf{C} \setminus T \to \mathbf{C}$, taking $\mathbf{C} \setminus T$ to the interior of the unit disc), then the Erokhin bases become the well known Faber bases (see the monograph by Suetin [1984]).

Thus the meaning of Vitushkin's theorem was revealed (see Theorem 24); that it discussed known special cases of Erokhin bases, which formed a chain: the classical bases (Taylor, Laurent, Fourier, Chebyshev), the Faber bases, the Erokhin bases. The author of this work has repeatedly posed the problem of construction of bases of Erokhin type for the space of functions analytic in the neighbourhood of certain continua. The generalization of Erokhin bases to multi-connected case was recently accomplished by Yu.A. Farkov [1984]. Let us mention several other important investigations where widths have been calculated by the method of bases: V.P. Zakharyuta Nguen Tkh'eu (1972) V.P. Zakharyuta and N.I. Skiba [1976], N.I. Skiba (1978).

3.4. Other Classes of Functions. By BH_p^r we denote the class of functions analytic in the unit disc, for which

$$\|x^{(r)}(\cdot)\|_{H_p} \leqslant 1,$$

where

$$\|x(\cdot)\|_{H_p} = \sup_{0<\rho<1} \|x(\cdot)\|_{H_{p,\rho}},$$

$$\|x(\cdot)\|_{H_{p,\rho}} = \begin{cases} \left((2\pi)^{-1} \int_{\mathbf{T}^1} |x(\rho e^{i\theta})|^p \, d\theta\right)^{1/p}, & 1 \leqslant p \leqslant \infty, \\ \|x(\cdot)\|_{H_{\rho,\infty}} =: \|x(\cdot)\|_{C(D_\rho)}. & p = \infty \end{cases}$$

Theorem 29.

$$d_n(BH_p^r, H_{p,\rho}) = \rho^n(n(n-1)\dots(n-r+1))^{-1} =: \alpha_{npr},$$

$$n > r, \quad r \in \mathbf{N}, \quad 1 \leqslant p \leqslant \infty, \quad \rho < 1.$$

Theorem 29 was proved by V.M. Tikhomirov (1960) for $p = \infty$. This was the first case of an exact calculation of the width for similar classes of analytic functions. For $1 \leqslant p < \infty$ this result is due to L.V. Taikov (1963). M.Z. Dveirin considered it for fractional (relative to Riemann-Liouville) r.

Theorem 29 follows from upper estimates of the quantity $d(BH_p^k, \mathscr{P}_n, H_{p,\rho})$ (K.I. Babenko, L.V. Taikov), lower estimates of $\beta(BH_p^r, \mathscr{P}_n, H_{p,\rho})$, from the Bernstein-Walsh inequalties, and theorems on the widths of spheres. Here no new apparatus arises; the best space is the space of polynomials.

We give some more formulae on widths.

Theorem 30. $/1 \leqslant p \leqslant q \leqslant \infty, \rho < 1/ \Rightarrow$

$$d_n(BH_p^r, H_{q,\rho}) \asymp n^{-r}\rho^n.$$

Let $A_p(G)$ be the closure of A^G in the metric of $L_p(G)$,

$$BA_p^r(G) = \{x(\cdot) \in A_p(G) \mid \|x^{(r)}(\cdot)\|_{L_p(G)} \leqslant 1\}, \quad r \in \mathbf{Z}.$$

Theorem 31. */∂T an analytic curve, $1 < p, q < \infty$/ $\Rightarrow$*

$$d_n(BA_p^r(G), L_q(G)) \asymp n^{-r+1/p-1/q}\rho^n,$$

where ρ is the conformal radius of the doubly-connected domain $G \setminus T$.

Let $\alpha(\cdot)$ be a finite Borel measure on $K_\rho = \{z: |z| = \rho\}$, $\rho < 1$.

Theorem 32.

$$d_n(BH_2, L_2(K_\rho, \alpha(\cdot))) \asymp (g(\alpha)\rho)^{1/2}\rho^n,$$

where

$$g(\alpha(\cdot)) = \exp\left((2\pi\rho)^{-1}\int_{K_\rho}\ln(d\alpha/|dz|)|dz|\right)$$

is the geometric mean of the measure $\alpha(\cdot)$.

Using the more precise method, described by V.M. Tikhomirov, of calculating ε-entropy for spaces with a basis, Tikhomirov [1976], it is possible to obtain the second term of the asymptotic form in the formulae for the ε-entropy.

Theorem 33.

$$\mathscr{H}_\varepsilon(BH_p^r, H_{p,\rho}) = \frac{(\log_2 1/\varepsilon)^2}{\log_2 1/\rho} - 2r\frac{(\log_2 1/\varepsilon)(\log_2 \log_2 1/\varepsilon)}{\log_2 1/\rho} + O(\log_2 1/\varepsilon).$$

We denote by $B^r(G)$ the class of functions $\{x(\cdot) \in A^G: |x^{(r)}(z)| \leqslant 1, z \in G\}$.

Theorem 34. $/\partial G \in C^{r+\alpha+1}, \partial T \in C^{1+\alpha}, \alpha > 1/ \Rightarrow$

$$\mathscr{H}_\varepsilon(B^r(G), C(T)) = \frac{(\log_2 1/\varepsilon)^2}{\log_2 1/\rho} - 2r\frac{(\log_2 1/\varepsilon)(\log_2 \log_2 1/\varepsilon)}{\log_2 1/\rho} + O(\log_2 1/\varepsilon),$$

where ρ *is the conformal radius of the doubly-connected domain* $G \setminus T$.

Theorems 30 and 34 were proved by Yu.A. Farkov, Theorem 31 by V.L. Oleinikov ($r = 0$, T and G concentric rings). O.G. Parfenov ($r = 0$, $\partial G \in C^{1+\varepsilon}$) and Yu.A. Farkov ($r \geqslant 1$), Theorem 33 was proved by M.Z. Dveirin.

The problems of best approximation for analytic functions are still far from completion.

§4. Exact Solutions of Approximation Problems

Exact solutions have always been highly valued in approximation theory, and we have already given a fair amount of attention to them. Here also we will compare classical and optimal methods. Two points of view are possible here. One of them is optimistic. According to this even the attempt to find an optimal method of approximation or recovery of functions usually leads to something new and interesting. But there are also followers of the classics, who believe and assume, that nothing essentially better, than the already approved methods, can be discovered. In the material to be discussed it is possible to find confirmation of either viewpoint.

The theme of this section is covered in detail in the monographs Akhiezer [1965], Kornejchuk [1976, 1984], Pinkus [1985], Tikhomirov [1976], and in a monograph by N.P. Kornejchuk "Exact constants in approximation theory" Moscow, Nauka [1987]. There the reader will find detailed bibliographic references.

4.1. Best Bernstein Constants for Trigonometric Polynomials and Splines and Bernstein Widths. The source of the results here is Bernstein's inequality (Theorem 36, Chap. 1).

Theorem 35. a) $/0 < p \leqslant \infty, r \in \mathbf{N}/ \Rightarrow$

$$b(BW_p^r(\mathbf{T}^1), \mathcal{T}_n, L_p(\mathbf{T}^1)) = n^{-r};$$

b) $/1 \leqslant q < \infty, r \in \mathbf{N}/ \Rightarrow$

$$b(BW_q^r(\mathbf{T}^1), \mathcal{T}_n, C(\mathbf{T}^1)) = n^{-r} \|\cos \cdot\|_{L_q(\mathbf{T}^1)}^{-1};$$

c) $/p = 1 \vee \infty, r \in \mathbf{N}/ \Rightarrow$

$$b(BW_p^r(\mathbf{T}^1), \hat{S}_{2n,r}, L_q(\mathbf{T}^1)) = b_{2n}(BW_p^r(\mathbf{T}^1), L_p(\mathbf{T}^1)) = K_r n^{-r}.$$

Relation a), for $p = \infty$, includes Bernstein's classical result. For $1 \leqslant p < \infty$ relation a) was proved by Zygmund. In the case $0 < p < 1$ Bernstein's inequality was extended by V.V. Arestov [1979]. Assertion b) was obtained by L.V. Taikov, assertion c) for $p = \infty$ by V.M. Tikhomirov (1960), for $p = 1$ by Yu.N. Subbotin (1976).

Result c) for $p = \infty$ showed for the first time that the space of trigonometric polynomials (in contrast to the space of splines $\hat{S}_{2n,r}$) is not optimal in the sense of the best Bernstein constants.

The classical Bernstein inequality has been extended to entire functions of exponential type (we will have occasion to speak in these in the next chapter). The class of functions of exponential type $\leqslant \sigma$, bounded on the real axis, is denoted B_σ (details in Akhiezer [1965]).

Theorem 36. $/x(\cdot) \in B_\sigma/ \Rightarrow$

a) $\|\dot{x}(\cdot)\|_{BC(\mathbf{R})} \leqslant \sigma \|x(\cdot)\|_{BC(\mathbf{R})}$;

b) $\|\dot{x}(\cdot)\|_{BC(\mathbf{R})} \leqslant \sigma(2\sin\sigma h)^{-1} \sup\{|x(t+h) - x(t-h)| \mid t \in \mathbf{R}\}, 0 \leqslant \sigma h < \pi$;

c) $\sup\{|x(t+i\tau)e^{-i\alpha} + x(t-i\tau)e^{i\alpha}| \mid t \in \mathbf{R}\} \leqslant 2(\operatorname{ch}^2 \sigma\tau - \sin^2 \alpha)^{1/2}$
$\times \|x(\cdot)\|_{BC(\mathbf{R})}, \alpha \in \mathbf{R}$;

d) $\sup\{|x(t+\tau) - x(t-\tau)| \mid t \in \mathbf{R}\} \leqslant 2\sin\sigma\tau \|x(\cdot)\|_{BC(\mathbf{R})}, 0 \leqslant \sigma\tau \leqslant \pi/2$.

All of these inequalities are exact. Inequalities a) and d) were obtained by S.N. Bernstein (special versions of b) were established by S.M. Nikol'skij and S.B. Stechkin), inequality c) by Boas. An interesting approach to inequalities of Bernstein type from the point of view of Banach algebras is contained in the article Gorin [1980], which has a detailed bibliography (see also Kamzolov [1974]).

4.2. Best Linear Methods of Summation of Fourier Series and Linear Widths. Here the starting point of the investigation is the theorems of Kolmogorov and Favard-Akhiezer-Krejn (Theorems 42 and 44 of Chap. 1).

Theorem 37. a) $/p = 2, r \in \mathbf{N}/ \Rightarrow$

$$\varphi(BW_2^r(\mathbf{T}^1), \mathscr{T}_{n-1}, L_2(\mathbf{T}^1)) = d_{2n-1}(BW_2^r(\mathbf{T}^1), L_2(\mathbf{T}^1))$$
$$= d_{2n}(BW_2^r(\mathbf{T}^1), L_2(\mathbf{T}^1)) = n^{-r};$$

b) $/p = 1 \vee \infty, r \in \mathbf{N}/ \Rightarrow$

$$\lambda(BW_p^r(\mathbf{T}^1), \mathscr{T}_{n-1}, L_p(\mathbf{T}^1)) = \lambda_{2n-1} = \lambda_{2n} = d_{2n-1}$$
$$= d_{2n}(BW_p^r(\mathbf{T}^1), L_p(\mathbf{T}^1)) = K_r \cdot n^{-r}.$$

Theorem 37 a) was obtained by A.N. Kolmogorov (1936). The best method of approximation turned out to be the Fourier method. (It can be shown, somewhat unexpectedly, that this is not the unique optimal linear method.)

The best linear method of approximation of the class $BW_\infty^r(\mathbf{T}^1)$ by the spaces $\mathscr{T}_n$ (clearly not the Fourier method; compare with Theorem 42 of Chap. 1) was found, we recall, by Favard and N.I. Akhiezer—M.G. Krejn (1936–37). The coincidence of approximations using this method with linear and Kolmogorov widths was established by V.M. Tikhomirov (1960). The linear deviations in the L_l metric by the method of duality, were found by S.M. Nikol'skij and then Yu.N. Subbotin (1971) proved their optimality in the sense of the widths λ_{2n-1} and d_{2n-1}.

Thus, here the classical methods turned out to be optimal. However, shortly afterwards it was shown that spline approximations also possess remarkable properties.

Theorem 38. a) $/1 \leqslant q \leqslant \infty, r \in \mathbf{N}/ \Rightarrow$

$$\lambda(BW_\infty^r(\mathbf{T}^1), \hat{S}_{2n,r-1}, L_q(\mathbf{T}^1)) = \lambda_{2n}(BW_\infty^r(\mathbf{T}^1), L_q(\mathbf{T}^1))$$
$$= d_{2n}(BW_\infty^r(\mathbf{T}^1), L_q(\mathbf{T}^1)) = \|\tilde{x}_{nr}(\cdot)\|_{L_q(\mathbf{T}^1)};$$

b) $/1 \leqslant p \leqslant \infty, r \in \mathbf{N}, m \geqslant r - 1/ \Rightarrow$

$$d(BW_p^r(\mathbf{T}^1), \hat{S}_{2n,m}, L_1(\mathbf{T}^1)) = d_{2n}((BW_p^r(\mathbf{T}^1), L_1(\mathbf{T}^1)) = \|\tilde{x}_{nr}(\cdot)\|_{L_{p'}(\mathbf{T}^1)};$$

c) $/1 \leqslant q \leqslant \infty, r \in \mathbf{N}, m \geqslant r - 1/ \Rightarrow$

$$d(BW_\infty^r(\mathbf{T}^1), \hat{S}_{2n,m}, L_q(\mathbf{T}^1)) = d_{2n}(BW_\infty^r, L_q(\mathbf{T}^1)) = \tilde{x}_{nr}(\cdot)\|_{L_q(\mathbf{T}^1)};$$

In 38 a) there is hidden the assertion that interpolation of functions from $BW_\infty^r(\mathbf{T}^1)$ by splines $\hat{S}_{2n,r-1}$ (relative to the uniform net) achieves the linear (and even the projective, for the definition see Tikhomirov [1976]) width of this class. For $q = \infty$ this result was obtained by V.M. Tikhomirov (1969), for $1 \leqslant q \leqslant \infty$ independently by A.A. Ligun [1980], Yu.I. Makovoz [1979] and Pinkus [1985] (in the latter case there is no analogue of Theorem 44 of Chap. 1). Interpolation by splines from $\hat{S}_{2n,r-1}$ provides a more convenient means of approximation, even than the linear method of Favard. Moreover, from c), proved by A.A. Ligun, (he also proved b)), it follows that there is an entire family of extremal subspaces, $\hat{S}_{2n,m}$ for $m \geqslant r-1$. In this connection the theorem of Favard-Akhiezer-Krejn deals with the "limit case" of spline approximation as $m \to \infty$ (V.L. Velikin [1981]).

4.3. Best Favard Constants for Trigonometric Polynomials and Splines and Gel'fand Widths. Here the source is the result of Bohr-Favard (Theorem 43 of Chap. 1).

Theorem 39 a). $/p = 1 \vee \infty, r \in \mathbf{N}/ \Rightarrow$

$$f(BW_p^r(\mathbf{T}^1), \mathcal{T}_{n-1}, L_p(\mathbf{T}^1)) = d^{2n-1}((BW_p^r(\mathbf{T}^1), L_p(\mathbf{T}^1)) = K_r \cdot n^{-r};$$

b) $/1 \leqslant p \leqslant \infty, r \in \mathbf{N}/ \Rightarrow$

$$f(BW_p^r(\mathbf{T}^1), \mathcal{T}_{n-1}, L_1(\mathbf{T}^1)) = d^{2n-1}((BW_p^r(\mathbf{T}^1), L_{1p}(\mathbf{T}^1)) = \|x_{nr}(\cdot)\|_{L_{p'}(\mathbf{T}^1)}.$$

From this theorem, due to L.B. Taikhov [1967], it follows that the space $\mathcal{T}_n$ is optimal in the sense of Gel'fand width.

But consideration of duality and Theorem 38 imply that equally the space $\hat{S}_{2n,r-1}$ is optimal.

4.4. Widths of Classes of Smooth Functions. We mention first two results. The following theorem of N.P. Kornejchuk was obtained in the development of his theory on the approximation of the classes $W^rH^\omega(\mathbf{T}^1)$ by the spaces $\mathcal{T}_n$ (Theorems 59, 60 of Chap. 1).

Theorem 40. *Let $\omega(\cdot)$ be a convex upwards modulus of continuity, $r \in \mathbf{Z}$. Then*

$$\begin{aligned} d(W^rH^\omega(\mathbf{T}^1), \mathcal{T}_{n-1}, C(\mathbf{T}^1)) &= d_{2n-1} \\ &= d_{2n}(W^rH^\omega(\mathbf{T}^1), C(\mathbf{T}^1)) = \|x_{nr\omega}(\cdot)\|_{C(\mathbf{T}^1)}. \end{aligned}$$

Theorem 41.

$$d_{2n-1}(B\mathring{W}_1^r(\mathbf{T}^1), L_2(\mathbf{T}^1) = \left(\sum_{k \geqslant 2n} k^{-2r}\right)^{1/2}, \quad d_0 = \infty.$$

This theorem is due to R.S. Ismagilov.

Here, as in the case $W^rH^\omega(\mathbf{T}^1)$, the classical methods turn out to be optimal.

We pass on now to the non-periodic case, where there is no longer such a natural and universal apparatus of approximation as $\mathcal{T}_n$.

The problem of Kolmogorov widths of the classes $W_p^r(\hat{I})$ in $L_q(\hat{I})$ for $q \leqslant p$ has turned out to be closely connected with a specific problem of variational calculus. We describe this problem.

Let $1 \leqslant s < \infty$ and let $x(\cdot)$ belong to $L_s(\hat{I})$, let $x_{(s)}(\cdot)$ denote the function $|x(\cdot)|^{s-1} \operatorname{sign} x(\cdot)$ and consider the system of differential equations:

$$x^{(r)} = y_{(p')}, \qquad y^{(r)} = (-1)^r \lambda^q x_{(q)} \tag{i}$$

with the boundary conditions

$$y^{(i)}(\pm 1) = 0, \qquad i = 0, 1, \ldots, r-1,$$

or, equivalently, the nonlinear differential equation

$$((x^{(r)})_{(p)})^{(r)} + (-1)^{r+1} \lambda^q x_{(q)} = 0, \qquad ((x^{(r)})_{(p)})^{(i)} (\pm 1) = 0,$$
$$i = 0, 1, \ldots, r-1. \tag{ii}$$

In (ii) $1 < q < \infty$, $1 < p \leqslant \infty$ (for $p = \infty$ it is necessary to consider the system (i)).

This equation is well known for $r = 1$, $p = q = 2$. Such an equation is called a *Sturm-Liouville equation.* Equations of Sturm-Liouville type have a "spectrum" of eigenvalues, that is, numbers $\{\lambda_k\}_{k \in \mathbf{Z}_+}$, which correspond to eigenfunctions $\{x_k(\cdot)\}_{k \in \mathbf{Z}_+}$:

$$x_k(t) = c_k \cos \pi(k + 1/2)t, \qquad n \in \mathbf{N}, \quad x_0(t) = c_0.$$

Similar facts are true in the general case.

Theorem 42. *Let $r \in \mathbf{N}$, $1 < p \leqslant \infty$, $1 < q < \infty$, k an integer greater than or equal to r. Then there is a function $x_k(\cdot) = x_{krpq}(\cdot)$ and a number $\lambda_k = \lambda_{krpq}$, satisfying equation* (ii) *(for $p = \infty$ $(x_k(\cdot), y_k(\cdot), \lambda_k)$ satisfying system* (i)*), $x^{(r)}(\cdot)$ having norm one in $L_p(\hat{I})$ and exactly k zeroes. In the case $p \geqslant q$ the function $x_k(\cdot)$ is given uniquely by these properties, and the set of numbers $\{\lambda_k\}_{k \geqslant r}$ and* 0 *exhaust set of numbers for which there is a solution of* (ii) *(for $p = \infty$* (i)*).*

A similar result is true in the periodic case.

Theorem 43. $/p \geqslant q/ \Rightarrow$

$$\lambda_k(BW_p^r(\hat{I}), L_q(\hat{I})) = d_k(BW_p^r(\hat{I}), L_q(\hat{I})) = \lambda_{krpq}^{-1},$$
$$k \geqslant r, \quad d_k = \lambda_k = \infty \quad \text{for} \quad 0 \leqslant k \leqslant r-1.$$

Here one of the extremal subspaces is the space of splines

$$\hat{S}_{k+r-1,r-1}(p, q) = \left\{ y(\cdot) | y^{(r)} = \sum_{i=1}^{k} c_i \delta(t - \tau_i) \right\},$$

where

$\tau_i = \tau_{krpqi}$, $i = 1, \ldots, k$ is the set of zeroes of $x_{krpq}(\cdot)$.

There is an analogue of this theorem in a periodic version (cases a) and b) of Theorem 40 are special cases of this analogue).

Theorem 43 has a long history. We describe some of its stages. The case $p = q = 2$ was investigated by A.N. Kolmogorov (1936). But as extremal subspaces he found harmonics forming the "major axes of ellipsoids". The case $p = q = \infty$ was investigated by V.M. Tikhomirov (1969) and there for the first time were found the functions $x_{kr\infty\infty}(\cdot)$ and the splines $\hat{S}_{k+r-1,r-1}(\infty, \infty)$. For

$r = 1$ and $p = q$ the theorem was proved by V.M. Tikhomirov and S.B. Babadzhanov, for $p = \infty$, $1 \leqslant q < \infty$ and $q = 1$, $1 \leqslant p < \infty$ it was proved by Miccelli and Pinkus, on the "diagonal", $p = q$, by Pinkus, in the general case was announced in A.P. Buslaev and V.M. Tikhomirov [1985].

Thus, "in the upper triangle", $p \geqslant q$, splines provide an extremal apparatus of approximation. In the lower triangle there are no exact results, if we ignore the result of R.S. Ismagilov (Theorem 41). We note further the long connecting thread from polynomials least deviating from zero (recall Chap. 1, § 2) and their equations of Sturm-Liouville type to Theorem 43. The Legendre polynomials and Chebyshev polynomials are initial functions in the series constructed ($x_{kk2\infty}(\cdot)$ and $x_{kk\infty\infty}(\cdot)$), but in the general case we are obliged to solve nonlinear equations of Sturm-Liouville type.

A comparison of splines and algebraic polynomials turns out to be in favour of the first. We give an example. According to the result of S.M. Nikol'skij (Chap. 1, Theorem 49), $d(BW_\infty^1(\hat{I}), \mathscr{P}_N, C(\hat{I})) \sim K_1 N^{-1} = \pi/2N$ and the approximation itself is not simple. If we divide the segment $\hat{I}$ into N equal parts and on each segment replace the function $x(\cdot)$ by a constant equal to the value of $x(\cdot)$ at the middle of the segment, we obtain a more precise and definitive) estimate ($\hat{L}_N$ is the space of piecewise constant functions)

$$d(BW_\infty^1(\hat{I}), \hat{L}_N, C(\hat{I})) = \frac{1}{N} \ll \frac{\pi}{2N}.$$

4.5. Appendix

4.5.1. Problems on Inequalities for Derivatives on Lines and Half-lines. By itself this theme is not directly related to approximation theory, but its contact with approximation problems is so vast, that we will nevertheless say something about it.

Let T be a line or a half-line. The problem of inequalities for derivatives consists in finding the least constant K in the relation

$$\|x^{(k)}(\cdot)\|_{L_q(T)} \leqslant K \|x(\cdot)\|^{\alpha}_{L_p(T)} \|x^{(n)}(\cdot)\|^{\beta}_{L_r(T)},$$

where $n, k \in \mathbf{N}$, $0 \leqslant k < n$, $1 \leqslant p, q, r \leqslant \infty$, $\alpha, \beta > 0$. From homogeneity considerations it follows that this inequality only be valid if $\alpha = (n - k - r^{-1} + q^{-1})/(n - r^{-1} + p^{-1})$, $\beta = 1 - \alpha$. Thus, the best constant depends on two integer parameters n and k and three real, parameters (p, q, r). We give a list of all the results, where for definite values of (p, q, r) the problem has been solved for any pairs n, k, $0 \leqslant k < n$.

Theorem 44. a) $T = \mathbf{R}$ *the problem is solved in four cases*:
1) $p = q = r = \infty$ (A.N. Kolmogorov, $n = 2$, $k = 1$-Hadamard).
2) $p = q = 2$ (Hardy-Littlewood-Pólya),
3) $q = \infty$, $p = r = 2$ (L.B. Taikov),
4) $p = q = r = 1$ (Stein).

b) *For* $T = \mathbf{R}_+$ there are two exact solutions: 1) $p = q = r = 2$ (Yu. I. Lyubich, N.P. Kuptsov),

2) $q = \infty$, $p = r = 2$ (V.N. Gabushin).

A detailed bibliography on inequalities for derivatives is contained in the commentary of G.G. Margaril-Il'yaev and V.M. Tikhomirov [1985].

4.5.2. Problems on Inequalities for Derivatives on a Finite Segment. The functions constructed in the preceding subsection, apparently, can turn out to be useful in many other problems of analysis.

We give two results, which develop Chebyshev's theorems on extrapolation for polynomials and Markov's theorems on inequalities for polynomials (Chap. 1, Theorems 12 and 15).

Theorem 45.

$$/x(\cdot) \in W_\infty^r(\mathbf{R}),\ \|x(\cdot)\|_{C(\hat{I})} \leqslant \|x_{nr\infty\infty}(\cdot)\|_{C(\hat{I})}/ \Rightarrow |\tau| > 1 \Rightarrow |x(\tau)| \leqslant |x_{nr\infty\infty}(\tau)|.$$

Theorem 46. *Under the conditions of Theorem* 45,

$$\|x^{(k)}(\cdot)\|_{C(\hat{I})} \leqslant \|x_{nr\infty\infty}^{(k)}(\cdot)\|_{C(\hat{I})}, \qquad 0 < k < r.$$

Theorem 45 was proved by V.M. Tikhomirov, Theorem 46 by A.I. Zvyagintsev.

§ 5. Exact Solutions of the Problem of Recovery

Side-by-side with approximation problems, much attention has been given in approximation theory to the problems of recovery. We touch on this theme in three aspects: the recovery of functionals, operators and elements. The latter has close connections with widths and cowidths. The main attention will be paid to exact solutions.

5.1. Recovery of Functionals. Best Quadratures. Let X be a normed space, W a subset of X and $f: W \to R$ a functional on W. It is required to "recover" $f(x)$ from certain "information" about x. It is usual to regard the information on x as given by some mapping $\varphi: W \to Z$, where Z is a "coding" set. The simplest problem on recovery of f from information, given φ, is to find the quantity

$$e(f, W, \varphi) = \inf \sup_{x \in W} |f(x) - \psi \circ \varphi(x)|,$$

where the inf is taken over all functions $\psi: Z \to R$. But usually we have not one coding mapping, but an entire class of them $\Phi = \{\varphi\}$, and then it is required to find the quantity

$$E(f, W, \Phi) = \inf\{e(f, W, \varphi) | \varphi \in \Phi\}.$$

The problem of recovery of functionals (and operators; on these next) was first posed by S.A. Smolyak (1965). S.A. Smolyak discovered an important and interesting phenomenon, that for the recovery of linear functionals relative to linear information it is sufficient to restrict oneself to linear means. Namely,

Theorem 47. *Let W be a convex centrally symmetric (with centre at zero) subset of a linear space X, $\varphi: X \to \mathbf{R}^n$ a linear mapping (that is, $\varphi(x) = (\langle x_1', x\rangle, \ldots, \langle x_n', x\rangle, x_i \in X')$, f a linear functional ($f(x) = \langle x', x\rangle$, $x' \in X'$). Then there is an optimal linear method of recovery, that is,*[3]

$$e(f, W, \varphi) = \inf\left\{\sup_{x \in W}\left|\langle x_0', x\rangle - \sum_{k=1}^{n} p_k\langle x_k', x\rangle\right| (p_1, \ldots, p_n) \in \mathbf{R}^n\right\}.$$

We consider a special case of the problem of recovery of functionals, the *problem of quadratures*, when W is a class of continuous (or smooth) functions given on a space T with a measure μ, the functional is the integral

$$\mathscr{I}(x(\cdot)) = \int_T x(t)\, d\mu,$$

and the information is given by the values of $x(\cdot)$ at n points

$$(\varphi(x(\cdot); \tau_1, \ldots, \tau_n)) = (x(\tau_1), \ldots, x(\tau_n))$$
$$\Phi = \{\varphi(x(\cdot), \tau_1, \ldots, \tau_n) | \tau_i \in T\}.$$

Thus the *problem of best quadratures* arises, that is (for a centrally symmetric class) the problem of finding the quantity

$$\kappa_n(W) = \inf\left\{\sup_{x(\cdot) \in W}\left|\int_T x(t)\, d\mu - \sum_{i=1}^{n} p_i x(\tau_i)\right|\right\},$$

where the infimum is taken over all possible nodes $\{\tau_1, \ldots, \tau_n\}$ and weights $\{p_1, \ldots, p_n\}$.

In his monograph [1974] S.M. Nikol'skij wrote: "The diversity of quadrature formulae is infinite. In this connection, depending on the class of functions to which the method is to be applied, one quadrature formula or another may have greater or lesser advantage compared with the rest". S.M. Nikol'skij stressed that the problem of finding quantities of the type $\varkappa$ was initiated by A.N. Kolmogorov. Therefore the problems of best quadrature in a class are called *Kolmogorov-Nikol'skij problems.* As in the case of approximations, the efforts of many were spent on finding optimal quadratures of Sobolev classes.

Theorem 48. *Let $r \in \mathbf{N}$, $1 \leqslant p \leqslant \infty$, $n \in N$. Then the quantity $\varkappa_n(BW_p^r(\mathbf{T}^1))$ is attained by the rectangles formula*

---

[3] The proof is given in [23]

$$\mathcal{I}(x(\cdot)) \cong n^{-1} \sum_{k=1}^{n} x(2\pi k/n).$$

For $1 < p < \infty$ the optimal formula is unique up to a rigid shift.

As in many earlier cases, this result was not obtained all at once. The optimality of the rectangles formula for small smoothness and for $p = \infty$ was discovered by S.M. Nikol'skij (see Nikol'skij [1974]). Then the results for small smoothness were obtained (N.E. Lushpai $r = 1, 2, 1 \leqslant p \leqslant \infty, r = 3, p = 2$, N.P. Kornejchuk $r = 3$, $p = 1$, T.N. Busarova $r = 3$, $p = \infty$). An essential movement on these problems was made by V.P. Motornyj [1974], who solved the problem for $p = \infty$ and for even n for $p = 1$. The case $p = 1$ was fully investigated by A.A. Ligun. The general result is due to A.A. Zhensykbaev [1976, 1981].

You may gain the impression, that the rectangles formula will be optimal for any shift invariant class of functions. Therefore the following result, due to K.I. Oskolkov [1979] was surprising to many.

Theorem 49. *There is a polynomial of second degree $\hat{P}(z) = z^2 + a_1 z + a_2$ such that for the class*

$$BW_{\infty}^{\hat{P}}(\mathbf{T}^1) =: \{x(\cdot) \mid \|\hat{P}(d/dt)x(\cdot)\|_{L_{\infty}(\mathbf{T}^1)} \leqslant 1\}$$

the rectangles formula (for some n) is not optimal.

It is natural to try to understand what the problem was here. It is clear that there is no analogue of Rollé's theorem for $\hat{P}(\cdot)$, this operator "increases oscillation". The experience of approximation theory showed that many results which used Rolle's theorem, will transfer to the case when there is some analogue of it.

Kernels, not increasing oscillation, are a special theme in analysis, having very close contact with approximation theory. This theme was developed by M.G. Krejn, F.R. Gantmakher, Shoenberg, Karlin and many others (see, for example, Karlin [1968], Karlin & Studden [1972], Pinkus [1985]).

We state one result.

Let $K \subset \mathbf{Z}$ be a subset not containing all the integers k lying between $-n$ and n. Denote by L_K the linear hull of $\{\exp(ik\cdot), k \in K\}$. We say that a function $Q \in L_K^{\perp} \cap L_1(T^1)$ is a kernel *not increasing the 2n-oscillation* (and we will write $\Omega \in \mathcal{A}_{2n}(K)$), if for any functions $p(\cdot) \in L_K$ and $v(\cdot) \in L_K^{\perp} \setminus \{0\}$ such that the number of sign changes $\nu(v(\cdot))$ of $v(\cdot)$ does not exceed $2n$, then $\nu(p(\cdot) + (\Omega * v)(\cdot)) \leqslant \nu(v(\cdot))$. We denote by $BW_p^{\Omega}(\mathbf{T}^1)$ the class of functions $x(\cdot)$, representable in the form

$$p(\cdot) + (\Omega * v)(\cdot), \quad p(\cdot) \in L_K, \quad v(\cdot) \in L_K^{\perp}, \quad \|v(\cdot)\|_{L_p(\mathbf{T}^1)} \leqslant 1.$$

Theorem 50. *If $\Omega(\cdot) \in \mathcal{A}_{2n}(K)$, then the value $\varkappa_n(BW_p^{\Omega}(\mathbf{T}^1))$ is attained by the rectangles formula*

$$\mathcal{I}(x(\cdot)) \cong n^{-1} \sum_{k=1}^{n} x(2\pi k/n), \qquad \text{if} \quad \dim L_K > 1,$$

and

$$an^{-1} \sum_{k=1}^{n} x(2\pi k/n), \qquad \textit{if} \quad \dim L_K = 1,$$

where a is the solution of an extremal problem.

Special cases of this result were obtained by M.A. Chakhiev, the general case by Nguen Tkhi Tkh'eu Khoa [1986].

In the non-periodic case, in the general situation, only existence and uniqueness theorems have been proved (Boyanov, A.A. Zhensykbaev [1976]) An explicit form of optimal quadratures can be obtained only for small smoothness.

Apparently, the following ideology is justified: in the analytic case it is reasonable to use Gaussian quadratures, for small smoothness formulae of rectangles, trapezia etc. The question of intermediate smoothness is at present open. The study of the asymptotic form of optimal quadratures for the classes $BW(\hat{\Gamma})$ as $n \to \infty$ and $r \to \infty$ is incomplete.

We are not able to dwell on the problems of many-dimensional quadratures. There are virtually no exact solutions of the Kolmogorov-Nikol'skij problem here. The basic ideology is Gaussian (see Mysovskikh [1981]). The presence of an invariant structure allows us to look for invariant formulae. This idea has been widely applied by S.L. Sobolev and his school.

5.2. Best Recovery of Operators. Let X and Y be normed spaces, $F: X \to Y$ a linear, although, generally speaking, unbounded operator, $W \subset X$ a subset lying in the domain of definition of F. We put the question of best approximation on W of F by a bounded operator. For this we introduce the quantity

$$E_N(F, W, X, Y) =: \inf\left\{\sup_{x \in W} \|Fx - \Lambda x\| \,|\, \Lambda \in \mathscr{L}(X, Y), \|\Lambda\| \leqslant N\right\}.$$

This quantity is of the type $E(f, W, \varphi)$, the topic of the preceeding subsection. The problem of best approximation on the class of an unbounded operator by bounded operators was first posed by S.B. Stechkin [1967]. Problems of this kind (recovery of operators with errors) were then studied in Micchelli-Rivlin [1976].

We illustrate the results obtained on one of the most significant examples; the approximation of the differentiation operator $D = d/dt$ in the classes BW_∞^r in the spaces $X = Y = C(\mathbf{R})$. We begin with the simplest case: $r = 2$. The quantity $E_N(D, BW_\infty^2(\mathbf{R}), C(\mathbf{R}), C(\mathbf{R}))$ characterises the best possibility of approximation of D on the class $BW_\infty^2(\mathbf{R})$ using operators of norm $\leqslant N$.

Theorem 51.

$$E_N(D, BW_\infty^2(\mathbf{R}), C(\mathbf{R}), \mathbf{C}(\mathbf{R})) = (2N)^{-1},$$

and the extremal operator has the form

$$(\Lambda_N x)(t) = (x(t+h) - x(t-h))/2h, \qquad h = N^{-1}.$$

The upper estimate for the explicit representation of the operator is trivial, and the lower estimate follows from the following inequality, due to S.B. Stechkin:

$$E_N(F, W, X, Y) \geqslant \sup_{M>0} (\Phi_M(F, W, X, Y) - NM),$$

$$\Phi_M(F, W, X, Y) =: \sup\{\|Fx\|_Y | x \in W, \|x\| \leqslant M\}.$$

The value $\Phi_M(D, BW_\infty^2(\mathbf{R}), C(\mathbf{R}), C(\mathbf{R})) = (2M)^{1/2}$ as immediately follows from Hadamard's inequality, quoted at the end of the preceeding subsection, and this gives an estimate from below and the proof of Theorem 51. Thus the Stechkin problem occurs in connection with "inequalities for derivatives".

We now pose a question. Suppose that a function $x(\cdot)$ belongs to the class $\alpha BW_\infty^2(\mathbf{R})$ (that is $\|\ddot{x}(\cdot)\|_{C(\mathbf{R})} \leqslant \alpha$) and is known with error $\leqslant \varepsilon$ ($\|\tilde{x}(\cdot) - x(\cdot)\|_{C(\mathbf{R})} \leqslant \varepsilon$). The question is: with what accuracy can we then approximate its derivative? Let Λ_N be a linear approximating operator with norm $\leqslant N$. Then we obtain:

$$\min_{N>0} \|\dot{x}(\cdot) - \Lambda_N \tilde{x}(\cdot)\|_{C(\mathbf{R})} \leqslant \min_{N>0} (\alpha(2N)^{-1} + \varepsilon N) = \sqrt{2\alpha\varepsilon},$$

where, a consequence of the same Hadamard inequality, the constant not may be reduced. Thus the problem of approximation of operators turns out to be connected with incorrectly posed problems. In addition it turns out to be connected with problems of linear and nonlinear smoothing of functions (see Theorem 52). In short, this domain also forms its own special world, closely bordering on many problems of approximation theory.

Theorem 52. a) *The rational function* $S_{2m-1} = \left(z\frac{d}{dz}\right)^{2m-1}\left(\frac{2z}{1+z^2}\right)$ *has* $m-1$ *distinct roots* $\{z_j\}_{j=1}^{m-1}$ *and a simple zero* $z = 0$ *in the interval* (0, 1).

b) *The best approximation of the differentiation operator* $D = d/dt$ *on the class* $BW_\infty^{2m}(\mathbf{R})$ *is attained by the operator*

$$(D_{N,2m}x)(t) = \sum_{j=0}^{\infty} c_{j,2m}(x(t + (2j+1)\pi/2) - x(t - (2j+1)\pi/2),$$

where

$$c_{k-1,2m} = (-1)^{k-1} 4(2m-1)\pi^{-1} \sum_{j=1}^{m-1} R_{2m-2}(z_j) z_j^{2k-2} / S_{2m-1}^1(z_j),$$

$$R_{2m}(z) = \left(z\frac{d}{dz}\right)^{2m}\left(\frac{1-z^2}{1+z^2}\right), \qquad N = \left(\frac{2m-1}{2m}\right)\frac{K_{2m-1}}{K_{2m}}.$$

This theorem (proved by A.P. Buslaev [1981]) contains the solution to the problem of finding $E_N(D, BW_\infty^r(\mathbf{R}), C(\mathbf{R}), C(\mathbf{R}))$. The odd case, $r = 2m+1$, is formulated in a similar way. Finding $E_N(D^k, BW_\infty^r(\mathbf{R}), C(\mathbf{R}), C(\mathbf{R}))$ is reduced to finding $E_N(D, BW_\infty^r(\mathbf{R}), C(\mathbf{R}), C(\mathbf{R}))$.

This theorem was preceded by the results of S.B. Stechkin ($r = 2, 3$) and V.V. Arestov ($r = 4, 5$ [1967]). We have already mentioned that for $r = 2$ the extremal operator has the form $(D_{N1}x)(t) = (x(t + h) - x(t - h)(2h)^{-1}, h^{-1} = N$. For $r = 3$ also a finite-difference operator is obtained. But for $r = 4$ the extremal operator has the form

$$(D_{N,4}x)(t) = -24\pi^{-1}(1 + \alpha)^{-1} \sum_{i=0}^{\infty} \alpha^i (x(t + (2i + 1)\pi/2) - x(t - (2i + 1)\pi/2)),$$

where α is the solution of the equation $\alpha^2 - 22\alpha + 1 = 0$ with modulus less than one.

The optimal operator turns out not to be a finite-difference and this indicates (in my opinion) that the statement itself has not yet attained a completely final form. But the attempt to solve the Stechkin problem has led to the establishment of very interesting connections with other extremal problems. Put

$$K(n, k, p, q, r) = \sup\{\|x^{(k)}(\cdot)\|_{L_q(\mathbf{R})} / \|x(\cdot)\|^{\alpha}_{L_p(\mathbf{R})} \|x^{(n)}(\cdot)\|^{1-\alpha}_{L_r(\mathbf{R})}\},$$
$$\alpha = (n - k + q^{-1} - r^{-1})/(n - r^{-1} + p^{-1}).$$

This is the exact constant in the Kolmogorov inequality (see §4).

Theorem 53. a) $/1 < p, q, r \leqslant \infty, p^{-1} + p'^{-1} = q^{-1} + q'^{-1} = r^{-1} + r'^{-1} = 1/ \Rightarrow$

$$E_N(D^k, BW_p^n(\mathbf{R}), L_r(\mathbf{R}), L_q(\mathbf{R})) = d(BW_{q'}^{n-k}(\mathbf{R}), NBW_{r'}^n(\mathbf{R}), L_{p'}(\mathbf{R}));$$

b) $/1 \leqslant p, q, r \leqslant \infty/ \Rightarrow$

$$\beta\alpha^{\alpha/\beta} N^{-(\alpha/\beta)} K(n, k, r, q, p) = \lambda(BM_{q'}^{n-k}(\mathbf{R}), NBW_{r'}^n(\mathbf{R}), L_{p'}(\mathbf{R})).$$

Theorem 53 was proved by V.V. Arestov, and in the periodic case a result similar to b) was obtained by B.E. Klots [1975, 1977]. Thus, the problem of linear smoothing turns out to be equivalent to the problem of inequalities for derivatives, and the Stechkin problem is the problem of nonlinear smoothing.

5.3. Optimal Recovery of Smooth and Analytic Functions. This is an interesting theme and not everything has been finally cleared up. In the development of this circle of questions an active part has been taken by de Boor, Boyanov, Winograd, Micchelli, Rivlin, K.Yu. Osipenko and others. The collection Micchelli & Rivlen (Eds) [1976] is devoted to the problem of recovery. We mention also the important article Micchelli, Rivlin & Winograd [1976], and the book Widom [1972], which contains a detailed bibliography.

We will pause briefly on these problems, stressing the analogy between the smooth and the analytic cases. We limit ourself to the parallel consideration of the problem of recovery of the values of functions from the classes $BW_{\infty}^r(\hat{I})$ and

$B\mathscr{A}_l^D$ (the latter consists of functions analytic in the unit disc D satisfying $|x(z)| \leqslant 1$, $z \in D$, whose restriction to the interval $(-l, 0)$ is real). Functions from $B\mathscr{A}_l^D$ will be investigated on the interval $[-l, 0]$, $0 < l < 1$.

For the class $BW_\infty^r(\hat{I})$ we discuss the problem of extrapolation (see above, theorem 45) Exactly the same problem can be posed for $B\mathscr{A}_l^D$ (it is called the *Heins problem*)):

$$x(\tau) \to \sup, \qquad \|x(\cdot)\|_{C([-l,0])} \leqslant \gamma, \qquad x(\cdot) \in B\mathscr{A}_l^D.$$

And just as in the application to the class $BW_\infty^r(\hat{I})$ the completion of the splines of Chebyshev and Zolotarev type are extremal, so in the Heins problem special Blaschke products, having the necessary alternation, have an analogous role. The Chebyshev splines we denoted above by $x_{nr\infty,\infty}(\cdot)$, the "Chebyshev" Blashke products we denote by $x_{n,D,l\infty\infty}(\cdot)$.

The functions $x_{nr\infty\infty}(\cdot)$ play a decisive role in the question of recovery of functions from $BW_\infty^r(\hat{I})$ from linear information with $n+r$ parameters, the functions $x_{n,D,l,\infty,\infty}(\cdot)$ play an analogous role for the recovery of functions from $B\mathscr{A}_l^D$ on $[-l, 0]$. In other words, there are the theorems.

Theorem 53.

$$2^{-1}\lambda^{n+r}(BW_\infty^r(\hat{I})) = d^{n+r}(BW_\infty^r(\hat{I})) = \|x_{nr\infty\infty}(\cdot)\|_{C(\hat{I})}$$

and the optimal linear coding function is the association of a function $x(\cdot)$ *with the vector* $(x(\tau^i_{nr\infty\infty}))$, $i = 1, \ldots, n+r$, *where the* $\{\tau^i_{nr\infty\infty}\}$ *are the zeroes of* $x_{nr\infty\infty}(\cdot)$.

Theorem 54.

$$2^{-1}\lambda^n(B\mathscr{A}_l^D) = d^n(B\mathscr{A}_l^D) = \|x_{n,D,l,\infty,\infty}(\cdot)\|_{C([-l,0])}$$

and the optimal coding function is the association of a function $x(\cdot)$ *with the vector* $(x(\tau^i_{n,D,l,\infty,\infty}))$, $i = 1, \ldots, n$ *where the* $\{\tau^i_{n,D,l,\infty,\infty}\}$ *are the zeroes of the function* $x_{n,D,l,\infty,\infty}(\cdot)$. (It is possible to give an explicit expression for this quantity in terms of elliptic functions, since the Heins problem is closely connected with one of the problems solved by E.I. Zolotarev.)

Theorem 53 is due to Miccelli, Rivlin and Winograd [1976] (however, this result is implicitly contained in the work of the author (1969), where the quantity $d^n(BW_\infty^r(\hat{I})))$, is calculated, theorem 54 is due to K.Yu. Osipenko [1976, 1985]. We mention also the problem of recovery of a function from its Fourier coefficients.

Theorem 55 (Boyanov).

$$\begin{aligned}\Phi(BW_\infty^r(\mathbf{T}^1), \mathscr{T}_N, C(\mathbf{T}^1)) &\overset{\text{def}}{=} \sup_{x(\cdot) \in BW_\infty^r(\mathbf{T}_1)} \{\operatorname{Diam}(\varphi^{-1}(\varphi(x(\cdot))|\varphi(x(\cdot)) \\ &= ((a_k, b_k), k = 0, \ldots, n)\} \\ &= 2K_r(N+1)^{-r}\end{aligned}$$

and here the recovering operator has the form

$$x(t) \mapsto \frac{a_0}{2} + \sum_{k=1}^{N} \lambda_k (a_k \cos kt + b_k \sin kt),$$

where (a_k, b_k) *are the Fourier coefficients, and* λ_k *are the multipliers in the convolution operator in Theorem* 44.

Chapter 4
Approximation Theory and its Connections with Neighbouring Domains of Mathematics

The theme of this chapter is so broad that it is not easy to give even a rough outline. We briefly discuss the following fragments: connections with the theory of extremal problems, harmonic analysis, and also with certain special questions in the theory of functions, functional analysis and geometry. At the very end of the chapter we endeavour to summarise. One of the aims of this chapter to note new problems and to try to look into the future.

§ 1. Approximation Theory and the Theory of Extremal Problems

In this article a lot of space has been devoted to exact solutions, we have touched on them in each of the previous chapters. The interest in exact solutions has not weakened throughout the development of approximation theory.

We can pose the question: what is the reason for such a stable interest? Experience shows that exactly solved problems often lead to some kind of new method of approximation or recovery, to new nonstandard special functions, polynomials, splines, which then obtain application in other areas.

But apart from this there is another reason why any exact solution has definite interest. The fact is that the set of completely solved problems can be perceived as a testing ground for the theory of extremal problems.

In many domains of mathematics, the natural sciences, economics and engineering there is a need to search for the maxima and minima of certain quantities. The problem of finding extrema has stood and been solved during the whole history of mathematics. Recently the efforts of many researchers has been directed to surveying all the various extremum problems from one theoretical position.

Each concrete completely solved extremal problem occupies in relation to the general theory one of two positions. Either it can be solved "standardly" or there is no such solution. In the latter case there are also two possibilities: either the researcher has not managed to use the theory or the theory is insufficiently developed for the solution of the given problem. The latter case is more interesting since it does not exclude the possibility of supplementing the theory by new methods with which the given problem can be solved.

Here we wish to briefly survey the material on exact solutions in approximation theory from the point of view of the theory of extremal problems.

But first let us say a few words about the current situation with the theory. First of all, we now have many general methods, permitting us to prove the existence of solutions. Furthermore, there is a fairly universal principle—the Lagrange principle—which permits us to describe necessary conditions, which solutions must satisfy. An application of the Lagrange principle leads to some equations. There are then several general principles, regarding sufficient conditions, and methods which make it easier to solve the equations which describe the solution (The reader may read about all of this in the books Alekseev, Galeev & Tikhomirov [1984], Tikhomirov [1976].)

However, one cannot hope that as time passes all extremal problems will be exactly solved. Explicitly written exact solutions can only be obtained for a very small proportion of all extremal problems. From this point of view, possibly, it is useful to mention certain situations in which we can hope for success. We single out three such situations: a) convexity of the problem, b) the presence of an invariant structure and c) the presence of a special method of solution of the equations.

We pass on now to the discussion of concrete material.

1) Problems on polynomials least deviating from zero (see Theorems 1–3) are convex problems. Here the necessary conditions are written in the form of integrable differential equations. In the case of the Chebyshev polynomials they lead to the equations for trigonometric functions. For more details see Tikhomirov [1976].

2) The theorem on alternation is a direct consequence of the refinement theorem (mentioned in the part "Convex analysis") (see Tikhomirov [1976]). However, it must be said that the refinement theorem crystallized largely as a consequence of the necessity to solve problems on Chebyshev approximations.

3) Theorems 8–11 of Chap. 1 (Chebyshev and Zolotarev on exact solutions) admit a standard solution, thanks to a necessary condition for an extremum obtained by an application of the refinement theorem, which leads to the equation for elliptic functions. Theorems 12 and 15 immediately follow from the refinement theorem (see Tikhomirov [1976]). Theorems 13 and 14 can be proved, using the fact that the corresponding problems reduce to a one-parameter family of convex problems, smoothly depending on the parameter. Here it turned out to be possible to apply a second order variation (see Tikhomirov [1976]).

4) The Bernstein-Zygmund inequality (Theorem 36 of Chap. 1 and 35 of Chap. 3) can be proved because of the convex structure of the problem and the presence of invariance. This leads to the Riesz identity as a criterion for extremality, which also decides the matter (see Tikhomirov [1976]).

5) For the proof of the Bohr-Favard and Favard-Akhiezer-Krejn theorems (43 and 44 of Chap. 1) the decisive circumstances are: convexity and invariance of the problem, and the possibility of using Rolle's theorem.

6) Problems on splines, discussed in Chap. 2, are convex optimal control problems, which reduce to so-called Lyapunov problems, which gives the possibility of obtaining standard solutions.

7) The problems from Chap. 3, discussed in Theorems 37–39 and 41, 45, are convex problems of optimal control, or their duals, whose theory is very strongly developed.

8) The discussions which led to the proofs of Theorems 42, 43, did not go outside the standard methods of variational calculus.

9) It would be interesting to survey all the material connected with the theme "inequalities for derivatives on $\mathbf{R}$ or $\mathbf{R}_+$". We limit ourselves to comments on Theorem 44. The solutions a) of A.N. Kolmogorov, b) of Hardy-Littlewood, c) of L.V. Taikov, d) of Yu.I. Lyubich-N.P. Kuptsov and e) of V.N. Gabushin follow in a standard way from the fact that the corresponding problem reduces a) to convex invariant problem of optimal control, b) to a problem of nonlinear programming, c) to a Lyapunov problem, d) to a quadratic problem of variational calculus, e) to convex and quadratic problem of variational calculus.

But apart from the problems listed, there remain a number of problems, for which standard solutions have not yet been found. We name some of them. These are Theorems 58–60 (V.K. Dzyadyk-Sun Yun Shen and N.P. Kornejchuk) from Chap. 1, the theorem of V.V. Arestov (the case $0 < p < 1$ in Theorem 35), the inequality of Bernstein type on $\mathbf{R}$ (Theorem 36) the inequality of V.N. Konovalov [1980], the theorem of Stein (the case a 4) in Theorem 44) and a number of theorems on the recovery of analytic functions from Chap. 3.

It is possible to the hope that mulling over these and a number of other results will enrich the theory of extremal problems to the same extent as mulling over Chebyshev's theorem on alternation enriched convex analysis.

§ 2. Approximation Theory and Harmonic Analysis

2.1. Harmonic Analysis on Homogeneous Spaces and Approximation Theory. In Chap. 2 we gave a lot of attention to questions on approximation of functions by trigonometric polynomials. We posed the question: why do trigonometric polynomials have such complete and universal approximative properties? Could we expect, that in some other situations there might be found an equally effective apparatus of approximation?

We make an attempt to answer this question. Trigonometric polynomials are the eigenfunctions of the shift operator on the circle. From this point of view they have a set of "close relatives". Apart from the circle, there are other homogeneous spaces, on which a group of motions acts. On such homogeneous spaces there are special functions, similar to the harmonics on the circle. Their study also is harmonic analysis. In many cases one can expect such functions to retain effective approximation properties.

There is one class of homogeneous spaces, where productive work on the investigation of the approximative methods of harmonic analysis was done a long time ago. These are the n-dimensional spheres $\mathbf{S}^n$ and more generally compact globally symmetric spaces of rank 1. There is a classification of such spaces. They form four series—the spheres $\mathbf{S}^n$, the real, complex and quaternionic spaces ($P^n(\mathbf{R})$, $P^n(\mathbf{C})$ and $P^n(\mathbf{H})$)—and one special space of dimension 16, the Cayley's elliptic plane $P^{16}(\mathrm{Cay})$. The index n here always denotes the dimension of the manifold, where dimension of the spheres and $P^n(\mathbf{R})$ can be any number, $\dim P^n(\mathbf{C})$ is even, $n \geqslant 2$, and $\dim P^n(\mathbf{H})$ is a multiple of four, $n \geqslant 4$. The listed manifolds of dimension n we denote by M^n. Each of the manifolds M^n can be considered as the orbit space of some compact subgroup of the orthogonal group $\mathscr{G}$, that is $M^n = \mathscr{G}/\mathscr{H}$, $\mathscr{H}$ a subgroup. We denote by π the natural mapping of $\mathscr{G}$ to $\mathscr{G}/\mathscr{H}$ and put $o = \pi(e)$, where e is the identity of G. This point is invariant for all the motions of H. It is called the *pole* of M^n (sometimes the north pole).

On any M^n there is an invariant Riemannian metric ρ, an invariant Haar measure and a *Laplace-Beltrami operator*, which can be written as:

$$(\Delta x)(\xi) = \lim_{\gamma \to 0} 2n\left(\int_{\rho(\xi,\tau)=\gamma} x(\tau)\,d\tau - x(\xi)\right)^{\gamma-2},$$

where $d\tau$ is the invariant normalized unit measure on the submanifold $\{\tau \in M^n: \rho(\xi, \tau) = \gamma\}$.

A function $x(\cdot)\colon M^n \to \mathbf{C}$ is called *zonal*, if $(T_h x)(\xi) = x(h^{-1}\xi) = x(\xi)\ \forall h \in H$. For a space of rank 1 the zonal functions depend only on $\rho(o, \xi)$. A function $\varphi\colon M^n \to C$ is called *spherical* if it is zonal, an eigenfunction of the Laplace-Beltrami operator and is normalized by the condition $\varphi(o) = 1$.

The set of eigenvalues of the Laplace-Beltrami operator forms its spectrum. It is discrete, real and situated on the non-negative axis. We write it in increasing order and denote by $\mathscr{H}_k$ the eigenspace of Δ corresponding to the eigenvalue λ_k. $\mathscr{H}_k$ is shift invariant and $L_2(M^n)$ decomposes into a direct sum of the subspaces $\mathscr{H}_k$:

$$L_2(M^n) = \sum_{k=0}^{\infty} \mathscr{H}_k.$$

In spaces of rank 1 this generates a natural family of approximating subspaces (analogues of the trigonometric polynomials):

$$\{\mathscr{T}_N(M^n)\}_{N \geqslant 0}, \qquad \mathscr{T}_N(M^n) = \mathscr{H}_0 + \cdots + \mathscr{H}_N.$$

Using shifts it is possible to define a subspace of Nikol'skij type $H^{\omega}(M^n)$, and the presence of the Laplace-Beltrami operator gives the possibility to define classes of Sobolev type

$$BW_p^{2r}(M^n) \Leftrightarrow BW_p^{\Delta^r} = \{x(\cdot) \mid \|\Delta^r x(\cdot)\|_{L_p(M^n)} \leqslant 1\}.$$

The theory of spherical functions (on $\mathbf{S}^2$) was founded in the eighteenth century by Laplace and Legendre. The general theory of spherical functions was formed in our time, although, for $\mathbf{S}^n$ a lot of research was done at the beginning of this century (Fejér, Kogbetliantz and others). There are not always complete analogues with the case $\mathbf{T}^1$. Here are two examples.

Theorem 1. *The Fourier series of a zonal function from $L_p(\mathbf{S}^n)$ converges to the function in the $L_p(\mathbf{S}^n)$ metric only for* $(2n/(n+1)) < p < (2n/(n-1))$.

The case $n = 1$ was investigated, as we recall, by M. Riesz, $n = 2$, by Pollard, Newman and Rudin, the general case by Mackenhoupt and Stein.

Theorem 2. *For any $p \neq 2$ there is a function $x(\cdot) \in L_p(\mathbf{S}^n)$, whose Fourier series diverges in $L_p(S^n)$, $n \geqslant 2$ (Bonami, Clerk* [1973]).

But, nevertheless, on manifolds of rank 1 there is an analogue of the theorem of Littlewood-Paley, Marcinkiewicz and others (Strichartz [1972], Bonami-Clerk).

On the manifolds M^n the beginnings of a constructive theory of functions has been constructed (G.G. Kushnirenko, Ragozin, S.M. Nikol'skij and P.I. Lizorkin and others), and a variety of versions of the Bernstein inequality proved.

Theorem 3.
$$b(BW_p^{2r}(M^n), \mathscr{T}_N(M^n), L_p(M^n)) \leqslant nN^2.$$
For $p = \infty$ the inequality is sharp.

Various special cases and versions of this theorem have been discussed in many works (A.P. Shaginyan, G.G. Kushnirenko, E.G. Gol'shtein, S.V. Topuriya, A.S. Dzhafarov, A.I. Kamzolov, Pawelke, Stein, Butser-Ionen, Ragozin and others). In the form given it is contained in Kamzolov [1974].

The first results on the deviations from the subspaces $\mathscr{T}_N(M^n)$, quantities of the form $f(BW_p^{2r}(M^n), \mathscr{T}_N(M^n), L_q(M^n))$, $\varphi(BW_p^{2N}(M^n), \mathscr{T}_N(M^n), L_q(M^n))$, and the widths d_N and λ_N etc. have been obtained (A.I. Kamzolov, [1974, 1982a, 1982b], V.A. Ivanov). These results show that the spaces $\mathscr{T}_N(M^n)$ in fact have universal and complete properties similar to the polynomials $\mathscr{T}_N \Leftrightarrow \mathscr{T}_N(\mathbf{S}^1)$.

It is possible to hope that we stand on the threshold of a broader and more extensive attack on these problems. In this connection it will be necessary to broach the non-compact case. But before that we are engaged in understanding the simplest object the line R.

2.2. Approximation of Functions on the Whole Line. We have said several times that S.N. Bernstein came to study approximation by entire functions of exponential type and, in particular, introduced the space B_σ. It is natural to think about the question of what kind of object this is or what kind of analogue of it can be compared with it in relation to "massiveness".

S.N. Bernstein arrived at B_σ (in calculating the asymptotic form of $d(|t|, \mathscr{P}_N, C(\hat{I}))$ by way of a limit process relative to polynomials. In this connection it was clear that the spaces B_σ generalize trigonometric polynomials. Then he began a broad programme of approximation of functions on the real axis by functions of exponential type. We give two of his theorems.

Theorem 4 (An analogue of Weierstrass' theorem). *In order that* $d(x(\cdot), B_\sigma, C(\mathbf{R})) \to 0$ *as* $\sigma \to 0$ *it is necessary and sufficient that* $x(\cdot)\colon \mathbf{R} \to \mathbf{R}(\mathbf{C})$ *be uniformly continuous.*

Theorem 5 (An analogue of Jackson's theorem). *Let* $x(\cdot)$ *be uniformly continuous on R. Then*

$$d(x(\cdot), B_\sigma, C(\mathbf{R})) \leqslant C\omega(\sigma^{-1}, x(\cdot)).$$

We have already mentioned the analogue of the Bernstein inequality. We could give analogues of the Favard-Akhiezer-Krejn theorem (see Krejn [1938]) and many other results. The theory of approximation by functions from B_σ was elaborated by M.G. Krejn, Nagy, S.M. Nikol'skij and many others.

So what are these spaces B_σ? They in fact occur in natural correspondence with the spaces of trigonometric polynomials. In the same way as $\mathscr{T}_N(\Leftrightarrow \mathscr{T}_N(\mathbf{S}^1))$ is the linear hull of the eigenfunctions of the shift operator (with periodic conditions) with eigenvalues satisfying $|\lambda_k| \leqslant N$, so is B_σ the linear hull of the harmonics with eigenvalues $|\lambda| \leqslant \sigma$. In other words, the functions with bounded spectrum. We denote by $B_{\sigma p}$ the class of functions of exponential type with exponent σ (or with spectrum, distributed on $[-\sigma, \sigma]$) and belonging to the space $L_p(\mathbf{R})$, $B_{\sigma\infty} =: B_\sigma$.

The following theorem holds

Theorem 6. *Every function from* $B_{\sigma 2}$ *admits a representation*

$$x(t) = \sum_{k \in \mathbf{Z}} x(k\pi/\sigma) \frac{\sin \sigma(t - k\pi/\sigma)}{\sigma(t - k\pi/\sigma)}.$$

A similar kind of formula was obtained in the thirties by Whittaker, Cartwright, L. Chakalov and others, and the information theoretic meaning of this result was discovered by V.A. Kotel'nikov [1933] (the theorem is sometimes called the *Kotel'nikov theorem*). Shannon wrote on the information theoretic meaning of Theorem 6 in his seminal work on information theory (1948). The crux of Theorem 6 is that the function $x(\cdot) \in B_{\sigma 2}$ is recovered from its values on an arithmetic progression with step π/σ. This situation prompted Shannon to introduce the notion of "rate of number measurement", which in the words of Shannon "is the mean number of measurements required per second to specify some term of an ensemble". Shannon dealt with random objects. A.N. Kolmogorov introduced the idea of mean ε-entropy of a class of functions or subspaces. Not being able to give exact definitions here, we proceed rather descriptively. The *mean ε-entropy* $\bar{\mathscr{H}}_\varepsilon(C, X)$ of some (non-compact) class C of functions on the line $\mathbf{R}$ is the limit of the ratio $\mathscr{H}_\varepsilon(C^T, X^T)/2T$ as $T \to \infty$, where C^T is the restriction of the functions of C to $[-T, T]$ in the metric generated by X on $[-T, T]$.

Theorem 7. a) *Let $BC(\mathbf{R})$ be the unit sphere of the space $C^b(\mathbf{R})$. Then*

$$\bar{\mathcal{H}}_\varepsilon(B_\sigma \cap BC(\mathbf{R}), C^b(\mathbf{R})) \asymp (2\sigma/\pi) \log 1/\varepsilon.$$

b) *Let $1 < p < \infty$, $A \subset R$ a Jordan measurable bounded set. $B_{A,p}$ the set of functions from $L_p(\mathbf{R})$, whose Fourier transform has a carrier which is situated in the set A. Then*

$$\bar{\mathcal{H}}_\varepsilon(B_{A,p} \cap BL_p(\mathbf{R}), L_p(\mathbf{R})) = (\pi)^{-1} \text{ meas } A \log 1/\varepsilon.$$

Theorem 7 a) was proved by V.M. Tikhomirov (Kolmogorov & Tikhomirov [1959]), Theorem 7b) by Din' Zung.

The idea of mean ε-entropy and the similar idea of mean ε-dimension (see Din' Zung [1980]) allows us to pose the same problems for approximations on the line and $\mathbf{R}^n$, as for the circle or torus. In other words, it is possible to pose the problem on subspaces of given mean dimension with the best approximation properties (an analogue of the problem on Kolmogorov widths), on best linear approximation processes (an analogue of the problem of Fourier widths and linear and harmonic widths, in the latter case the role of the number of measurements in Theorem 7 is played by the measure of the carrier) etc. Recently interesting results in this direction were obtained by Magaril-Il'yaev. The idea to go outside the limits of Euclidean spaces and consider non-compact homogeneous spaces (say, Lobachevskii spaces) suggests itself.

§ 3. Approximation Theory and Functional Analysis

Here we will touch on three themes: Banach algebras, superposition, and dimension of function spaces.

In the first chapter we covered the theorems of Stone-Weierstrass and Wiener. Their role in the creation of the theory of Banach algebras was enormous. The works of M.A. Lavrent'ev, M.V. Keldysh and S.N. Mergelyan, cited in the first chapter, promoted the general statements on the coincidence of the algebras $C(T)$ and $\bar{P}(T)$ not only for subsets of functions of one variable, but also for many variables. In the investigation of these problems a major role was played by notions and results from the theory of Banach algebras (see Gonchar (ed.) [1963]). It could be that many facts of the classical approximation theory obtain their development in contiguity with the theory of Banach algebras. An example of similar interaction is the interpretation of Bernstein's inequality mentioned in Chap. 3.

We pass on to work on superpositions. Is there a notion of: the number of variables of a function? Is it possible to reduce any function to functions of two variables? The very formulation of such kind of questions was touched with blasphemy, and this prompted Hilbert to pose his *thirteenth problem* in the sharpest possible form: "show that the equation of seventh degree $f^7 + xf^3 + yf^2 + zf + 1 = 0$ is unsolvable using continuous functions, depending on only two variables". On the one hand a concrete algebraic function, on the other any continuous functions of two variables. However, the general hope was not justified.

Theorem 8. *Any continuous function* $f: I^n \to R$ $(I = [0, 1])$ *is representable as a superposition* $f(x_1, \ldots, x_n) = \sum_{s=1}^{2n+1} \chi_s(\sum_{k=1}^{n} \varphi_{sk}(x_k))$, *where the functions* $\chi_k(\cdot)$ *are continuous, and the functions* $\varphi_{sk}(\cdot)$ *are continuous, monotone and may be chosen once and for all.*

This theorem is due to A.N. Kolmogorov [1985]. It preceded results of A.N. Kolmogorov himself, which represented every function of $n \geqslant 4$ variables in the form of a superposition of functions of three variables and V.I. Arnol'd, who proved that any continuous function of $n \geqslant 3$ variables is representable in the form of a superposition of functions of two variables (that disproves the Hilbert conjecture). The survey Arnol'd [1985] is devoted to this theme.

The idea of "number of variables", undoubtedly, must be subjected to a thorough analysis in approximation theory. It could be that if instead of the word "continuous" Hilbert had inserted "continuously differentiable", it would have turned out to be true (the *Kolmogorov conjecture*).

"Metric massivity" of smooth functions of many variables was investigated by A.G. Vitushkin. He proved (see Kolmogorov & Tikhomirov [1959]).

Theorem 9. *There is a function of class* $C^r(I^n)$, *which is not representable as a superposition of s times continuously differentiable functions of m variables if* $sm^{-1} > rn^{-1}$.

This result (side-by-side with the ideas of information theory) served as a spur for the introduction by A.N. Kolmogorov of the notion of ε-entropy. The notion of ε-entropy and width allow us to then define the idea of an approximative dimension (see Kolmogorov [1985] No. 57), which allowed the proof of the non-isomorphism of the spaces A^{G_1} and A^{G_2}, where G_1 is a domain in $\mathbf{C}^n$, G_2 in $\mathbf{C}^m$, $m \neq n$. But for continuous functions the matter gets more complicated.

Theorem 10. *For any uncountable metric compact T the space C(T) is linearly isomorphic to the space C(I)* (A.A. Milyutin).

There is an interesting collection of results (A.G. Vitushkin, A.G. Vitushkin-G.M. Khenkin, Pel'chinskij and others) on the non-representability of functions by superpositions, isomorphisms and non-isomorphisms. But the central problems still await their resolution.

§ 4. Approximation Theory and Geometry

Many times both here and in the first part of this volume, we have spoken on the close contact of approximation theory and geometry. Who should investigate the general properties of solutions of the problem $\|x - y\| \to \inf$, $y \in A$ in a normed space? Geometer or approximation theorist? We have repeatedly spoken out on the conviction that "this is not approximation theory".

But we show caution. P.L. Chebyshev discovered the uniqueness of rational approximations in $C(\Delta)$ (Chap. 1, Theorem 6). This, certainly, is approximation theory. The analogous question in $L_2(\Delta)$ is, without doubt, also approximation theory. But the answer to the question is this: in Hilbert space under certain compactness requirements (which the set $\mathcal{R}_{mn}$ satisfies) a non-convex set cannot be a set of uniqueness. This is a purely geometric fact. Does it not follow from this example that a complete separation of geometry from approximation theory is not reasonable?

We throw light on several individual questions.

Convexity of Chebyshev sets. This was the subject of the article "Convex analysis". There we gave the Bunt-Motskin theorem on convexity of Chebyshev sets (the sets for which the solution of the problem of least distance always exists and is unique) in finite-dimensional Euclidean spaces. This result gave rise to a lifetime of scientific direction. First the finite-dimensional problem was studied and at the beginning of fifties the "result in infinite-dimensional space" was completed.

Theorem 11. *In a finite-dimensional Banach space the class of Chebyshev sets coincides with the class of closed convex sets if and only if the sphere of the space is convex and smooth.* (N.V. Efimov, S.B. Stechkin).

Theorem 12. *In a smooth space*[1] *every Chebyshev space is convex* (V. Klee).

Here apparently for the first time, an infinite-dimensional result was obtained. The further development of this theme is in the important papers of N.V. Efimov and S.B. Stechkin. They introduced some interesting new ideas. The first of these was related to the notion of approximative compactness. A set is called *approximatively compact*, if every minimizing sequence in the problem of least distance contains a convergent subsequence. Spaces in which all subspaces are approximatively compact came to be called *Efimov-Stechkin spaces.*

Theorem 13 (N.V. Efimov, S.B. Stechkin). *Let X be a smooth uniformly convex Banach space. For a Chebyshev set to be uniformly convex it is necessary and sufficient that it be approximatively compact.*

It follows from Theorem 13, in particular, that $\mathcal{R}_{mn}$ is not a Chebyshev set in $L_2(\Delta)$, for $n \geqslant 1$ this was mentioned in the first section. In the development of this direction much was done by L.P. Vlasov, I.G. Tsar'kov et al. However, up to now the question of whether every Chebyshev set in a Hilbert space is convex has not been solved.

Research into the connectedness of Chebyshev sets was begun by Wulbert. He proved

[1] that is, a space with a smooth sphere.

Theorem 14. *A Chebyshev set with a continuous metric projection is connected.*

A very paradoxical example in approximative geometry is due to Dunham who constructed an example of a Chebyshev set in $C(I)$ with an isolated point.

S.B. Stechkin [1963] began to develope the new area of *approximative properties of arbitrary sets.* Let X be a normed space and $A \subset X$. Denote by $E(A)$ the set of elements of X with a non-empty metric projection on A, and by $U(A)$ the set x for which there is either no closest element in A, or it is unique. We give some results about the sets $E(A)$.

Theorem 15. *Let X be a uniformly convex Banach space and A a closed set ($A \in \mathrm{Cl}(x)$). Then $X \setminus E(A)$ is of first category.*

Theorem 16. *The following conditions on the space X are equivalent:*

a) $A \in \mathrm{Cl}(X) \Rightarrow \mathrm{Cl}\, E(A) = X$,
b) $A \in \mathrm{Cl}(X) \Rightarrow X \setminus E(A)$ *has first category,*
c) $A \in \mathrm{Cl}(A) \Rightarrow E(A)$ *is connected,*
d) *is an Efimov-Stechkin space.*

Theorem 17. *In any infinite-dimensional Banach space there is a closed set A such that $E(A)$ is not Borel.*

Theorem 15 was proved by S.B. Stechkin, Theorem 16 by Lau and S.V. Konyagin, Theorem 17 by S.V. Konyagin.

We move on to questions of uniqueness.

Theorem 18 (S.B. Stechkin). *Let X be not strictly convex. Then there is a hyperplane $H \subset X$ such that $U(H) = H$. If X is strictly convex, then $U(A)$ is dense in X $\forall A \subset X$.*

The result here had already been used in concrete problems of analysis after the discovery of the non-uniqueness of the solutions of differential equations (A.V. Fursikov). It is thought that it will be applied to similar questions many more times.

Another interesting cycle of problems bordering geometry and approximation theory, is *the description of the geometric properties of spaces via their approximation properties.* Here are some examples.

Theorem 19. *If for some $N \in \mathbf{Z}_+$ the equality $d_N(A, X) = d_N(A, M)$, holds for any closed linear manifold M and $A \subset M$, then $X(\dim X > N)$ is a Hilbert space.*

Theorem 20. *If there is an $N \in \mathbf{Z}_+$ such that for any A the inequality $b_N(A, X) \leqslant a_N(A, X)$ holds, then $X(\dim X > N)$ is a Hilbert space.*

The first principle in Theorems 19 and 20 is the Kakutani geometric criteria for Hilbertness. Theorem 19, in the case $N = 0$, was proved by Klee, the general case was proved in the papers of V.M. Tikhomirov, R.S. Ismagilov, S.B. Babadzhanov. Theorem 20 is due to M.I. Stesin. A number of interesting results on this theme are due to A.L. Garkavi.

The geometric problems of approximation theory have occupied a large number of mathematicians. We mention two surveys: Vlasov [1973] and Garkavi [1969].

§ 5. Some Results and Thoughts About the Future

Each survey is subjective and, naturally, so is this one. We have tried to cover a great deal of material and have presented before the reader close to two hundred and fifty results. But the material from which we have selected is huge, containing possibly tens of thousands of theorems relations and formulae... It is difficult to look at this huge collection with the eye of a future investigator. What will remain, what will disappear, who knows?

The fate of each concrete result is whimsical. We have seen how much effort was spent on the solution of the Vallée-Poussin problem on the estimate of the size of the best approximation of $|t|$ on the interval $[-1, 1]$ by algebraic polynomials and by rational functions. The function $|t|$ is so simple, that no approximation is required. However, S.N. Bernstein was occupied with the solution of the problem of approximation of $|t|$ by polynomials for several decades. Expending a huge amount of effort he proved that $\lim_{n\to\infty} nd(|t|, P_n, C(\hat{I})) = \mu$ and calculated it to two places (Theorem 34). Now in a few minutes on a computer it is possible to find this number with much greater accuracy. What remains of all this effort? A highly unexpected and unpredictable thing: the results S.N. Bernstein's research on the approximation of $|t|$ turned out to be a new approximative tool; entire functions of finite order. And this apparatus has remained for a long time.

General methods and general ideas are the least subject to the destructive influence of time. But, being radiated, they are, as it were, dissolved into our consciousness. The author has tended everywhere to draw attention to general methods and concepts, which are stated and take shape using approximation theory. We recall some of them. The connection of approximations to special functions; the idea of best approximation; "refinement"; smoothing; intermediate approximations; saturation; the connection between the structural properties of functions and characteristics of approximation; the connection between approximation and the algebraic structure of polynomials and, between the theory of approximations and Banach algebras; universality of the methods of harmonic analysis for compatible metrics and its asymptotic ineffectiveness in a series of cases of mismatched metrics; the ideas of trigonometric widths and new opportunities for approximation by sums of a given number of harmonics; the connection between approximation by polynomials and complex analysis (the complex significance of real results of the constructive theory of functions); the "limit passage" from harmonics to functions of exponential type; the ideas of mean entropy and dimension; the need for a specific optimal apparatus of approximation; the optimality of splines and the need for "special splines"; connections and analogies between splines and rational functions, methods of selecting harmonics for the approximation of functions of many variables; the connection between widths and harmonic analysis, nonlinear analysis and the nonlinear theory of differential equations. This is an incomplete list of the questions we have discussed here. Many ideas are now complete, but a significant proportion of the most important problems have simply been remarked.

Let us try to present some perspective on this theory.

Certainly, the number of problems in the traditional channel of approximation theory is innumerable. There are a huge number of most absorbing problems in bordering territories, in the theory of extremal problems, harmonic, functional and nonlinear analysis and in geometry, where the methodology and ideology of approximation theory has the power to give rise to a lifetime of new interesting directions.

And together with this we must think about the extension of the theme of "going back to the roots". In our introduction we said that we can expect an attack on the fourth stage of approximation theory when alongside accuracy as the criterion of quality of approximation, the idea of complexity of approximation will be exposed to a similar scrupulous analysis. How should we interpret this opinion?

From the very sources of analysis came two interpolation formulae, Newton and Lagrange. They can be compared, which is preferable? Both lead to one and the same result. But the calculator, it seems, places his choice on the Newton formula. One reason for this is that the Newton formula is simpler in respect of number of operations.

But number of operations is not the unique characteristic of complexity. It was said above that the method of Fourier summation proposed by Favard (for the class $BW_\infty^1(\mathbf{T}^1)$), and the method, of associating a function with a piecewise constant function, equal in value to the function at the middle of each segment, in the uniform partition give the same accuracy of approximation. But intuitively it is clear that the second method is simpler. How can we give meaning to this expression?

Mathematicians have still to take great pains to become familiar with the idea of complexity.

Apparently, it will be necessary once again to return to the starting points. P.L. Chebyshev was interested in the question of how to extract roots. Now to extract roots it is sufficient to press the button on any microcomputer. The growth of opportunities presented by electronic calculators as a method will reach unprecedented proportions. But they are not unlimited. Electronic calculators are limited for physical reasons (in particular, the finite speed of propogation of electromagnetic waves). At the same time it is already possible to pose problems whose solutions cannot be calculated (for example, long term weather forecasting). This means that there is a set of problems, whose solution is possible because of skill alone, that is, in the final analysis, because of the theory

(An active propagandist of the ideology of numerical methods, based on trial by approximation theory, is K.I. Babenko. Not being able to comment on this in detail, we refer the reader to Babenko [1979, 1985]).

But theory is required essentially to complement. And, again and again, conforming with a new level of development of technique, we must pose fundamental questions of the kind: what is the best way to extract a root? What is the best way to multiply, evaluate functions, interpolate, smooth etc etc.

Alongside these questions, researchers will certainly be enticed to theoretical questions, perhaps, far from practical aims. How complicated is the number π? Which is more complicated to calculate, π or e? The ideology of approximation theory allows us to attach meaning to these questions and they, certainly, will eventually be raised and solved, as will many other problems connected with "finitization" of the idea of number, function, set and mapping.

Undoubtedly also the ideology of approximation theory (on optimality in relation to accuracy and complexity, approximation and recovery) must be further largely extended to random objects. Here the very ideas of approximation theory such as, deviation, width etc., will be "averaged" relative to some measure, concentrated on classes of approximable objects.

And, finally, some words in conclusion. We must not forget another reason why mathematical results, as other creations of the human spirit, are decreed a long life. The reason is aesthetic. There is the opinion that beauty does not die. And I would like to express the hope that the many results of approximation theory, which blend elegant ideas from diverse branches of mathematics: classical complex and functional analysis, the theory of extremal problems, theory of functions of a real variable, geometry and topology, will not immediately forgotten and will give life to other beautiful and fruitful directions in mathematical thought.

Summary of the Literature

There are many comments of a historical nature in the text of the article. Here we give general bibliographical indications and references to works, where the various theorems are proved.

In the introduction we quoted the most important monographs and survey articles on approximation theory.

Chapter 1. We mention first the basic monographs, in which are expounded the problems of classical approximation theory: Akhiezer [1965], Dzyadyk [1977], Goncharov [1954], Korovkin [1959], Natanson [1949], Timan [1960], Tikhomirov [1976], Borel [1905], Cheney [1966], Collatz & Krabs [1973], Lorentz [1966], Rice [1964]. The majority of the theorems cited in Chapter 1 are proved in the monographs Akhiezer [1965], Dzyadyk [1977], Natanson [1949], Timan [1960], Tikhomirov [1976]. In addition to this, we mention the foundational works: Bernstein [1937, 1952], Zolotarev [1877], Markov [1948], Chebyshev [1947], Jackson [1911], Shapiro [1970], Vallée-Poussin [1919], Weierstrass [1885] and the surveys Akhiezer [1945], Goncharov [1945], Dzyadyk [1977], Nikol'skij [1948], Nikol'skij [1961].

For § 1.1 see Archimedes et al. [1911]. Shanks & Wretch [1962]. § 2 on orthogonal polynomials and systems of functions see Akhiezer [1945], Badkov [1983], Kashin & Saakyan [1984], Suetin [1979], Jackson [1941], Legendre [1795], Szegö [1921]. On Gaussian quadratures, Gauss [1986]. § 3, Basic literature: Akhiezer [1965], Dzyadyk [1977], Natanson [1949]. Tikhomirov [1976]. Theorem 5 and the extremal problems see Alekseev, Galeev & Tikhomirov [1984]. Exact solutions Akhiezer [1965], Zolatarev [1877], Chebyshev [1947]. Extremal properties of polynomials: Markov [1948], Tikhomirov [1976], problems of uniqueness: Akhiezer [1965], Tikhomirov [1976], Haar [1918], Mairhuber [1956]. § 4. The Weierstrass theorem is proved in Weierstrass [1885], Lebesgue [1898]. See also Akhiezer [1965], Korovkin [1959], Natanson [1949] and others. For § 4.3.2 see Dzyadyk [1977], Keldysh [1945], Lavrent'ev [1934], Mergelyan [1951], Runge [1885], Walsh [1926]. The Stone-Weierstrass theorem is proved in many books on approximation theory, for example, Dzyadyk [1977]. Muntz's theorem Akhiezer [1965]. To § 5. Foundational works Bernstein [1937], Müntz [1914], Bohr [1935], Borel [1905], Hadamard [1892], Jackson [1911], Lebesgue [1909], Lebesgue [1910], the monographs Akhiezer [1965], Goncharov [1954], Dzyadyk [1977], Natanson [1949], Timan [1960], etc. We mention further the survey articles of S.M. Nikol'skij [1945, 1948, 1958, 1961, 1969], and also Kornejchuk [1985].

For a refinement of Jackson's theorem see Kornejchuk [1976]. On approximations by polynomials Dzyadyk [1977], Timan [1960]. On theorem 58 see Dzyadyk [1953], Theorems 59 and 60 see Kornejchuk [1976, 1984, 1985].

Chapter 2. Theorems 1 and 2 are well-known, Theorem 3 is in Hadamard [1892], Fabry [1898], Theorem 5 in Montessus de Ballore [1902], Theorems 6–11 in Akhiezer [1965]. On splines see Zav'yalov, Kvasov & Miroshnichenko [1980], Kornejchuk [1984], Stechkin & Subbotin [1976], Ahlberg, Nilsen & Walsh [1967], Boor [1974, 1978], Karlin [1975]. On interpolation see Goncharov [1954], Davis [1965]. On smoothness and classes of smooth functions: Besov, Il'in & Nikol'skij [1975a], Galeev [1985], Magaril-Il'yaev & Tikhomirov [1984], Nikol'skij [1969]. For the basic theorems of harmonic analysis see Besov, Il'in & Nikol'skij [1975a], Temlyakov [1985].

The proof of Theorem 18a) is in the monograph Nikol'skij [1969]. An equivalent description of spaces of type $H_p^\alpha(\mathbf{T}^n)$ and similar was carried out in Bakhvalov [1963], Nikol'skij [1969] and others. This has led to the creation of a single concept of smoothmess, developed in detail in the monograph Triebel [1978] (see also Besov, Il'in & Nikol'skij [1975a], Nikol'skij [1969]). For details regarding harmonc analysis see the monograph of E. Stein and D. Weiss "Introduction to harmonic analysis in Euclidean spaces". Moscow: Mir 1974. On harmonic analysis and approximations see Butzer & Nessel [1971].

In Theorems 19 and 21 sufficiency for $\mathbf{T}^n$ was proved by O.V. Besov (Besov, Il'in & Nikol'skij [1975b], necessity by E.M. Galeev, in $\mathbf{R}^n$ the theorem was proved by G.G. Magaril-Il'yaev ([1979]). The proof of Theorem 22 is in Tikhomirov [1976], Theorem 23 in Galeev [1978]. The problem of approximation of Sobolev classes on $\mathbf{T}^n$ was first considered by K.I. Babenko [1960]. There hyperbolic crosses appeared for the first time. B.S. Mityagin [1962] brought the apparatus of harmonic analysis on L_p to bear on the problem of approximation of Sobolev classes. In the form stated Theorems 24 and 25 were proved by E.M. Galeev [1978], Theorems 26, 27 by G.G. Magaril-Il'yaev [141]. Here we must also mention the works Din' Zung [1984], Telyakovskij [1964], Besov, Il'in & Nikol'skij [1975b]. See also the surveys in the recently published monograph V.N. Temlyakov [1986]. The upper estimate in Theorem 29 was proved by K.I. Babenko [1960], the lower by S.A. Telyakovskij [1964].

To § 5. Theorems 47, 48 and 50 are due to A.A. Gonchar, 49 to S.N. Bernstein, 51, 52 and 54 to E.P. Dolzhenko, 53 to V.V. Peller, 54a) to Boehm, 54b) to A.L. Levin and V.M. Tikhomirov, 57 to S.P. Suetin, 58 to Saff and A.A. Gonchar. The upper estimate in Theorem 59 was obtained in the thirties by Walsh, the lower estimate under various assumptions on T and G is contained in the works of A.L. Levin-V.M. Tikhomirov, Bagby, Boyadzhiev and others.

Chapter 3, The monographs Kornejchuk [1976], Kornejchuk [1984], Tikhomirov [1976], Lorentz [1966], Pinkus [1985], V.M. Temlyakov [1986] are devoted to the theme of this chapter. We mention also the important works Babenko [1958], Erokhin [1968], Ismagilov [1968, 1974, 1977], Karlin & Studden [1972], Makovoz [1972], Boor, de Vore & Hollig [1980], Golitschek [1979], Melkman

[1976], Melkman & Miccelli [1978], Mergelyan [1951], Miccelli & Pinkus [1978]. For the definitions of widths see Aleksandrov [1978], Kolmogorov & Tikhomirov [1959], Temlyakov [1982], Tikhomirov [1976], Uryson [1951].

Some commentaries to individual results.

Theorem 1: for the proof of a) and b) see Tikhomirov [1976], c) follows from the definition (this was noted by S.V. Pukhov), d) was proved by S.A. Smolyak and then repeatedly generalized, e) was proved by R.S. Ismagilov. The notion of absolute width is due to R.S. Ismagilov, although the first prototype was suggested by A.G. Vitushkin. Assertion a) of Theorem 2 was proved by A.G. Vitushkin, b) by V.M. Tikhomirov, c) by R.S. Ismagilov. Theorem 3: inequalities a) follow immediately from the definitions; a major role in all these questions was played by the first inequality of b), it first appeared in the work of V.M. Tikhomirov 1960 (although in fact an equivalent result was proved by M.A. Krasnosel'skij, M.G. Krejn and D.D. Mil'man in 1947). V.M. Tikhomirov in 1960 was the first to relate the problems of approximation theory with topological ideas (namely, Borsuk's theorem on antipodes). Subsequently analogous methods came to be used in a very large number of works (Yu.I. Makoboz [1972], A.A. Zhensykbaev [1981], V.I. Ruban [1974], B.L. Velikin, Micelli, Pinkus, see Pinkus [1985] and others) The remaining inequalities in b) were proved by V.M. Tikhomirov, c) by Helfrich and Ha, d) by A.N. Kolmogorov. Theorem 4: a) was proved by A.N. Kolmogorov, b) is in Archimedes et al [1911] and the note of A.I. Mal'tsev, included in Kolmogorov, Petrov & Smirnov [1947]. The duality formula in c) was proved by S.A. Smolyak (1965), d) was proved by S.V. Pukhov (except for the case $d_1(\Sigma^{2s}, l_2^{2s+1})$, where S.V. Pukhov reduced the problem to an extremum problem, solved by A.P. Buslaev). The majority of these facts are proved either in Tikhomirov [1976] or in Pinkus [1985].

Theorem 5: a) was proved by M.I. Stesin, Stesin [1975], b) by Pietch (1974) and M.I. Stesin (1975), c) by V.M. Tikhomirov. Theorem 6: the most important of the relations of the theorem is the equivalence c). The definitive result was not proved all at once. In the "upper triangle" it is trivial. The point $(0, \frac{1}{2})$ was mentioned above. The "segment" $[(1, 1), (1, \frac{1}{2})]$ was investigated by M.Z. Solomyak and V.M. Tikhomirov (1967) and they also discovered, that there were some anomalies in the lower triangle. The definitive result was obtained by B.S. Kashin (1975). It is desirable to note here the important role in this cycle of the papers R.S. Ismagilov [1974, 1977] and B.S. Kashin [1977]. The methods developed by Kashin led to a vigorous development of the entire theme of widths. Relation b) was obtained by V.E. Majorov [1980] and Hollig [1979].

Theorem 7 was proved by E.D. Gluskin [1983].

We touch on the question of the history of Theorems 8–11. The first result was obtained by A.N. Kolmogorov [1985, No. 28] (1936), where he investigated the points $(\frac{1}{2}, \frac{1}{2})$ in Theorem 10. Later results on Kolmogorov widths developed as follows: Rudin [1952]—$(1, \frac{1}{2})$, S.B. Stechkin—(0, 0), V.M. Tikhomirov—(0, 0), Yu.I. Makovoz—I (Fig. 2), M.Z. Solomyak—V.M. Tikhomirov—$[(1, 1), (1, \frac{1}{2})]$, R.S. Ismagilov—II (Fig. 2). Here as we said the Fourier method is optimal (as also is spline-interpolation relative to the uniform partition). It also seemed that the formula $d_n(BW_p^\alpha(\mathrm{T}^1), L_q(\mathrm{T}^1)) \asymp d(B\mathring{W}_p^\alpha(\mathrm{T}^1), L_q(\mathrm{T}^1))$ holds everywhere. An important breakthrough was the result of R.S. Ismagilov, where it was discovered that this is not the case. Then B.S. Kashin [1977] made remarkable progress in estimating the widths of finite-dimensional sets, obtaining definitive results. A major role in the development of the many-dimensional theme was played by the works of V.N. Temlyakov [1979, 1982, 1985].

Theorem 9 in II (Fig. 2) was proved by V.E. Majorov [1980] (independently and slightly earlier by Hollig [1979]). Theorem 9 follows immediately from this result, Theorem 10 and the duality relation for linear widths. Theorem 8 was proved (when $p \leqslant q$) by V.M. Temlyakov [1982], Theorem 11a) by M.I. Stesin [1975] and M.Sh. Birman-M.Z. Solomyak [1967] proved Theorem 11b).

Theorem 12 was proved by I.G. Tsar'kov (for $1/p \leqslant 1/q$) and by V.E. Majorov (in the remaining cases). Theorems 14 and 16 were proved by V.E. Majorov, we spoke on Theorem 15 in the text. Theorem 17 for an A consisting of one point was proved by V.N. Temlyakov [1985], for the general case (when $p \leqslant q$) by E.M. Galeev, Theorem 18 was proved by E.M. Galeev.

Theorem 19, as a one-dimensional result, was not proved all at once. The first to investigate the problem was K.I. Babenko (1960) (see Babenko [1958]), then came the work of B.G. Mityagin 1962

($p = q$, $\alpha \in N$), E.M. Galeev 1977, 1978 ($p = q, \alpha \in R$; $2 \leqslant p \leqslant q, \alpha \in R$), B.M. Temlyakov 1980 ($p \leqslant q \leqslant 2$), 1982 ($p \leqslant q$) (essentially shifting the whole theme) and finally E.M. Galeev 1984 ($q \leqslant p$).

Theorem 20 is due to E.M. Galeev. In Theorem 21 the case $q \geqslant \max(p, 2)$ was proved by B.N. Temlyakov (1982), $2 \leqslant p \leqslant q$ by Din' Zung (1983), $q \leqslant 2 \leqslant p$ by E.M. Galeev (1984).

Theorem 22 was proved by M.Sh. Birman and M.Z. Solomyak [1967].

We spoke on the results of §§ 3–5 in the text. We mention again Magaril-Il'yaev [1979] and Badkov [1983], relating to § 4.5.1.

Chapter 4. We mention a number of works, included in the bibliography, on theme of this chapter. § 1—Alekseev, Galeev & Tikhomirov [1984], Tikhomirov [1976], Tikhomirov & Boyanov [1979]. § 2—Din' Zung [1980], Din' Zung & Magaril-Il'yaev, Kamzolov [1982b, 1982c], Kolmogorov & Tikhomirov [1959], Bonami & Clerk [1973], Weierstrass [1885], § 3—Arnol'd [1985], Kolmogorov & Tikhomirov [1959], § 4—Vlasov [1973], Garkavi [1969], Stechkin [1963], § 5—Babenko [1979, 1985].

References

For the convenience of the reader, references to reviews in Zentralblatt für Mathematik (Zbl.), compiled using the MATH database, and Jahrbuch über die Fortschritte der Mathematik (Jrb.) have, as far as possible, been included in this bibliography.

Ahlberg, J.H., Nilson, E.N., Walsh, J.L.

[1967] The theory of splines and their applications. New York: Academic Press. Zbl.158.159

Ahlfors, L.V.

[1947] Bounded analytic functions. Duke Math. J. *14*, 1–11. Zbl.30,30

Akhiezer, N.I.

[1938] On best approximations of a class of analytic functions, Dokl. Akad. Nauk SSSR *18*, 241–244. Zbl.19,13

[1945] General theory of Chebyshev polynomials, In: Scientific heritage of P.L. Chebyshev. Moscow-Leningrad: Izdat. Akad. Nauk SSSR, 5–42.

[1965] Lectures on the theory of approximations. Moscow: Nauka. German transl.: Berlin, Akatemic-Verlag (1967). Zbl.152,253

Akhiezer, N.I., Krejn, M.G.

[1937] On the best approximation by trigonometric sums of differentiable periodic functions. Dokl. Akad. Nauk SSSR *15*, 107–111. Zbl.16,300

Aleksandrov, P.S.

[1978] On Urysohn widths. In: Dimension theory and neighbouring questions. Articles of a general nature. Moscow: Nauka, 155–165

Alekseev, V.M., Galeev, E.M., Tikhomirov, V.M.

[1984] A collection of problems on optimization. Moscow: Nauka

Andrienko, V.A.

[1967] Embedding of certain classes of functions. Izv. Akad. Nauk SSSR, Ser. Mat. *31*, 1311–1326. Zbl.176,433. English transl.: Math. USSR, Izv. 1, 1255–1270 (1967) Zbl.176,433

Archimedes, Huygens, Legendre and Lambert

[1911] On the quadrature of the circle. Odessa: Mathesis

Arestov, V.V.

[1967] On best approximation of differentiation operators. Mat. Zametki *1*, 149–154. Zbl.168,122. English transl.: Math. Notes 1, 100–103, 1967. Zbl.168,122

[1979] On the Bernstein inequalities for algebraic and trigonometric polynomials. Dokl. Akad. Nauk SSSR *246*, 1289–1292. English transl.: Sov. Math., Dokl. 20, 600–603, 1979. Zbl.433.41004

Arnol'd, V.I.
[1985] Superposition. In: Kolmogorov A.H., Selected works. Mathematics and mechanics. Moscow: Nauka, 455–451. Zbl.566.01012
Babenko, K.I.
[1958] a) On best approximations of a class of analytic functions. Izv. Akad. Nauk SSSR, Ser. Mat. *22*, 631–640. Zbl.87,71
b) On the ε-entropy of a class of analytic functions. Nauchn. Dokl. Vyssh. Shkol'., Ser. Fiz-Mat. *2*, 9–16. Zbl.124,37
[1960] Approximation of periodic functions of many variables by trigonometric polynomials. Dokl. Akad. Nauk SSSR *132*, 247–250. English transl.: Sov. Math., Dokl. 1, 513–516, 1960. Zbl.99,283. [*132*, 982–985. Zbl.102,53. English transl.: Sov. Math., Dokl. 1, 672–675]
[1979] (Ed) Theoretical principles and construction of numerical algorithms of mathematical physics, Moscow: Nauka. Zbl.464.65082
[1985] Some problems in approximation theory and numerical analysis. Usp. Mat. Nauk *40*, No. 1, 3–27. English transl.: Russ. Math. Surv. 40, No. 1, 1–30 (1985). Zbl.618.41035
Babenko, V.F.
[1983] Non-symmetric extremal problems in approximation theory. Dokl. Akad. Nauk SSSR *269*, No. 3, 521–524. English transl.: Sov. Math., Dokl. 27, 346–349 (1983). Zbl.568.41032. English transl.: Sov. Math., Dokl. 27, 346–349 (1983)
Badkov, V.M.
[1983] Uniform asymptotic representations of orthogonal polynomials. Tr. Mat. Inst. Steklova *164*, 6–36. English transl.: Proc. Steklov Inst. Math. 164, 5–41 (1985). Zbl.583.42014
Bagby, T.
[1967] On interpolation by rational functions. Duke Math. J. *36*, 95–104. Zbl.223.30049
Bakhvalov, N.S.
[1963] Imbedding theorems for classes of functions with several bounded derivatives. Vestn. Mosk. Univ., Ser. I. 18, No. 3, 7–16. Zbl.122,113
[1971] On optimal linear methods of approximation of operators in convex classes of functions. Zh. Vychisl. Mat. Mat. Fiz. *11*, 1014–1018. Zbl.252.41024
Barrar, R.B., Loeb, H.L.
[1982] The fundamental theorem of algebra and the interpolating envelope for totally positive perfect splines. J. Approximation Theory *34*, 167–186. Zbl.489.41013
Bernstein, S.N.
[1937] Extremal properties of polynomials. Moscow-Leningrad: ONTI
[1952] Complete collected works I, II. Moscow: Izdat. Akad. Nauk SSSR. Zbl.47,73
Besov, O.V., Il'in V.P., Nikol'skij, S.M.
[1975] a) Multiplicative estimates for integral norms of differentiable functions of several variables. Tr. Mat. Inst. Steklova *131*, 3–15 English transl.: Proc. Steklov Inst. Math. 131, 1974, 1–14 (1975). Zbl.309.46020
b) Integral representations of functions and imbedding theorems, Moscow: Nauka, 1975. Zbl.352.46023
Birman, M.Sh., Solomyak, M.Z.
[1967] Piecewise-polynomial approximations of the classes W_p^α. Mat. Sb., Nov. Ser. *73*, 331–355 English transl.: Math. USSR, Sb. 2 (1967), 295–317 (1968). Zbl.173,160
Boehm, B.
[1964] Functions whose best rational Chebyshev approximations are polynomials. Numer. Math. *6*, 235–242. Zbl.127,291
Bohr, H.
[1935] Un théorème générale sur l'integration d'un polynóme trigonométrique. C.R. Acad. Sci., Paris *200*, 1276–1277. Zbl.11.110
Bonami, A., Clerk, J.-L.
[1973] Sommes de Cesaro et multiplicateurs des dévelopoements en harmoniques sphériques. Trans. Amer. Math. Soc. *183*, 223–263. Zbl.278.43015

Boor, C.de,
[1974] A remark concerning perfect splines. Bull. Amer. Math. Soc. *80*, 724–727. Zbl.286.41010
[1978] A practical guide to splines. New York: Springer-Verlag. Zbl.406.41003
Boor, C. de, Vore, R. de, Höllig, K.,
[1980] Mixed norm n-widths. Proc. Amer. Math. Soc. 80, 577–583. Zbl.458.41018
Borel, E.
[1905] Leçons sur les fonctions de variables réelles et les développements en séries de polynömes. Paris: Gauthier-Villars. Jrb.36,435
Bulanov, A.P.
[1968] Asymptotic behaviour of the least deviation of $|x|$ from the rational functions. Mat. Sb., Nov. Ser. *76*, 288–303. Zbl.169,393
[1969] On the order of approximation of convex functions by rational functions. Izv. Akad. Nauk SSSR, Ser. Mat. *33*, 1132–1148. Zbl.194,94
Buslaev, A.P.
[1981] Approximation of a differentiation operator. Mat. Zametki *29*, 731–738. English transl.: Math. Notes 29, 372–378, 1981. Zbl.488.41022
Buslaev, A.P., Tikhomirov, V.M.
[1985] Some questions of nonlinear analysis and approximation theory. Dokl. Akad. Nauk SSSR *283*, 13–18. Zbl.595.34018, English transl.: Sov. Math., Dokl. 32, 4–8, 1985
Butzer, P.L., Nessel, R.J.
[1971] Fourier analysis and approximation. Vol. I: One-dimensional theory. Basel-Stuttgart: Birkhäuser-Verlag. Zbl.217,426
Chebyshev, P.L.
[1947] Complete collected works. Moscow-Leningrad: Izdat. Akad, Nauk SSSR. Zbl.33,338
Cheney, E.W.
[1966] Introduction to approximation theory. New York: McGraw Hill. Zbl.161,252
Ciesielski, Z.
[1981] The Franklin orthogonal system as unconditional basis in ReH^1 and BMO. Func. Anal. Approx. Proc. Conf. Oberwolfach, Aug. 9–16 1980, 117–125. Zbl.472.42013
Ciesielski, Z., Figiel T.
[1983] Spline bases in classical function spaces on compact C^∞-manifolds, I, II. Stud. Math. *76*, 1–58, *76*, 96–136. Zbl.599.46041, Zbl.599.46042
Collatz, L., Krabs, W.
[1973] Approximationstheorie. Tschebyscheffsche Approximation mit Anwendungen. Stuttgart: Teubner. Zbl.266.41019
Davis, P.J.
[1963] Interpolation and approximation. New York: Blaisdell. Zbl.111,60
DeVore, R.A., Scherer, K.
[1980] Quantitative approximation. New York: Acad. Press. Zbl.453.00027
Din' Zung
[1980] Mean ε-dimension of the class of functions $B_{G,p}$. Mat. Zametki *28*, 727–736. English transl.: Math. Notes 28, 818–823 (1981). Zbl.462.42014
[1984] Approximation of classes of smooth functions of many variables. Tr. Semin. In. I.G. Petrovskogo 10, 207–226. Zbl.574.41033
Din' Zung, Magaril-Il'Yaev, G.G.
[1979] Problems of Bernstein and Favard type and mean ε-dimension of some function classes. Dokl. Akad. Nauk. SSSR *249*, 783–786. English transl.: Sov. Math., Dokl. 20, 1345–1348. Zbl.459.46028
Dolzhenko, E.P.
[1962] Speed of approximation by rational fractions and properties of functions. Mat. Sb., Nov Ser. *56*, 403–432. Zbl.115,57
[1967] Comparison of rates of rational and polynomial approximations. Mat. Zametki *1*, 313–320. English transl.: Math. Notes 1 (1967), 208–212 (1968). Zbl.201,76

Domar, Y.

[1968] An extremal problem related to Kolmogorov's inequality for bounded functions. Arch. Math. 7, 433–441. Zbl.165,488

Dunford, N. Schwartz, J.T.

[1958] Linear operators. Part 1. General theory. New York-London: Interscience. Zbl.84,104

Dunham, C.B.

[1975] Chebyshev sets in $C[0, 1]$ which are not sums. Can. Math. Bull. *18*, 35–37. Zbl.309.41025

Dzyadyk, V.K.

[1953] On best approximation in the class of periodic functions, having bounded s-th derivative $(0 < S < 1)$. Izv. Akad. Nauk SSSR, Ser. Mat. *17*, 135–162. Zbl.50,71

[1977] Introduction to the theory of uniform approximation of functions by polynomials. Moscow: Nauka. Zbl.481.41001

Efimov, A.V.

[1960] Approximation of continuous periodic functions by Fourier sums. Izv. Akad. Nauk SSSR, Ser. Mat. *24*, 243–296. Zbl.96,44

[1966] On best approximation of classes of periodic functions by trigonometric polynomials. Izv. Akad. Nauk SSSR, Ser. Mat. *30*, 1163–1178. Zbl.171,314.

Erokhin, V.D.

[1968] On best linear approximation of functions, analytically continued from a given continuum to a given domain. Usp. Mat. Nauk *23*, No. 1, 91–132. English transl.: Russ Math. Surv. 23, No. 1, 93–135 (1968). Zbl.184,100

Faber, G.

[1914] Uber die interpolatorische Darstellung stetiger Funktionen. Jahresber. Dtsch. Math.-Ver. *23*, 192–210

Fabry, E.

[1898] Sur les séries de Taylor qui ont une infinité de points singulières. Acta Math. *22*, 65–87

Farkov, Yu.A.

[1984] Faber-Erokhin basic function in a neighbourhood of several continua. Mat. Zametki *36*, No. 6, 883–892. English transl.: Math. Notes 36, 941–946. Zbl.578.30002

Favard, J.

[1936] Sur l'approximation des fonctions périodiques par des polinômes trigonométriques. C.R. Acad. Sci., Paris 203, *203*, 1122–1124. Zbl.16,59

[1937] Sur les meilleurs procédés d'approximation de certaines classes de fonctions par des polinômes trigonométriques. Bull. Sci. Math., II. Ser *61*, 209–224; 243–256. Zbl.17,251

Fejér, L.

[1904] Untersuchungen über Fouriersche Reihen. Math. Ann. *58*, 501–569

Fisher, S.D., Miccelli, Ch.A.

[1980] The n-width of sets of analytic functions. Duke Math. J. *47*, 789–801. Zbl.451.30032

Gaffney, P.W., Powell, M.J.D.

[1976] Optimal interpolation. Lect. Notes Math. *506*, 90–99. Zbl.317.65002

Galeev, E.M.

[1978] Approximation by Fourier sums of classes of functions with several bounded derivatives. Mat. Zametki 23, 197–212 English transl.: Math. Notes 23, 109–117 (1978). Zbl.402.42003

[1981] Kolmogorov diameters of the intersection of classes of periodic functions and finite-dimensional sets. Mat. Zametki *29*, 749–760. English transl.: Math. Notes 29, 382–388. Zbl.465.41017

[1985] Kolmogorov widths of classes of periodic functions of many variables $\tilde{W}_{\bar{p}}^{\bar{\alpha}}$ and $\tilde{H}_p^{\alpha}$ in the space $\tilde{L}_q$. Izv. Akad. Nauk, Ser. Mat. *49*, No. 5, 916–934. Zbl.626.41018

Ganelius, T.

[1957] On one-sided approximation by trigonometrical polynomials. Math. Scand. *4*, 247–258. Zbl.77,70

Garkavi, A.L.
[1969] Theory of best approximation in linear normed spaces. Itogi Nauki Tekh. Ser. Mat. Anal, 1967, 75–132, English transl.: Prog. Math. 8, 83–150 (1970). Zbl.258.41019
Gauss, C.F.
[1986] Methodus nova intergralinm valores per approximationen inveniendi. Werke, Göttingen 3, 163–196
Gel'fand, I.M., Raikov, D.A., Shilov, G.E.
[1960] Commutative normed rings. Moscow: Fizmatgiz. Zbl.134,321
Gluskin, E.D.
[1983] Norms of stochastic matrices and diameters of finite-dimensional sets. Mat. Sb., Nov. Ser. *120*, 180–189. Zbl.472.46012
Golitschek, M. von
[1972] On the convergence of interpolating periodic spline functions of high degree. Numer. Math. *19*, 146–154. Zbl.222.65011
[1979] On n-widths and interpolation by polynomial splines. J. Approximation Theory *26*, 132–141. Zbl.428.41015
Golovkin, K.K.
[1964] Imbedding theorems for fractional spaces. Tr. Mat. Inst. Steklova *70*, 38–46. Zbl.119,106
Gonchar, A.A.
[1963] (Ed) Some questions of approximation theory. Moscow: Inost, Lit.
[1968] Speed of approximation by rational fractions and properties of functions. In: Tr. Mezhd. Kongr. Mat., Moskva, 1966. Moscow: Mir, 329–356. Zbl.193,34
[1984] On the rate of rational approximation of analytic functions. Tr. Mat. Inst. Steklova *166*, 52–60. English transl.: Proc. Steklov Inst. Math. 166, 52–60 (1986). Zbl.575.30035
Goncharov, V.L.
[1945] Theory of best approximation of functions. In: Scientific heritage of P.L. Chebyshev. Moscow-Leningrad: Izdat. Akad. Nauk SSSR, 5–42
[1954] Theory of interpolation and approximation of functions. Moscow: Gostekhizdat. Zbl.57,298
Gorin, E.A.
[1980] Bernstein inequalities from the point of view of operator theory. Vestn. Khar'k. Univ *205*, Prikl. Mat. Mekh. 45, 77–105. Zbl.487.41028
Haar, A.
[1917] Die Minkowskische Geometric und die Annäherung an stetige Funktionen. Math. Ann. *78*, 294–311. Jrb.46,418
Hadamard, J.
[1992] Essai sur l'étude des fonctions données par leur dévelopeement de Taylor. Jaivn. de Math. *4*, No. 8, 101–186. Jrb.24,359
Holladay, J.C.
[1957] Smoothest curve approximation. Math. Tales Aids Comput. *11*, 233–243. Zbl.84,349
Höllig, K.
[1979] Approximationszahlen von Sobolev-Einbettungen. Math. Ann. *242*, 273–281. Zbl.394.41011
Ismagilov, R.S.
[1968] On n-dimensional diameters widths of compact sets in Hilbert space. Funkts. Anal. Prilozh. *2*, No. 2, 32–39. English transl.: Funct. Anal. Appl. 2, 125–132, 1968. Zbl.179,190
[1974] Diameters of sets in linear normed spaces and approximation of functions by trigonometric polynomials. Usp. Mat. Nauk *29*, No. 3, 161–178. English transl.: Russ. Math. Surv. 29, No. 3, 169–186, 1974. Zbl.303.41039
[1977] Width of compacta in normed linear spaces. In: Geometry of linear spaces and operator theory. Yaroslavl', 75–113. Zbl.418.41017
Jackson, D.
[1911] Über die Genauigkeit der Annäherung stetige Funktionen durch ganze rationale Functionen gegebenen Grades und trigonometrische Summen gegebener Ordnung. Göttingen: Diss.

[1930] The theory of approximation. New York: Amer. Math. Soc. *11*. Jrb.56,936
[1941] Fourier series and orthogonal polynomials. Oberlin, Ohio: Amer. Math. Soc. Zbl.60,169
Johnson, R.S.
[1960] On monosplines of least deviation. Trans. Amer. Math. Soc. *96*, 458–477. Zbl.94,39
Kakutani, S.
[1939] Some characterisations of Euclidean spaces. Jap. J. Math. *16*, 93–97. Zbl.22,150
Kamzolov, A.I.
[1974] a) On the Riesz interpolation formula and the Bernstein inequality for functions on uniform spaces. Mat. Zametki *15*, 967–978. English transl.: Math. Notes 15, 576–582 (1974). Zbl.305.41001
[1982] b) On approximation of smooth functions on the sphere by the Fourier method. Mat. Zametki *31*, 847–853, 1982. English transl.: Math. Notes 31, 428–431, 1982. Zbl.514.42036
c) On best approximation of classes of functions by polynomials with spherical harmonics. Mat. Zametki *32*, 285–293, 1982. English transl.: Math. Notes 32, 622–626 (1983). Zbl.514.41031
Karlin, S.
[1968] Total positivity. I. Stanford: Calif. Univ. Press. Zbl.219.47030
[1975] Interpolation properties of generalised perfect splines and the solution of certain extremal problems. Trans. Amer. Math. Soc. *206*, 25–66. Zbl.303.41011
Karlin, S., Studden, W.J.
[1972] Tchebycheff systems: with applications in analysis and statistics. New York: Wiley. Zbl.153,389
Kashin, B.S.
[1977] Widths of certain finite-dimensional sets and classes of smooth functions. Izv. Akad. Nauk SSSR, Ser. Mat. *41*, 334–351. English transl.: Math. USSR, Izv. 11, 317–333 (1977). Zbl.354.46021
[1981] On the diameters of Sobolev classes of small smoothness. Vestn. Mosk. Univ., Ser. I 1981, No. 3, 50–54. English transl.: Mosc. Univ. Math. Bull. 36, No. 5, 62–66. Zbl.474.46025
Kashin, B.S., Saakyan, A.A.
[1984] Orthogonal series. Moscow: Nauka. Zbl.632.42017
Keldysh, M.V.
[1945] Determination of a function of a complex variable by series of polynomials in closed domains. Mat. Sb., Nov. Ser. *16*, (58), 249–258.
Kirchberger, P.
[1903] Über Tchebycheffsche Annäherungsmethoden. Math. Ann. *57*, 505–540. Jrb.34,438
Klots, B.E.
[1975] Linear deviations of classes *W* and approximation in multiplier spaces. Mat. Zametki *18*, 97–108. English transl.: Math. Notes 18, 640–646 (1976). Zbl.329.41023
[1977] Approximation of differentiable functions by functions of greater smoothness. Mat. Zametki *21*: 1, 21–32. English transl.: Math. Notes 21, 12–19. Zbl.346.41015
Knuth D.E.
[1969] The art of computer programming. v. 2. Addison-Wesley. Zbl.191,180
Kolmogorov, A.N.
[1985] Selected works. Mathematics and mechanics. Moscow: Nauka. Zbl.566.01012
Kolmogorov, A.N., Petrov, A.A., Smirnov, Yu.M.
[1947] A formula of Gauss from the theory of the method of least squares. Izv. Akad. Nauk SSSR, Ser. Mat. *11*, 561–566. Zbl.29,405
Kolmogorov, A.N., Tikhomirov, V.M.
[1959] ε-entropy and ε-capacity of sets in function space. Usp. Mat. Nauk *14*, No. 2, 3–86. Zbl.90,335
Konovalov, V.N.
[1980] Supplement to Kolmogorov's inequalities. Mat. Zametki *27*, 209–215. English transl.: Math. Notes 27, 100–104. Zbl.432.26003

[1981] An approximation theorem of the Jackson type for functions of several variables. Ukr. Mat. Zh. *33*, 757–764. English transl.: Ukr. Math. J. 33, 570–575, 1982. Zbl.492.41013

Konyagin, S.V.

[1978] Bounds on the derivatives of polynomials. Dokl. Akad. Nauk SSSR *243*, 1116–1118. English transl.: Sov. Math., Dokl. 19, 1477–1480, 1978. Zbl.418.41002

[1981] On a problem of Littlewood. Izv. Akad. Nauk SSSR, Ser. Mat. *45*, 243–265. English transl.: Math. USSR, Izv. 18, 205–225 (1982). Zbl.493.42004

Kornejchuk, N.P.

[1976] Extremal problems of approximation theory. Moscow: Nauka

[1984] Splines in approximation theory. Moscow: Nauka. Zbl.532.41009

[1985] S.M. Nikol'skij and the development of research on approximation theory in the USSSR. Usp. Mat. Nauk *40*, No. 5, 71–131. English transl.: Russ. Math. Surv. *40*, No. 5, 83–156. Zbl.601.01018

Kornejchuk, N.P., Ligun, A.A., Doronin, V.G.

[1982] Approximations with constraints. Kiev: Naukova Dumka. Zbl.531.41001

Korovkin, P.P.

[1959] Linear operators and approximation theory. Moscow: Nauka. Zbl.94,102

Kotel'nikov, V.A.

[1933] Material for the first all-union congress on question of reorganisation of telecommunications and the development of the communications industry. Moscow: Izdat. PKKA

Krejn, M.G.

[1938] On best approximation of continuous differentiable functions on the whole real axis. Dokl. Akad. Nauk SSSR *18*, 619–623. Zbl.19,14

Krylov, V.I.

[1959] Approximate calculation of integrals. Moscow: Fizmatgiz. Zbl.86,329

Kulanin, E.D.

[1983] Estimates of diameters of Sobolev classes of small smoothness. Vestn. Mosk. Univ., Ser I 1983, No *2*, 24–30. English transl.: Mosc.Univ. Math. Bull. 38, No. 2, 29–35, 1983. Zbl.532.46016

Lavrent'ev, M.A.

[1934] Towards a theory of conformal mappings. Tr. Inst. Fiz-Mat Inst. im V.A. Steklova, 159–245. Zbl.10,264

Lebed', G.K.

[1958] Inequalities for polynomials and their derivatives. Dokl. Akad. Nauk SSSR *117*, No. 4, 570–572. Zbl.127,292

Lebesgue, H.

[1998] Sur l'approximation des fonctions. Bull. Sci. Math. Fr., *22*, 278–287. Jrb.29,352

[1909] Sur les intégrales singullières. Ann. Toulouse *1*, 27–117

[1910] Sur la representation trigonometriques approachée des fonctions satisfaisant à une condition de Lipshitz. Bull. Math. Soc. Fr.

Legendre, A.M.

[1795] Mémoires de mathemtiques et de physique présentes à l'Academic royale des science par divers savants, **10**, 411–434

Levin, A.L., Tikhomirov, V.M.

[1967] On the approximation of analytic functions by rational functions. Dokl. Akad. Nauk SSSR *174*, 279–282. English transl.: Sov. Math., Dokl. 8, 622–626 (1967). Zbl.184,299

[1968] On a problem of V.D. Erokhin, Usp. Mat. Nauk *23*, No. 1 (139) 119–132. English transl.: Russ. Math. Surv. *23*, No. 1, 121–135. Zbl.184,100

Ligun, A.A.

[1973] Sharp inequalities for the suprema of seminorms in classes of periodic functions. Mat. Zametki *13*, 647–654. English transl.: Math. Notes 13, 389–393 (1973). Zbl.258.42003

[1980] On the diameter of certain classes of differentiable periodic functions. Mat. Zametki *27*, 61–75. English transl.: Math. Notes 27, 34–41 (1980). Zbl.447.46025

Lorentz, G.G.
[1966] Approximation of functions. New York: Holt Rinehart Winston. Zbl.153,389
Lorentz, G.G., Schumaker, L.L.
[1972] Saturation of positive operators. J. Approximation Theory *5*, 413–424. Zbl.233.41007
Lozinskij, S.M., Natanson, I.P.
[1959] Metric and constructive theory of functions of a real variable. In: 40 Years of Mathematics in the USSR. Moscow: Fizmatgiz, 295–380
Magaril-Il'yaev, G.G.
[1979] Problem of mean derivatives. Mat. Zametki *25*, 81–96. English transl.: Math. Notes 25, 43–51 (1979). Zbl.409.26009
[1983] Inequalities for derivatives and duality. Tr. Mat. Inst. Steklova *161*, 183–194. English transl.: Proc. Steklov Inst. Math. 161, 199–212, 1984. Zbl.555.26006
[1986] Inequalities of Bernstein-Nikol'skij type and approximations of generalized Sobolev classes, Tr. Mat. Inst. Steklova **173**, 190–212, 204. English transl.: Proc. Steklov Inst. Math. 173, 205–220 (1987). Zbl.622.41009
Magaril-Il'yaev, G.G., Tikhomirov, V.M.
[1984] On some problems of harmonic analysis on $\mathbf{T}^{n'} \times \mathbf{R}^{n''}$, In: Some problems of modern analysis. Moscow: Izdat. Misk. Univ. 57–82. Zbl.631.00003
Mairhuber, J.C.
[1956] On Haar's theorem concerning Chebyshev approximating problems having unique solutions. Proc. Math. Soc. *7*, 609–615. Zbl.70,291
Majorov, V.E.
[1975] Discretization of problem on widths. Usp. Mat. Nauk *30*, No. 6, 179–180. Zbl.319.46024
[1980] On linear widths of Sobolev classes and chains of extremal subspaces. Mat. Sb., Nov. Ser. 113 (155), 437–463. English transl.: Math. USSR, Sb. 41, 361–382 (1982). Zbl.458.41019
[1981] On a modification of the Bernstein-Nikol'skij inequality for trigonometric polynomials. Dokl. Akad. Nauk SSSR *258*, 23–26. English transl.: Sov. Math., Dokl. 23, 471–474. Zbl.509,42001
Makovoz, Yu.I.
[1972] On a method of lower estimation of widths of sets in a Banach space. Mat. Sb., No. Ser. *87* (129), 136–142. English transl.: Math. USSR, Sb. 16, 139–146. Zbl.247.46036
[1979] Widths of Sobolev classes and splines, least deviating from zero. Mat. Zametki *26*, 805–812. English transl.: Math. Notes 26, 897–901 (1980). Zbl.428.46026
Markov, A.A.
[1948] Selected works. Moscow-Leningrad: GITTL. Zbl.34,37
McGehee, O.C., Pigno L., Smith, B.
[1981] Hardy's inequality and the L^1-norm of exponential sums. Ann. Math., II. Ser. *113*, 613–618. Zbl.473.42001
Melkman, A.A.
[1976] n-width under restricted approximation. In: Approximation Theory II. Proc. Int. Symp., Austin 1976, 463–468. Zbl.361.41012
Melkman, A.A., Micchelli, Ch. A.
[1978] Spline spaces are optimal for L^2 n-width. Ill. J. Math *22*, 541–564. Zbl.384.41005
Mergelyan, S.N.
[1951] Some problems of the constructive theory of functions. Tr. Mat. Inst. Steklova *37*, 1–92. Zbl.45,353
Micchelli, Ch.A., Pinkus, A.
[1978] Some problems in the approximation of functions of two variables and n-widths of integral operators. J. Approximation Theory *24*, 51–77. Zbl.356.41014
Micchelli, Ch.A., Rivlin, Th.J., Winograd, S.
[1976] The optimal recovery of smooth functions. Numer. Math. *26*, 191–200. Zbl.335.65004
Micchelli, Ch.A., Rivlin, Th.J. (Eds)
[1976] Optimal estimation in approximation theory. New York: Plenum Press. Zbl.378.00010

Mityagin, B.S.

[1962] Approximation of functions in the spaces C and L_p on the torus. Mat. Sb., Nov. Ser. *58*, 387–414. Zbl.125,280

de Montessus de Ballore, R.

[1902] Sur les fractions continues algébriques. Bull. Soc. Math. Fr. *30*, 28–36. Jrb.33,227

Motornyi, V.P.

[1974] On the best quadrature formula of the form $\sum_{k=1}^{n} p_k f(x_k)$ for some classes of periodic differentiable functions. Izv. Akad. Nauk SSSR, Ser. Mat. *38*, 583–614. English transl.: Math. USSR, Izv. 8, 591–620 (1975). Zbl.306.41013

Müntz, Ch.H.

[1914] Über den Approximationssatz von Weierstrass. Schwarz-Festschrift, 303–312

Mysovskikh, I.P.

[1981] Interpolated cubature formulae. Moscow: Nauka. Zbl.537.65019

Nagy, S.

[1938] Über gewisse Extremalfragen bei transformierten trigonometrishen Entwicklungen. Ber. Sächs. Akad. Wiss. Leipzig *90*, 103–130. Zbl.21,401

Natanson, I.P.

[1949] Constructive theory of functions. Moscow-Leningrad: GITTL. Zbl.41,186

Newman, D.J.

[1964] Rational approximation to $|x|$. Mich. Math. J. *11*, 11–14. Zbl.138,44

Nguen Tkhi Tkh'eu Khoa

[1986] Best quadrature formula and the method of recovery of functions in classes, defined by kernels not increasing oscillation. Mat. Sb., Nov. Ser. 130, 105–119. English transl.: Math. USSR, Sb. 58, 101–117 (1987). Zbl.613.41026

Nikol'skij, S.M.

[1945] Approximation of periodic functions by trigonometrical polynomials. Tr. Mat. Inst. Steklova *15*, 76 p. Zbl.60,168

[1948] Approximation by polynomials of functions of a real variable, In: 30 years of mathematics in the USSR. Moscow: OGIZ, 288–318. Zbl.41,28

[1958] Approximation of functions of many variables by polynomials. Tr. III All-Union Mat. Congr. *3*, 226–231. Zbl.124,285

[1961] Some problems of approximation of functions by polynomials. Proc. Internat. Congr. Math. 1954, Amsterdam 1, 371–382. Zbl.79.91

[1969] Approximation of functions of many variables and the theory of embeddings. Moscow: Nauka. Zbl.185,379

[1974] Quadrature formulae, 2nd ed. Moscow: Nauka. Zbl.298.65017

Osipenko, K.Yu.

[1976] Best approximation of analytic functions from information about their values at a finite number of points. Mat. Zametki *19*, 29–40. English transl.: Math. Notes 19, 17–23 (1976). Zbl.328.30031

[1985] A problem of Heins and optimal extrapolation of analytic functions given with an error. Mat. Sb., Nov. Ser. *126*, 566–575. English transl.: Math. USSR, Sb. 54, 551–559 (1986). Zbl.578.30034

Oskolkov, K.I.

[1975] Estimates of approximation of continuous functions by sequences Fourier sums. Tr. Mat. Inst. Steklova *134*, 240–253. English transl.: Proc. Steklov Inst. Math. 134, 273–288 (1977). Zbl. 359.42002

[1979] On optimality of quadrature formula with equidistant nodes in classes of periodic functions. Dokl. Akad. Nauk SSSR *249*, 49–51. English transl.: Sov. Math., Dokl. 20, 1211–1214 (1979). Zbl.441.41018

Peetre, J.

[1966] Espaces d'interpolation et théorème de Soboleff. Ann. Inst. Fourier *16*, No. 1, 279–317. Zbl.151,179

Pietsch, A.
[1974] s-number of operators in Banach spaces. Stud. Math. *51*, 201–223. Zbl.294.47018
Pinkus, A.
[1985] n-widths in approximation theory. Berlin: Springer-Verlag. Ergebnisse der Mathematik und ihrer Grenzgebiete. 3. Folge. Vol. 7. Zbl.551.41001
Popov, V.A., Petrushev, P.P.
[1977] Exact order of best uniform approximation of a convex function by rational functions, Mat. Sb., Nov. Ser. *103*, 285–292. English transl.: Math. USSR, Sb. 32, 245–251. Zbl.363.41012
Potapov, M.K.
[1974] Imbedding of classes of functions with dominating mixed modulus of smoothness. Tr. Mat. Inst. Steklova *131*, 199–210. English transl.: Proc. Steklov Inst. Math. 131, 206–218 (1975). Zbl.316.46024
[1980] Imbedding theorems in a mixed metric. Tr. Mat. Inst. Steklov *156*, 143–136. English transl.: Proc. Steklov Inst. Math. 156, 155–171 (1983). Zbl.454.46030
Rice, J.R.
[1969] The approximation of functions. Vol 1. Linear theory. Addison-Wesley. Zbl.114,270
Rivlin, T.
[1974] An introduction to the approximation of functions. New York-London: Wiley Interscience. Zbl.189,66
Ruban, V.I.
[1974] Even widths of classes $W^{(r)}H^{\omega}$ in the space $C_{2\pi}$. Mat. Zametki *15*, 387–392. English transl.: Math. Notes 15, 222–225, 1974. Zbl.297.41023
Rudin, W.
[1952] L^2-approximation by partial sums of rothogonal developments. Duke Math. J. *19*, 1–4. Zbl.46,66
Runge, C.
[1985] Zur Theorie der eindeutigen analytischen Funktionen. Acta Math. *6*, 229–244. Jrb.17,379
Rvachev, V.L., Rvachev, V.A.
[1979] Non-classical methods of approximation theory in boundary value problems. Kiev: Naukova Dumka
Schoenberg, I.J.
[1946] Contribution to the problem of approximation of equidistant data by analytic functions. Q. Appl. Math. *4*, 45–99; 112–141. Zbl.61,288
[1951] On Pólya frequency functions I. The totally positive functions and their Laplace transform. J. Anal. Math. *1*, 331–374. Zbl.45,376
[1964] Spline interpolation and best quadrature formulae. Bull. Amer. Math. Soc. *70*, 143–149. Zbl.136,362
[1965] On monosplines of least deviation and best quadrature formulae, SIAM J. Numer. Anal. *34*, 144–170. Zbl.136,362
[1972] Notes on spline functions I. The limits of the interpolating periodic spline functions as their degree tends to infinity. Nederl. Akad. Wet., Proc., Ser. A 75, 412–422. Zbl.271.41004
Schumaker, L.L.
[1981] Spline functions: basic theory. New York: John Wiley & Sons. Zbl.449.41004
Sendov, B.
[1979] Hausdorff approximations. Sofia: BAN. Zbl.416.41002
Shanks, D., Wretch Jr., J.W.
[1962] Calculation of π to 10^5 decimals. Math. Comput. *16*, 76–99. Zbl.104,360
Shapiro, H.S.
[1970] Topics in approximation theory. Berlin: Springer-Verlag. Zbl.213,85
Sharma, A., Meir, A.
[1966] Degree of approximation os spline interpolation. J. Math. Mech. *15*, 759–767. Zbl.158,307

Singer, I.
[1970] Best approximation in normed linear spaces by elements of linear subspaces. Bucharest: Acad RSR, Berlin: Springer-Verlag. Zbl.197,386
[1984] Best approximation and optimization. J. Approximation Theory *40*, 274–284. Zbl.552.41020
Smirnov, V.I., Lebedev, N.A.
[1964] Constructive theory of functions of a complex variable. Moscow-Leningrad: Nauka. Zbl.122,315
Sobolev, S.L.
[1938] On a theorem of functional analysis. Mat. Sb., Nov. Ser. *4*, 471–497. Zbl.22,148
[1974] Introduction to the theory of cubature formulae. Moscow: Nauka. Zbl.294.65013
Stechkin, S.B.
[1951] On the order of best approximations of continuous functions. Izv. Akad. Nauk SSSR, Ser. Mat. *15*, 219–242. Zbl.42,300
[1953] An estimate of the remainder of the Taylor series for certain classes of analytic functions. Izv. Akad. SSSR, Ser. Mat. *17*, 461–472. Zbl.52,79
[1963] Approximative properties of sets in normed linear spaces. Acad. Republ. popul. Roumaine, Rev. Math. Pures Appl. *8*, 5–18. Zbl.198,162
[1967] Best approximation of linear operators. Mat. Zameki *1*, 137–148. English transl.: Math. Notes 1 (1967), 91–99 (1968). Zbl.168,122
[1980] An estimate of the remainder term of Fourier series for differentiable functions. Tr. Mat. Inst. Steklov *145*, 126–151. English transl.: Proc. Steklov Inst. Math. 145, 139–166 (1981). Zbl.442.42004
Stechkin, S.B., Subbotin, Yu.N.
[1976] Splines in numerical mathematics. Moscow: Nauka. Zbl.462.65007
Stepanets, A.I.
[1981] Uniform approximations by trigonometrical polynomials. Kiev: Naukova Dumka. Zbl.481.42001
Stesin, M.I.
[1975] Aleksandrov diameters of finite-dimensional sets of classes of smooth functions. Dokl. Akad. Nauk SSSR *220*, 1278–1281. English transl.: Sov. Math., Dokl. 16, 252–256 (1975). Zbl.333.46012
Stone, M.H.
[1937] Applications of theory of Boolean rings to general topology. Trans. Amer. Math. Soc. *41*, 375–481. Zbl.17,135
Strichartz, R.S.
[1972] Multipliers for spherical harmonic expansions. Trans. Amer. Math. Soc. *167*, 115–124. Zbl.246.42003
Subbotin, Yu.N.
[1971] a) The application of spline functions and estimates of widths. Tr. Mat. Inst. Steklova *109*, 35–60. English transl.: Proc. Steklov Inst. Math. 109, 39–67 (1974). Zbl.256.41011
b) The connection of spline approximations with the problems of the approximation class of a class. Mat. Zametki. *9*, 501–510. English transl.: Math. Notes 9, 289–294. Zbl.218.41003
Suetin, P.K.
[1979] Classical orthogonal polynomials. Moscow: Nauka. Zbl.449.33001
[1984] Series of Faber polynomials. Moscow: Nauka
Szegö, G.
[1921] Über orthogonale Polynome, die zu einer gegebenen Kurve der komplexen Ebene gehören. Math. Z. *9*, 218–270. Jrb.48,374
[1939] Orthogonal polynomials. New York: Amer. Math. Soc. Coll. Publ. 23. Zbl.23,215
Taikov, L.V.
[1967] On best approximation in the mean of certain classes of analytic functions. Mat. Zametki *1*, 155–162. English transl.: Math. Notes 1, 104–109 (1968). Zbl.168,109

Tamrazov, P.M.

[1975] Smoothness and polynomial approximations. Kiev: Naukova Dumka. Zbl.351.41004

Telyakovskij, S.A.

[1960] On approximation of differentiable functions by linear means of Fourier series. Izv. Akad. Nauk SSSR, Ser. Mat. *24*, 213–242. Zbl.116,276

[1961/3] On norms of trigonometric polynomials and approximation of differentiable functions by linear means of their Fourier series. [Tr. Mat. Inst. Steklova *62* (1961) 61–97;] Izv. Akad. Nauk. SSSR, Ser. Mat. *27* (1963) 253–272. Zbl.106,275. Zbl.122,310

[1964] Some estimates for trigonometric series with quasi-convex coefficients. Mat. Sb., Nov. Ser. *63*, 426–444. Zbl.134,286

Temlyakov, V.N.

[1979] Approximation of periodic functions of several variables with bounded mixed derivative. Dokl. Akad. Nauk SSSR *248*, 527–531. English transl.: Sov. Math., Dokl. 20, 1032–1035 (1979). Zbl.468.42003

[1982] Approximation of functions with bounded mixed difference by trigonometric polynomials and diameters of some classes of functions. Izv. Akad. Nauk SSSR, Ser. Mat. *46*, 171–186. English transl.: Math. USSR, Izv. 20, 173–187 (1983). Zbl.499.42002

[1985] Approximation of periodic functions of several variables by trigonometric polynomials and widths of some classes of (certain) functions. Izv. Akad. Nauk. SSSR, Ser. Mat. *49*, No. 5, 986–1030. English transl.: Math. USSR, Izv. 27, 285–322 (1986). Zbl.608.42005

[1986] Approximation of functions with bounded mixed derivative. Tr. Mat. Inst. Steklov., Tom 178, Moscow: Nauka, 1986. Zbl.625.41028

Tikhomirov, V.M.

[1976] Some questions of the theory of approximations. Moscow: Izdat. Mosk. Gos. Univ.

Tikhomirov, V.M., Boyanov, B.D.

[1979] On certain convex problems of approximation theory. Serdica *5*, 83–96. Zbl.437.41034

Timan, A.F.

[1960] Theory of approximations of functions of a real variable. Moscow: Fizmatgiz. English transl.: Oxford-London-Paris: Pergamon Press, 1963. Zbl.117,290

Traub, J.F., Wozniakowski, H.

[1980] A general theory of optimal algorithms. New York: Academic Press. Zbl.441.68046

Triebel, H.

[1978] Interpolation theory. Function spaces. Differential operators. Berlin: Dtsch. Verl. Wiss. Zbl.387.46033

Ul'yanov, P.L.

[1968] Embedding of certain classes of functions. Izv. Akad. Nauk SSSR, Ser. Mat. *32*, 649–688. English transl.: Math. USSR, Izv. 2, 601–637. Zbl.176,433

[1970] Embedding theorems and the relation between best approximation with moduli of continuity in various metrics. Mat. Sb., Nov. Ser. *81* (123) 104–131. English transl.: Math. USSR, Sb. 10, 103–126. Zbl.196,83

Ural Scientific Centre

[1984] Splines and Numerical methods in the theory of approximations of functions. Table. Sverdlovsk: Ural'skij Nauchn. Tsentr

Uryson, P.S.

[1951] Works on topology and other domains of mathematics I, II. Moscow-Leningrad: Gostekhizdat. Zbl.44,193

Vallée-Poussin, Ch.-J.

[1911] Sur les polinômes d'approximation a une variable complexe. Bull. Acad. R. Belg. Ci. Sc., 199–211

[1919] Leçons sur l'approximation des fonctions d'une variable réels. Paris: Gauthier-Villars. Jrb.47,908

Velikin, V.L.

[1981] On the limit connection between approximations of periodic functions by splines and by trigonometric polynomials. Dokl. Akad. Nauk SSSR *258*, 525–529. English transl.: Sov. Math., Dokl. 23, 540–544 (1981). Zbl.523.41007

Vitushkin, A.G.

[1959] Estimates of complexity for the problem of tabulation. Moscow: Fizmatgiz. Zbl.122,119

[1967] Analytic capacity of sets in problems of approximation theory. Usp. Mat. Nauk *22*, No. 6, 141–199. English transl.: Russ. Math. Surv. 22, No. 6, 139–200. Zbl.157,354

Vlasov, L.P.

[1973] Approximative properties of sets in linear normed spaces. Usp. Mat. Nauk *28*, No. 6, 3–66. English transl.: Russ. Math. Surv. 28, No. 6, 1–66 (1973). Zbl.291.41028

Voronovskaya, E.V.

[1932] The determination of the asymptotic form of the approximation of functions by Bernstein polynomials. Dokl. Akad. Nauk SSSR, A, No. 4, 79–85. Zbl.5,12

Walsh, J.L.

[1926] Über der Grad der Approximation einer analytischen Funktion. Sitzungsberichte Munchen 223–229. Jrb.52,299

[1935] Interpolation and approximation by rational functions in the complex domain. New York: Amer. Math. Soc. Zbl.13,59

Weierstrass, K.

[1985] Über die analytische Darstellung sogenannter willkürlicher Funktionen einer reellen Veränderlichen. Sitzungsberichte der Akad. zu Berlin, 633–640; 789–806. Jrb.17,384

Weil, H.

[1917] Bemerkungen zum Begriff des Differentialquotienten gebrochener Ordnung. Vierteljahreschrift. Naturw. Ges. Zürich *62*, 296–302

Whitney, H.

[1957] On functions with bounded n-th differences. J. Math. Pure Appl., IX, Sér. *36*, 67–95. Zbl.77,69

Widom, H.

[1972] Rational approximation and n-dimensional diameter. J. Approximation Theory 5, 343–361. Zbl.234.30023

Zakharyuta, V.P., Skiba, N.I.

[1976] Estimates of n-diameters of some classes of functions, analytic on Riemannan surfaces. Mat. Zametki *19*, 899–911. English transl.: Math. Notes 19, 525–532 (1976). Zbl.375.46028

Zav'yalov, Yu.S., Kvasov, B.I., Miroshnichenko, V.L.

[1980] Spline function methods, Moscow: Nauka. Zbl.524.65007

Zhensykbaev, A.A.

[1981] Monosplines of minimal norm and best quadrature formulae. Usp. Mat. Nauk *36*, No. 4, 107–159. English transl.: Russ. Math. Surv. 36, No. 4, 121–180. Zbl.504.41024

[1976] On best quadrature formula in the class W^rL_p. Dokl. Akad. Nauk SSSR *227*, 277–279. English transl.: Sov. Math., Dokl. 17, 377–380 (1976). Zbl.354.65010

Zhuk, V.V., Natanson, G.I.

[1983] Trigonometric Fourier series and elements of approximation theory. Leningrad: Izdat. Leningrad. Univ. Zbl.535.42010

Zolotarev, G.

[1977] Application of elliptic functions to questions on functions, least deviating from zero. Abh. St. Petersb. XXX, No. 5, Jrb.9,343

Zygmund, A.

[1945] Smooth functions. Duke. Math. J. *12*, 47–76. Zbl.60,138

Author Index

Ahlberg, J.H. 178, 229, 231
Ahlfors, L.V. 126, 231
Akhiezer, N.I. 86, 97, 99, 116–118, 135, 205, 228, 229, 231
Akilov, G.P. 85, 86, 89
Aleksandrov, A.D. 15, 21, 73, 86
Aleksandrov, P.S. 188, 230, 231
Alekseev, V.M. 9, 58, 68, 69, 73, 85, 86, 109, 217, 229, 231
Andrienko, V.A. 161, 231
Archimedes, 4, 86, 104, 229–231
Arestov, V.V. 204, 214, 218, 231
Aristotle 59
Arkin, V.I. 67, 69, 73, 85, 86
Arnol'd, V.I. 223, 231, 232
Asplund, E. 22, 86
Astaf'ev, N.N. 57, 59, 85, 88
Aubin, J-P. 86

Babadzhanov, S.B. 208, 225
Babenko, K.I. 157, 196, 200, 202, 228–232
Babenko, V.F. 140, 186, 232
Badkov, V.M. 229, 231, 232
Bagby, T. 229, 232
Bakhvalov, N.S. 155, 229, 232
Banach, S. 80
Barrar, R.B. 176, 232
Belyi, V.I. 181
Berger, M. 15, 61, 85, 86
Bernoulli, I. 98, 106
Bernoulli, J. 98
Bernstein, S.N. 83, 96, 97, 104, 121, 122, 127, 130–132, 136, 165, 167, 168, 184, 198, 204, 220, 226, 228, 229, 232
Besicovitch, A.S. 76
Besov, O.V. 229, 232
Bieberbach, L. 20, 60, 86
Birman, M.Sh. 230–232
Bishop, E. 83
Blagodatskikh, V.I. 3
Blaschke, W. 13, 20, 60, 63, 86
Boas, R.P. Jr. 204
Boehm, B. 229, 232
Bogolyubov, N.N. 21, 66, 85, 87
Bohr, H. 135, 229, 232
Boltyanskii, B.G. 60, 73, 85, 87, 91, 92
Bonami, A. 220, 231, 232
Bonnice, W.E. 80, 87
Bonnesen, T. 15, 20, 85, 87
Boor, C.de 97, 176, 178, 184, 214, 229, 233
Borel, E. 97, 116, 126, 228, 229, 232
Bouligand, G. 76, 87
Bourbaki, N. 37, 40, 85, 87
Boyadzhiev, P.G. 229
Boyanov, B. 176, 212, 214, 215, 231, 242
Bronsted, A. 21, 53, 87
Brouncker, W. 105, 106
Brouwer, A. 63
Brown, A. 62
Brudnyj, Yu.A. 138, 182
Brunn, H. 15, 63
Bulanov, A.P. 172, 232
Bunt, L.N. 63, 224
Busarova, T.N. 211
Buseman, H. 15, 87
Buslaev, A.P. 98, 208, 213, 230, 232
Butzer, P.L. 97, 220, 229, 233

Carathéodory, C. 11, 20, 87
Cartier, P. 83
Cartwright, M. 221
Castaing, Ch. 22, 55, 187
Cauchy, A. 13–15, 87, 106, 142
Chakalov, L. 221
Chakhiev, M.A. 212
Chebyshev, P.L. 53, 87, 96–99, 104, 107, 108, 112, 113, 115–119, 127, 142, 172, 176, 182, 184, 224, 226, 228, 229, 233
Cheney, E.W. 97, 228, 233
Chernikov, S.N. 74, 87
Chernykh, N.I. 137
Choquet, G. 10, 21, 82, 83, 86, 87
Cieselski, Z. 175, 233
Clarke, F.H. 22, 76, 79, 86, 87
Clerk, J.-L. 220, 231, 232

Collatz, L. 97, 178, 228, 233
Cotes, R. 106
Courant, R. 18, 87

D'Alembert, Jean de Rond 106
Dantzig, G.B. 59, 85, 88
Danzer, L. 11, 61, 85, 88
Davis, P.J. 229, 233
De Leeuw, K. 83
Debrunner, H. 89
Dem'yanov, V.F. 78, 88
Din' Zung 159, 160, 222, 229–231
Dini, U. 74
Dirichlet, L. 70
Dolzhenko, E.P. 168, 186, 229, 233
Domar, Y. 234
Doronin, V.G. 140, 186, 229, 233
Doronin, Yu.A. 167
Dubois, Reymond P. 132
Dubovitskii, A.Ya. 17, 21, 52, 81, 88
Dunford, N. 234
Dunham, C.B. 225, 234
Dveirin, M.Z. 202, 203
Dzhafarov, A.C. 220
Dzyadyk, V.K. 97, 133, 136, 138, 139, 164, 166, 180, 181, 218, 228, 229, 234

Efimov, A.V. 164, 165, 224, 234
Eggleston, H. 61, 85, 88
Ekeland, I. 58, 70, 85, 86, 88
Eremin, I.I. 57, 59, 85, 88
Erokhin, V.D. 182, 200, 201, 229, 234
Euclid 14
Euler, L. 19, 74, 98, 106, 146

Faber, G. 184, 234
Fabry, E. 143, 171, 229, 234
Fan Ky 61, 73, 74
Farkas, J. 73
Farkov, Yu.A. 98, 202, 203, 234
Favard, J. 135, 176, 234
Fedorov, E.S. 14, 88
Fejer, L. 123, 162, 220, 234
Fell, J.M.G. 83
Fenchel, W. 8, 15, 19–21, 43, 46, 85, 87, 88
Fermat, P.de 109
Figiel, T. 175, 233
Filippov, A.F. 3
Fisher, S.D. 201, 234
Fomin, A.B. 137
Fomin, S.V. 9, 40, 58, 68, 69, 73, 86, 90
Fourier, J. 162

Frechet, M. 75
Freud, G. 138
Frobenius, F. 142
Fuchssteiner, B. 83, 86, 88
Fursikov, A.B. 225

Gabushin, V.N. 209, 218
Gaffney, P.W. 234
Galeev, E.M. 58, 85, 86, 98, 109, 196, 217, 229–231, 234
Galkin, P.V. 133
Gamkrelidze, R.V. 85, 88, 91
Ganelius, T. 186, 201, 234
Gantmakher, F.R. 211
Garkavi, A.L. 225, 226, 231, 235
Gass, S.L. 59, 85, 88
Gauss, C.F. 98, 114, 229, 235
Gavrilyuk, V.T. 166
Gavurin, M.K. 56, 59, 85, 88
Gel'fand, I.M. 124, 188, 235
Giles, J.B. 76, 86, 88
Girsanov, I.V. 85, 88
Glicksberg, I. 61
Glushkin, E.D. 195, 230, 235
Gol'shtein, E.G. 21, 59, 85, 89, 220
Golitschek, M. von 178, 229, 234
Golovkin, K.K. 161, 179, 235
Gonchar, A.A. 118, 168, 201, 229, 235
Goncharov, V.L. 97, 99, 228, 229, 235
Gorin, E.A. 204, 235
Gregory, J. 98
Grunbaum, B. 11, 61, 85, 88

Ha, C.W. 230
Haar, A. 120, 229, 235
Hadamard, J. 126, 143, 208, 229, 235
Hadwiger, H. 15, 85, 89
Hahn, H. 80
Hardy, G.H. 73, 89, 153, 208, 218
Hausdorff, F. 13, 186
Helfrich, H-P. 230
Helly, E. 12, 13, 20, 85, 89
Hermite, C. 106
Hilbert, D. 18, 87
Hille, E. 162
Holder, E. 73
Holladay, J.S. 176, 235
Hollig, K. 229, 230, 233, 235
Holmes, R. 12, 42, 85, 89
Hörmander, L. 9, 17, 21, 89
Huygens, Ch. 105, 231

Il'in, V.P. 229, 232

Ioffe, A.D. 3, 21, 48, 49, 53, 54, 60, 69, 78, 79, 85, 86, 89
Ismagilov, R.S. 97, 190, 195, 206, 208, 225, 229, 230, 235
Ivanov, V.A. 98, 220

Jackson, D. 96, 97, 104, 112, 127–129, 132, 162, 198, 228, 229, 235
James, R.S. 61, 89, 120
Jensen, J.L.W.V. 19, 89
John, F. 21, 61, 89
Johnson, R.S. 176, 235

Kakutani, S. 74, 89, 236
Kamzolov, A.I. 98, 163, 180, 204, 220, 231, 236
Kantorovich, L.V. 20, 21, 53, 56, 71, 80, 85, 86, 89, 90
Karlin, S. 86, 90, 97, 120, 211, 229, 236
Karmanov, V.G. 59, 85, 90
Kashin, B.S. 97, 98, 194, 195, 229, 230, 236
Keldysh, M.V. 124, 222, 229, 236
Kelly, J.L. 63
Kepler, J. 98, 106
Kharatishvili, G. 88
Kharshiladze, F.I. 167
Khavinson, S.Ya. 53
Khenkin, G.M. 223
Khovanskij, A.G. 74, 85, 90
Kirchberger, P. 116, 236
Klee, V. 11, 12, 61, 85, 89, 90, 225
Klots, B.E. 214, 235
Knuth, D. 96, 236
Kogbetliantz, E. 220
Kolmogorov, A.N. 40, 76, 85, 90, 96, 97, 99, 104, 134, 138, 164, 166, 168, 188, 199, 200, 205, 207, 208, 218, 221–223, 230, 231, 236
Konovalov, V.N. 98, 183, 218, 236
Konyagin, S.V. 3, 98, 196, 225, 237
Korkin, A.N. 108, 112
Kornejchuk, N.P. 97, 98, 136, 137, 139, 140, 164, 166, 178, 186, 189, 204, 206, 211, 218, 229, 237
Korovkin, P.P. 84, 97, 179, 180, 228, 229, 237
Kotel'nikov, V.A. 221, 237
Krabs, W. 97, 228, 233
Krasnoselskij, M.A. 85, 90, 230
Krejn, M.G. 10, 20, 42, 81, 86, 90, 120, 135, 205, 211, 230, 231, 237
Krylov, V.I. 184, 237
Kuhn, H. 21, 58
Kulanin, E.D. 194, 237
Kuptsov, N.P. 209, 218
Kushnirenko, G.G. 220
Kusraev, A.G. 3, 22, 59, 82, 85, 86, 90
Kutateladze, S.S. 22, 40, 42, 81, 83–86, 90
Kvasov, B.I. 178, 229, 243

Lagrange, J.L. 74, 98, 106
Lambert, J.H. 231
Landau, E. 127, 131, 139
Laplace, P.S. 106
Lau, K-S. 225
Laurent, P-J. 85, 90
Lavrent'ev, M.A. 124, 222, 229, 237
Lebed', G.K. 138, 237
Lebedev, N.A. 97, 241
Lebesgue, H. 96, 122, 126, 138, 229, 237
Legendre, A.M. 19, 106, 229, 231, 237
Leibniz, G.W. 19, 74, 106, 146
Leichtweis, K. 11, 12, 15, 20, 60, 61, 85, 90
Levin, A.L. 201, 229, 237
Levin, V.L. 3, 21, 22, 53, 55, 67, 69, 71, 73, 81, 85, 86, 89–91
Levitin, E.S. 85, 91
Ligun, A.A. 137, 140, 177, 186, 205, 206, 211, 237
Lipshitz, R. 99
Littlewood, J. 73, 89, 193, 208, 218, 220
Lizorkin, P.I. 151, 220
Loeb, H. 176, 232
Lorentz, G.G. 97, 189, 228, 229, 238
Lozinskij, S.M. 167, 228, 238
Lushpai, N.E. 211
Lusky, W. 83, 86, 88
Lyapunov, A.A. 21, 64, 91
Lyubich, Yu.I. 209, 218
Lyusternik, L.A. 61, 85, 91

Machin, J. 105
Mackenhoupt, B. 220
Mackey, G. 37, 38
MaClaurin, C. 106
Magaril-Il'yaev, G.G. 3, 98, 101, 209, 229, 231, 233, 238
Mairhuber, I. 120, 229, 238
Majorov, V.E. 98, 195, 230, 238
Makarov, V.L. 57, 91
Makovoz, Yu.I. 205, 229, 230, 238
Mal'tsev, A.I. 230
Malozemov, B.N. 59, 85, 88
Marcinkewicz 153, 195, 220
Markov, A.A. 119, 172, 228, 229, 238
Markov, V.A. 119
McGehee, O.C. 196, 238
McShane, E.J. 85
Meir, A. 240
Melkman, A.A. 229, 230, 238

Mendeleev, D.I. 119
Mercator, N. 98, 105
Mergelyan, S.N. 124, 168, 180, 222, 229, 230, 238
Meyer, P. 83
Micchelli, Ch.A. 201, 208, 212, 214, 215, 230, 234, 238
Michael, E. 74
Mil'man, D.P. 10, 20, 42, 90, 230
Milyutin, A.A. 17, 21, 52, 71, 81, 85, 91, 223
Minkowskii, H. 9, 10, 11, 13, 15, 17–19, 73, 85, 91
Miroshnichenko, V.L. 178, 229, 243
Mishchenko, E.F. 91
Mityagin, B.S. 196, 229, 230, 239
Monge 71
Montessus de Ballore, R. de 143, 171, 229, 239
Mordukhovich, B.Sh. 3, 78, 85, 86, 91
Moreau, J-J. 8, 17, 21, 43, 46, 52, 85, 91
Mosolov, P.P. 71, 86, 91
Motornyj, V.P. 140, 211, 239
Motskin, T.S. 63, 224
Muntz, Ch.H. 125, 229, 239
Myasnikov, B.P. 71, 86, 91
Mysovskikh, I.P. 184, 212, 239

Nagy Sz., B. 162, 166, 221, 239
Natanson, G.I. 243
Natanson, I.P. 97, 228, 229, 238, 239
Nessel, R.I. 97, 229, 233
Newman, D.J. 170, 220, 239
Newton, I. 74, 98, 105, 106
Nguen Tkhi Tkh'eu Khoa 212, 239
Nikaido, H. 91
Nikolesku, M. 61
Nikol'skij, S.M. 61, 62, 96, 97, 99, 127, 136, 137, 153, 154, 160, 162, 164–167, 204, 205, 208, 210, 211, 220, 221, 228, 229, 232, 239
Nilson, E.N. 178, 229, 231
Nudel'man, A.A. 90

Oldenburg, H. 105
Oleinikov, V.L. 203
Orlicz, W. 53
Oshman, E.V. 62
Osipenko, K.Yu. 98, 214, 215, 239
Oskolkov, K.I. 167, 211, 239
Osmolovskii, N.P. 85, 91

Padé, H. 142
Parfenov, O.G. 203
Pawelke, S. 220
Peetre, J. 239
Pel'chinskij, A. 223
Peller, V.V. 229
Penkov, B.I. 186
Petrov, A.A. 230, 236
Petrushev, P.P. 172, 240
Phelps, R.R. 21, 78, 86, 91
Pietsch, A. 230, 240
Pigno, L. 196, 238
Pinkevich, B.T. 164
Pinkus, A. 97, 189, 204, 205, 208, 211, 229, 230, 238, 240
Pogorelov, A.V. 15, 21, 62, 91
Pollard, H. 220
Pólya, G. 73, 89
Pontryagin, L.S. 21, 91
Popov, V.A. 172, 240
Potapov, M.K. 240
Powell, M.J.D. 234
Pshenichnyi, B.N. 21, 73, 85, 91
Pukhov, S.V. 230

Quade, E. 178

Rachev, S.T. 71, 85, 91
Radon, J. 12, 13, 20, 92
Ragozin, D.L. 220
Raikov, D.A. 124, 182, 235
Rice, J.R. 97, 228, 240
Riesz, F. 20, 92
Riesz, M. 134, 151, 153, 179, 220
Rivlin, Th.J. 97, 212, 214, 215, 238, 240
Robertson, A.P. 37, 40, 85, 92
Robertson, W. 37, 40, 85, 92
Rockafellar, R.T. 14, 17, 21, 28, 40, 52, 53, 55, 74, 85, 92
Rolle, M. 211
Ruban, V.I. 140, 230, 240
Rubinov, A.M. 57, 78, 83–86, 88, 90–92
Rubinshtein, G.Sh. 71, 85, 86, 90
Rudin, W. 220, 230, 240
Runge, C. 124, 183, 229, 240
Rutitskii, Ya.B. 85, 90
Rutman, M.A. 86, 90
Rvachev, V.A. 186, 240
Rvachev, V.L. 186, 240

Saakyan, A.A. 229, 236
Scherer, K. 223
Schoenberg, I.T. 120, 175, 176, 178, 211, 240
Schumaker, L.L. 237, 240
Schwartz, J.T. 234
Sendov, Bl. 97, 186, 240
Shaginyan, A.L. 220
Shanks, D. 229, 240

Shannon, C. 221
Shapiro, H.S. 228, 240
Sharma, A. 240
Sharygin, I.F. 188
Shevchuk, I.A. 181
Shilov, G.E. 124, 235
Shnirel'man, L.G. 53
Silverman, R.S. 80, 87
Simpson, Th. 106
Singer, I. 61, 97, 241
Skiba, N.I. 202, 243
Slater, M. 21
Smirnov, V.I. 97, 230, 236, 241
Smith, B. 196, 238
Smolyak, S.A. 188, 210, 230
Sobolev, S.L. 184, 212, 241
Solomyak, M. 230, 231, 232
Soltan, V.P. 84, 92
Stechkin, S.B. 53, 97, 98, 136, 137, 139, 164, 165, 178, 204, 212, 214, 224, 225, 229–231, 241
Steiltjes, Th. 143, 172
Stein, E.M. 208, 218, 220, 229
Steiner, J. 15
Stepanets, A.I. 97, 164, 166, 241
Stesin, M.I. 225, 230, 241
Stolz, O. 19, 74, 92
Stone, M. 124, 241
Strichartz, R.S. 220, 241
Studden, W. 97, 211, 229, 236
Subbotin, Yu.N. 97, 177, 178, 204, 205, 229, 241
Suetin, P.K. 112, 180, 201, 229, 241
Suetin, S.P. 229
Sun Yun Shen 139, 218
Szegö, G. 112, 229, 241

Taikov, L.V. 202, 204, 206, 208, 218, 241
Tamarkin, J.D. 162
Tamrazov, P.M. 181, 242
Taylor, B. 106
Telyakovskij, S.A. 157, 164–166, 229, 242
Temam, R. 58, 70, 85, 88
Temlyakov, V.N. 160, 161, 188, 196, 229–231, 242
Tiel, J. van 85, 92
Tikhomirov, V.M. 9, 21, 48, 49, 53, 54, 58, 60, 62, 68, 69, 73, 74, 85, 86, 89, 92, 97, 109, 120, 177, 188, 189, 199, 201–205, 207–209, 217, 221, 223, 225, 228–231, 233, 236, 237, 242
Timan, A.F. 97, 136, 138, 228, 229, 242
Tonelli, L. 65, 92
Topuriya, S.B. 220
Torricelli, B. 106
Traub, J. 97, 242
Treibel, H. 229, 242
Tsar'kov, I.G. 224, 230
Tucker, A.W. 21, 58, 90

Ul'yanov, P.L. 161, 242
Uryson, P.S. 188, 230, 242
Valadier, M. 22, 55, 87
Vallée-Poussin, Ch-J. 53, 96, 97, 117, 126, 130, 132, 137, 162, 228, 242
Vasil'ev, L.V. 88
Velikin, V.L. 140, 178, 206, 230, 242
Verchenko, N.Ya. 76
Vieta, F. 105
Vinter, R.B. 87
Vitushkin, A.G. 97, 126, 190, 199, 200, 223, 230, 243
Vlasov, L.P. 62, 63, 224, 226, 231, 243
Vore, R. de 229, 233
Voronoj, G.F. 73
Voronovskaya, E.V. 126, 243
Vyacheslavov, N.S. 172

Wallis, J. 98
Walsh, J. 97, 178, 229, 231, 243
Warga, J. 78
Weierstrass, K. 18, 21, 74, 96, 97, 104, 121, 137, 228, 229, 231, 243
Weil, H. 149, 243
Weiss, G. 229
Whitney, H. 182, 243
Whittaker, E.T. 221
Widom, H. 201, 214, 243
Wiener, N. 125
Winograd, S. 214, 215, 238
Wozniakowski, H. 97, 242
Wretch, J.W.Jr. 229, 240
Wulbert, D.E. 224

Yaglom, I.M. 60, 85, 92
Young, H.W. 20, 60, 74, 92
Young, L.C. 85, 92

Zakharyuta, V.P. 202
Zav'yalov, Yu.S. 178, 229, 243
Zhensykbaev, A.A. 140, 176, 177, 211, 212, 230, 243
Zhuk, V.V. 137, 243
Zolotarev, E.I. 108, 112, 113, 117, 118, 215, 228, 229, 243
Zukhovitskii, S.I. 53
Zvyagintsev, A.I. 209
Zygmund, A. 130, 132, 137, 204, 243

Subject Index

Alternation 115
Analysis, convex 4
— — global 82
— — local 82
— harmonic 218
— nonsmooth 76
Annihilator 33, 101
Apparatus of approximation 101
Approximation, Padé 142

B-splines 174
— normalised 174
Bilinear form 31

Calculus, convex 15
— subdifferential 17
Capacity, analytic 125
— logarithmic 200
— of a condensor 200
— of a condensor (T, F) 171
— of a pair (T, G) 171
Circle of ideas due to Chebyshev 115
Class, Besov 152
— Nikol'skij 104
— Nikol'skij $BH_p^\alpha(T^1)$ 104
— Nikol'skij $H_p^\omega(T^1)$ 151
— Sobolev 104
Closure 38
— convex 38
— of a function 39
Combination, affine 24, 27
— conical 24, 27
— convex 24, 27
— linear 24, 27
Condition, complementary slackness 58, 68, 69, 73
— non-negativity 68, 69, 73
— Slater 59
— stationarity 73
— transversality 69
— Weierstrass 72
Conjecture, Kolmogorov 223
— Littlewood 195

Cone 25
— conjugate 5, 33, 36
— contingent 76
— convex 5, 24
— — polyhedral 27
— right circular 25
— recessive 11
— tangent 76
Cones in general position 81
Constant, Lebesgue 133
— — of the operator I_τ 183
Constants, Bernstein 101
— Favard 100, 101, 135
— Kolmogorov-Nikol'skij 164
Constraint 55
Contingency 76
Continuum, admissible 181
Convex, curve 4
— function 28
— geometry 11
— hull 4, 27, 29
— — minimum 29
— — union 29
— *L*-integrand 54
— set 24
— surface 4
Convexity, in economics 19
— in mathematics 18
— in the natural sciences 18
Convolution, infimal 17
— integral 48
— of functions 29
— of sets 30
Cowidth 103
— Alexandrov 103, 189
— entropic 189
— Fourier 103
— linear 103
Criterion, for a space to be Asplund 76
— for ext A to be nonempty 11
— polyhedral 14
Criteria for solutions in linear programming 58

Cross hyperbolic 156
Curve, convex 4

Defect of a spline 146
Degree of a polynomial 141
Derivative, lower Dini 78
— Weil fractional 149
Deviation, Fourier 101
Direction recessive 11
Distance 101
— Hausdorff 13
Duality 4, 31
— in the problem of moments 62
— in variational problems 67

ε-entropy 188
— absolute 189
— mean 221
Epigraph of a function 6, 27
Equality, Parseval 144
Equation, Euler-Lagrange 69
— Legendre 110
— of Sturm-Liouville type 207
— Pearson 111
— Sturm-Liouville 207
Example, Runge 183
— Weierstrass 65
Extended number line 27

Figure, convex 1, 13
— polyhedral 14
Formalization 56
Forms, Canonical 31
Formula, Cauchy-Hadamard 142
— duality for least distance problems 62
— Dubovitskij-Milyutin 17
— for subdifferential of an integral 54
— Gauss 114
— inversion, for r-times differentiation 109, 141
— Moreau-Rockafellar 17
— quadrature 114
— rectangles 21
— Rodrigues 108
— — generalized 111
Fourier cowidth 103
— deviation 102
— N-width 187
— width 102
Function, affine 6, 28
— conjugate 6
— convex 4, 28
— — homogeneous 28
— Φ-convex 84
— Favard 100
— homogeneous of first degree 7
— improper 28
— indicator 28
— Lagrange 56, 59
— linear 28
— Minkowski of a set 28
— proper 28
— rational 142
— spherical 219
— strictly differentiable 77
— subdifferential at a point 51
— sublinear 7, 28
— support 7, 33, 36
— zonal 219
Functions, analytic 141
— elementary 33
— Favard 146
— generating 110
Fundamental theorems of algebra 177

Gauge 28
General position 17
— of cones 81
— of convex sets 81

Halfspace 33
— open 33
Hull, affine 27
— convex 27, 29
— convex conical 27
— linear 27
Hyperplane 8, 26, 33

Image, of a function under a linear mapping 29
— of a set under a linear mapping 36
Inequality, Bernstein's 131, 160
— Bernstein-Nikol'skij's 101
— Bernstein-Walsh's 181
— Bieberbach's 60
— Bohr-Favard's 101, 135
— Jensen's 28
— Lebesgue's 133
— Markov V.A. 119
— Markov A.A. 119, 160
— Young's 43, 82
Inequalities for derivatives of polynomials and functions 119
Interior 38
Intersection of sets 30

Inverse image of a, function under a linear mapping 29
— set under a linear mapping 30

Jacobian generalized 77

k-alternation 116
K-space 80
Kernel, Dirichlet 133, 145
— Fejér 123
— Jackson 128
— not increasing oscillation 211
— Poisson 113, 121

Lattice, order complete 79
— vector 80
Lemma, projection 190
Lifting 54

Manifold affine 24, 26, 27
Matrix regular 162
Maximum of functions 29
Means, Riesz 162
Method, Darboux 110
— of recovery, optimal linear 210
Modulus of continuity 100
Monospline 147
— Bernoulli 100, 147
Multipliers Lagrange 56, 73

N-width, Alexandrov 187
— Bernstein 187
— Gel'fand 188
— Kolmogorov 187
— linear 187
Nodes of a spline 146

Octahedron 25
Operator, Bernstein-Rogosinski 162
— Fejér 102, 145
— Fourier 102, 133, 145
— Jackson 102
— Korovkin 162
— Kutateladze canonical 81
— Laplace-Beltrami 219
— polynomial 179
— positive linear 179
— Vallée-Poussin 145
— Zygmund 162
Operators, basic, of convex analysis 4
Order of a spline 146

p-ellipsoid 190
p-sphere 25
Parallelopiped 25
Perturbation 57
Point, extreme 10, 42
— saddle 67
points, affinely independent 27
— linearly independent 27
Polar 5, 33, 35
Pole 219
Polyhedron, convex 13, 27
— Newton 74
Polynomial, Bernoulli 146
— Faber 180
Polynomials 136
— algebraic 141
— Chebyshev 112
— — of first kind 108, 111
— — of second kind 108, 111, 113
— Hermite 111
— interpolation 141
— Jacobi 111
— Lagrange 148
— Laguerre 111
— least deviating from zero 107
— Legendre 108
— Taylor 141
— trigonometric 143
— Zolotarev 118
Principle, Lagrange, for Lyapunov problems 67
— maximum 69, 72
— — for problems linear in the phase variables 68
— — Pontryagin's 72
— minimum 67
Problem, dual 57
— extension 66
— extrapolation, for polynomials 66
— extremal 55
— Favard's 175
— Heins 215
— Hilbert's thirteenth 222
— isoperimetric 59
— Kolmogorov-Nikol'skij 164, 210
— linear programming 56
— Lyapunov 67
— Monge-Kantorovich 71
— of classical variation calculus 64
— of convex programming 64
— of inequalities for derivatives 119
— of best quadratures 210
— of quadratures 210
— of recovery of functionals 210

— of recovery of operators 210
— optimal control 67
— transportation 56
Product Blaschke 201
Programming, convex 56
Projection 167
— metric 101
Property, isoperimetric, of the circle 59
— — of the sphere 60

Quasidifferential 89

Ray, extreme 12
Representation, integral, of the Fejér operator 145
— — regular 54
Rows of the Padé table 142

S-function 57
Semitangent 76
Separation 8
— proper 8
— strict 8
Series of classical polynomials 111
Set, absolutely convex 25
— approximately compact 224
— Chebyshev 62
— convex 24
— — in general position 81
— d-convex 84
— effective 27
— $\mathscr{T}$-set 25
— $\mathscr{H}$-convex 84
— of elements of best approximation 101
— of feasible elements 55
Simplex 27, 83
— Choquet 83
Smoothing 121, 141
Smoothness 150, 151
— elementary 151
— of a function 151
Sobolev space $W_p^\alpha(T^1)$ 103
— space $W_p^\alpha(T^n)$ 103
Solid, platonic 13
Space, Asplund 75
— Chebyshev 120
— Kantorovich 79
— linearly ordered 79
— Nikol'skij $H_p^\alpha(T^n)$ 104
— smooth 224
— Sobolev 103, 150
— vector ordered 79
Spaces, in duality 9, 31
— Efimov-Stechkin 224
— Nikol'skij 150
Spline 146
— Bernoulli 147
— complete 172
— Euler 100, 147
— ideal 100, 147
— perfect 147
— periodic 147
— natural 176
Subdifferential 36
— approximative 79
— Clarke 76
— Dini 78
— ε-Dini 78
— Ioffe-Mordukhovich 79
— of a function at a point 16, 34, 36
— of a homogeneous function 7, 33
— of a sublinear function 7
— of a sublinear operator 80
— of an operator 80
Subgradient 51
Subspace 24, 26, 27
Sum, Fejér 102, 123
— Fourier 102
— Jackson 102
— Kelly of function 29
— Kelly of sets 30
— of functions 29
— of sets 30
Supremum of a set 80
Surface, convex 4
System, Korovkin 84
— of orthogonal polynomials 111

$\mathscr{T}$-hull 25
Table, Padé 142
Theorem, Ahlfors' 126
— Alexandrov-Ky Fan 73
— Aumann's 64
— Banach's 63
— Banach-Alaoglu-Bourbaki 42
— Bernstein's 131
— — inverse 130
— Bernstein-Walsh direct and inverse 181
— Bishop-de Leeuw existence 83
— Blaschke's 60, 63
— Blaschke's compactness 13
— Bogolyubov's 66
— Bonenblast's 60
— Bonnice-Silverman-To 80
— Boyanov's 215
— Brunn's 63

Theorem (*continued*)
— Brunn-Minkowski's 15
— Bunt-Motskin's 63
— Carathéodory's 11
— Cauchy's on polyhedra 13, 14
— Chebyshev's for rational approximation 116, 118
— Chebyshev's on polynomials 112, 119
— Choquet's existence 83
— Choquet's uniqueness 83
— duality 69
— duality in linear programming 58
— Dubovitskij-Milyutin 52
— Dvoretskij's 64
— Efimov-Stechkin 224
— Fabry 156
— Fejér's 123
— Fenchel-Moreau 8, 43
— — an analogue 82
— Gauss' 114
— Haar's 120
— Hadamard's 143
— Hahn-Banach 40
— — in subdifferential form 40
— Hahn-Banach-Kantorovich 80
— Hardy-Littlewood 153
— Helly's 12
— Holmes' 12
— Hörmander's on sublinear functions 9
— Jackson's, an analogue 221
— — first 128
— — second 129
— James 120
— Kakutani-Tikhonov-Ky Fan 74
— Keldysh's 124
— Kelly's 63
— Klee's 11
— Kolmogorov's 134
— Kornejchuk's 140
— Korovkin's three functions 179
— Kotel'nikov's 221
— Krejn-Mil'man 42
— Kuhn-Tucker 58
— Lavrent'ev's 124
— Leichtweiss' 60
— Levin-Milyutin 71
— Littlewood-Paley 153
— Lyapunov's 64
— Mackey's 38
— Mairhuber's 120
— Michael's 74
— Marcinkewicz's 153, 195
— Markov A.A. 119
— Markov V.A. 119
— Mazur's 38, 42
— — on differentiability 75
— Mergelyan's 124
— Milyutin's 223
— Minkowski's on extreme points 10
— — on intersection of half-spaces 9
— — on polyhedra 15
— Minkowski-Farkas 73
— Montessus de Ballore 143
— Moreau-Rockafellar 52
— Muntz's 125
— Nikol'skij's 164, 165, 166, 167
— on Alexandrov widths and ε-entropy 193
— on alternation 115
— on approximation 14
— — by Fourier sums 136
— — by polynomial functions analytic in ellipses 135
— — of the conjugate class 135
— — of functions from $B\tilde{W}_\infty^r(\hat{I})$ 136
— — of functions satisfying a Lipschitz condition 136
— — of the Hölder class by Fejer sums 136
— on Bernstein widths 194
— on best approximation of the class $BW_\infty^r(T^1)$ 135
— on bipolars 7, 43
— on calculus of the Legendre-Young-Fenchel transform 46
— on compactness of the subdifferential 51
— on composition 52
— — finite increments and inverse functions 78
— — of linear operators, an analogue 81
— on conjugate cones 7, 74
— on existence in the small 66
— on existence of solutions to the linear programmimg problem 58
— on Fourier widths 193
— — of intersection of Sobolev classes 196
— on inverse functions 59
— on inversion of the operator ∂ 8
— on inversion of the operator s 8
— on involutiveness 43
— on Kolmogorov widths 193
— — of a Sobolev class 197
— on linear widths 193
— — of a Sobolev class 197
— on operators inverse to s and ∂ 43
— on the regular integral representation 54
— on saddle points 58
— on the subdifferential of the maximum 52

— on the subidfferential of a sum 52
— Rademacher's 76
— Radon 12
— refinement 17, 52
— Riesz M. 153, 179
— Runge's 124
— separation, first 8, 41
— — second 8, 41
— — strict 42
— Stechkin's 136, 165, 225
— Steiner's 15
— Stepanets' 166
— Stone-Weierstrass 124
— Telyakovskij's 165, 166
— Timan's 138
— Vallée-Poussin's 117
— Vitushkin's 126
— Weierstrass' 121, 143
— — an analogue 240
— Wiener's 125
— Young's 60
— Zolotarev's 117, 118
— Zygmund's 130, 132
Theorems, direct 128
— inverse 130
Theory, Choquet 83
— constructive functions 127
Topology 9
— compatible with duality 37
— Mackey 37
— weak 37
— weak* 37
Transform, Legendre 33
— Legendre-Young-Fenchel 6, 33, 34, 80
— multiplier 153
— (p, p)-multiplier 153
— Young 33, 80
— Young-Fenchel 33

up-function 203

Vector lattice 80
Volume mixed 15

Width 112
— absolute 190
— Alexandrov 102
— Bernstein 103
— entropy 103
— Favard 103
— Fourier 102
— Gel'fand 1103
— Kolmogorov 102
— linear 102
— Urysohn 188